W9-BWR-395

Naturalism and Pragmatism

Also by Jay Schulkin

ACTION, PERCEPTION AND THE BRAIN: Adaptation and Cephalic Expression (2012)

Naturalism and Pragmatism

Jay Schulkin
Georgetown University, USA

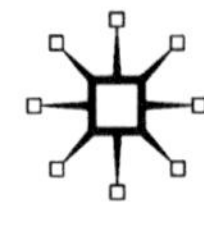

First published 2012 by
PALGRAVE MACMILLAN

Palgrave Macmillan in the UK is an imprint of Macmillan Publishers Limited, registered in England, company number 785998, of Houndmills, Basingstoke, Hampshire RG21 6XS.

Palgrave Macmillan in the US is a division of St Martin's Press LLC, 175 Fifth Avenue, New York, NY 10010.

Palgrave Macmillan is the global academic imprint of the above companies and has companies and representatives throughout the world.

Palgrave® and Macmillan® are registered trademarks in the United States, the United Kingdom, Europe and other countries.

ISBN: 978–1–137–02648–4

This book is printed on paper suitable for recycling and made from fully managed and sustained forest sources. Logging, pulping and manufacturing processes are expected to conform to the environmental regulations of the country of origin.

A catalogue record for this book is available from the British Library.

A catalog record for this book is available from the Library of Congress.

10 9 8 7 6 5 4 3 2 1
21 20 19 18 17 16 15 14 13 12

Transferred to Digital Printing in 2013

To Gordon Marino, Jonathan Moreno and Daniel Todes

Contents

Figures

Tables

Preface and Acknowledgments

I have been thinking about pragmatism since I was an undergraduate. My first academic teacher and friend Bob Neville introduced me to the "classical" pragmatists: I took immediately to Charles Sanders Peirce, William James, and John Dewey.

During this same period, now nearly 40 years ago, I spent a summer with Paul Weiss, student of Morris Cohen and Alfred North Whitehead, and one of the earliest editors of Peirce's work. It was a formidable experience. When I first proposed a tutorial on Peirce and James, Weiss replied in his characteristically blunt Lower East Side way, "Peirce I can understand, but why James?" I can't recall what I said; we read Peirce and no James, and, as Weiss rightly noted, "Peirce had size" (Weiss 1965). Weiss missed the profundity of James, but he understood Peirce at many levels, particularly at the level of speculative philosopher and generator of metaphysical categories. However, I never felt that he understood Peirce the experimentalist, despite the importance of scientific inquiry for Peirce and the pragmatist movement in general. Of course, Weiss was not blind to the scientific side of pragmatism. For instance, Weiss noticed something important about evolution in one of his early books: "nature is more an affair of fits and starts, of breaks and bumps, than of smooth and easy passage of continuity and harmony" (Weiss 1967 [1947]). Stephen Jay Gould, evolutionary biologist, noted something quite similar; he called it "punctuated equilibrium" – his now-famous theory of non-linear evolutionary change.

In graduate school, while I continued to study pragmatism with Betty Flower and Abe Edel in the philosophy department at the University of Pennsylvania, I eventually found my professional path in the behavioral sciences. I never lost my interest in classical pragmatism and my taste for its unique combination of naturalism with an adventurous sense of inquiry. But even as I returned again and again to the works of Peirce, James, and Dewey, I never stopped doing empirical science in the behavioral and neural sciences, either. This book, like my others, reflects both my philosophical roots in classical pragmatism and the scientific practice of an investigator.

In my experience, some philosophers are more in touch with the scientific side of pragmatism than others. I met Willard Van Orman Quine at the American Philosophical Society in Philadelphia in the early 1990s when my scientific advisor and senior colleague, Eliot Stellar, was President of that old society, founded by Benjamin Franklin in 1745. When I was introduced by Eliot as a philosopher, Quine took a step back; but when Eliot added, "in addition to a scientist," Quine stepped forward again, at least in my sense of it all. Perhaps Quine was too enthralled by science: the aesthetic and social dimensions of pragmatism were out of his purview, though he did consider himself to be within the pragmatic tradition – he exemplifies its commitment to naturalism and its love of science.

As befits its democratic image, pragmatism is a diverse movement that spans the full gamut of human interests and endeavors. In recent years, I have enjoyed the company of a diverse group of kindred pragmatic spirits and have taken a great pleasure in discovering that pragmatism attracts more than just white males. We pragmatists are a pluralistic group of individuals who enrich one another through mutual, respectful engagement – at least, that is the ideal – tempered by a humble sense of human progress.

In that spirit, this book offers reflections on the pragmatic tradition from what I hope is a fresh perspective: that of a working neuroscientist. Though naturalism and evolution are not the only topics of discussion, they are important themes of the following discussion. Both pragmatism and modern behavioral science grew up in the wake of Darwin's theory of evolution. Indeed, it is impossible to imagine either without evolutionary theory and the more general 19th century trend of naturalism from which modern evolutionary theory emerged. And yet, for a variety of reasons, these common origins have not ensured a close affinity between pragmatist philosophy and the behavioral sciences. Among the wide diversity of scientific theories of human cognition and its evolutionary origins, only a few are congenial to pragmatism in its original or "classical" form, which embraces the full range of human experience. Thus, this book presents not only a scientist's take on the pragmatic tradition, but also a pragmatist's take on the evolution of human problem solving.

In this regard, I am building on the pragmatist tradition from a scientist's perspective. Readers should not expect a thoroughgoing exegesis, however. Rather, in what follows are the consequences of various strains of thought within classical pragmatism that resonate today with what we now know from the inquiries that have continued to go on since the heyday of classical pragmatism. Furthermore, modern trends of enactive, embodied, and extended theories of mind and cognition link up with the pragmatist tradition.

Each chapter in this book builds upon classical pragmatism by examining how its characteristic orientation to inquiry has manifested itself in history. It engages historical figures as well as diverse contemporary contexts. It shows that the development of pragmatic inquiry is itself a process of evolution, as evidenced by its great diversity of expression and its ingenious modes of adaptation and extension – variation is a key theme in all forms of evolutionary process.

Unlike some depictions of evolution as unrelentingly progressive, this account presents both the evolution and the devolution of function, the lure towards integration, advance, and adaptation as well as the breakdown of function and coherence. Alongside evolutionary genius is the inevitable breakdown of functional relationships and devolution of expression – the converse of evolution. The downward spiral leads to extinction, a key concept in evolutionary theorizing.

Of course, if one is grounded in biological and evolutionary categories, naturalism has already been accepted as a background condition. Another premise is realism: not naïve realism, where things simply are, but critical realism. Objects are not simply given but are understood from a perspective – more precisely, from the midst of an ongoing search for a more stable perspective that enjoys fairly reliable powers of prediction. This is a search that is cephalic in origin (viz., originates in the whole of the brain, not any specific part) and is constantly working towards objects of various kinds.

For its classical progenitors, pragmatism was grounded in naturalism and realism, and was anchored to an evolutionary perspective. Evolution was the essential background, whether narrowly or broadly understood. My goal is to highlight the naturalist and evolutionary core of pragmatism while at the same time expanding its empirical horizons. This view is more enmeshed with the biological and cognitive sciences than that of Peirce, though Peirce is indispensible for understanding inquiry and constraints on hypothesis formation. James was actively engaged in psychobiology, as was Mead, but neither were sufficiently rooted in the experimental tradition. Of the four, Dewey is the most profound with regard to the connection between naturalism and pragmatism, and the importance of both for the enrichment of culture and the flourishing of participatory democracy: his capacious theory of inquiry underlies much of this book. But in each case I go beyond the purview of the classical pragmatists. I expand upon their core basic ideas and link them to contemporary science, drawing connections between the classical period and the modern period, and modern notions such as "embodied cognition," or what Dewey called "lived experience" or "enactive cognition" – the deep link between cognitive systems and action.

It should be noted at the outset that the story of pragmatic inquiry is deeply rooted in the American experience – the "Great American Experiment." America is ripe with contradictions: we suffered an original sin (slavery) at the same time that we reveled in a sense of innocence and adventure amidst the plenty of a seemingly unbounded nature – a new Eden. Natural piety, if not naturalism, is a deep American tradition (witness the great National Parks), yet we have been aggressive clearers of the wilderness and dominators of the new land. Nature was a prominent theme of early American thought, especially prominent in Emerson, Thoreau, and Margaret Fuller: after all, it was hard to avoid when so many 19th century Americans lived in rural settings. Natural piety could emanate quite readily in these circumstances. Of course, there was great variation in this natural piety as it interacted with an equally potent sense of social justice that prioritized upward mobility and expansion of opportunity.

Influenced by the American "frontier mentality," pragmatists moved away from entanglement with traditional philosophical conundrums while acknowledging that a good many matters of genuine philosophical importance will always remain partly unsettled. Decisive action requires that much is accepted on good faith, but intellectual honesty requires criticism. Doubts linger amidst an animal faith, let alone a human faith. But the backdrop is still naturalism and a modest realism. Putnam calls this "realism with a human face," or perhaps we could term it "naturalism with a human face" – and what other face could there be, since we constitute the frameworks of our own understanding, built on cephalic adaptive systems that evolved over millions of years.

In this book, Chapter 1 offers a perspective on the connection between naturalism and pragmatism. Chapter 2 outlines the work of Peirce, the person and the inquirer, especially as he fits with modern notions of inquiry. Chapter 3 focuses on a critical realism knotted to evolutionary considerations. Chapter 4 links a naturalized sense of inquiry and our evolutionary legacy to a broad sense of problem solving akin to Dewey's vision of inquiry. In particular, it examines medical decision making, its cultural evolution, and its integration with findings in cephalic capability and cultural expression. In Chapter 5, I present a Jamesian view of self-regulation and cephalic function. Chapter 6 considers social/moral smarts, evolution, and the philosophy of George Herbert Mead. Chapter 7 addresses important questions about human progress. The conclusion considers topics of well-being and social hope.

Years ago, when I was a graduate student in the late 1970s, a fellow graduate student (Dan Marino) would make fun of me for my interest in pragmatism, but in a serious gesture he wrote the following inscription to a book for me in 1979: "May the truth for you lie not in defunct black or white, may you have the courage to take it and its implications to heart. May you live it, not just see it." I hope I have done so.

I thank my family and friends and wonderful colleagues. I also thank Tibor Solymosi, who worked through the entire manuscript with me. To those not acknowledged directly, I apologize in advance.

1
Pragmatism and Naturalism amidst Social Sensibilities

Introduction

An abrupt beginning: Naturalism, like realism, has many meanings – a family of meanings, many of which are associated with diverse pragmatists. I have settled on critical realism. Why?

- Because the real is not something taken for granted.
- There are too many counter examples for what we take to be given, simply delivered to us with or without diverse forms of intellectual and technological lenses.
- We are cultural animals; our nature is bound to the cultures we live in, the diverse forms of natural phenomena that we have created and adapted to.

Both critical realism (e.g. realism not just taken for granted) and naturalism are at the heart of classical pragmatism;[1] and with that comes a sense, or senses of nature, not particularly sentimental or mystical, nor necessarily crudely practical and instrumental. Nature and the human condition, for Emerson, are rich in the contours of beauty, self-respect and respect for others, the cultivation of the good soul, a worthy individual with common virtues, including self-reliance.[2] It is a state in which choice predominates, an existential flower, along with existential angst.

The main themes of this book are a sense of nature and the predilection for inquiring into kinds of objects and events. These themes highlight the importance of an evolutionary framework for pragmatism. One forgotten proto-pragmatist, Chauncey Wright, embraced an evolutionary perspective, and the importance of what later became critical realism. I contrast this variety of realism with naïve realism.

Naïve realism, as I understand it here, holds that the world is simply given to a perceiver, regardless of the perspectival framework of that perceiver. Critical realism embraces the evolutionary fact that a significant amount of work has gone into our capacity for taking perspectives toward the world in order to act. This distinction is a central theme throughout the book.

Wright may be the forgotten one. Peirce, James, and Dewey, and to a lesser extent Mead, are not. One forged a deep sense of inquiry, one captured the human condition, the other understood what counted, and the final captured the social milieu.

This chapter sets the context and some of the historical routes of pragmatism and naturalism. The narrative is to weave a theme in which classical pragmatism, grounded in a sense of nature and evolutionary considerations and tied to the Enlightenment notion of human possibilities, is oriented to broad-based self-corrective investigation. But the investigation is towards facilitating socially cooperative experiences. The theme is how naturalism and critical realism are important factors in the social evolution of socially cooperative behaviors. This was a core theme for Dewey; the narrative theme is science without scientism, amidst a sense for social impact and human well-being – social hope.

Born cataloging and testing: Aristotle

Aristotle, like Kant and Peirce, saw human beings as born categorizers. We automatically taxonomize the kinds of objects that we encounter in nature.[3] We look to logic, logic of what is in the class of objects, what constitutes the class, and what core features are presupposed. What dominates in Aristotle is the life trajectory. Aristotle asserts the logic of development, along with an analytic capability of endless classification and logic, somewhat like Darwin himself so many centuries and continents later. Core features (or essences) and the logic of causation are common preoccupations for Aristotle in the classification of things. Central questions for him include: what is potential, what is actual, and what is the developmental trajectory of a plant, a platypus, or a person?

Aristotle suggests that "in general, the intellect in activity is its objects."[4] Developmental propensities predominate the intellectual landscape; "What is a thing's trajectory?" is a core question.

Enlightenment theory and Jefferson

Of course, the fact that there are many distinguishing and overlapping features of naturalism is a recurring theme in all cultures, but in the West it took on theological proportions as the grand theologies were minimized and nature's elegance emphasized. Thomas Jefferson held such a conception of nature, steeped in this Enlightenment sensibility.[5]

Nature permeates the cortical limbs of the Enlightenment thinker.[6] A conception of nature is also part of the conception of progress. The normative goal is an enlightened trajectory by effort. The beckoning, harnessing, and development of the new sciences mapped terrains through diverse instruments, which traversed the heavens and the territorial landscapes. Such knowledge gathered in glorious and evolving libraries.[7]

Thomas Jefferson would survey, build, and describe these natural capabilities. He was a true representative of his age, with diverse forms of Enlightenment sensibilities, along with minimalist expectations of gender, race, and ethnic capabilities. For Jefferson, devolution of expression existed along with a cultural evolution, but nature was at the forefront. Though a student of the Enlightenment, for Jefferson it was not a radical enlightenment in which liberty of thought and freedom was extended to African Americans; he was not as enlightened as other Enlightenment thinkers.[8] At the University of Virginia, a school built by Jefferson, the original curriculum reveals these concerns, as do Jefferson's notes on Virginia, the building of the University, and his statements about nature and religion, although his works also demonstrate his misguided ideas about race and racial superiority.[9]

There is an agrarian component in the roots of pragmatism.[10] It is rooted in the land, soil, nature; a philosophical context in which nature is palpable.

Jefferson begins his University of Virginia curriculum with history, but distinguishes human from natural history. He does the same for philosophy before going on to link history to memory, philosophy to reason, and fine arts to imagination. Jefferson's natural history is about nature, and his civil history is about culture. Philosophy is both moral and mathematical (including the sciences), and the fine arts go from literature and languages to gardening, painting, and sculpture. For an overview of Jefferson's integration of nature into education see Table 1.1.

Jefferson was fixed on understanding nature; for instance, he obsessively cataloged geography and species.[11] Part of that fixation was good and refreshing and bold, other parts narrow, self-serving and utterly mistaken (e.g. his thoughts on slavery). The enlightenment sensibility of his colleague Humboldt had a much more liberal view of human

Table 1.1 Jefferson and scientific education

Nature in Higher Education[13]

- *Plan for General Schools (1814)*
  - Department of Mathematics
    - Pure mathematics
    - Physio-mathematics (including astronomy and geography)
    - Physics or Natural Philosophy
    - Chemistry
    - Natural history, to wit; mineralogy
    - Botany
    - Zoology
    - Anatomy
    - Theory of Medicine
- *Professors Proposed in the Rockfish Gap Report (1818)*
  - Physio-mathematics (including astronomy and geography)
  - Physics or Natural Philosophy
  - Chemistry (including agriculture)
  - Mineralogy (including geology)
  - Botany
  - Zoology
- *University of Virginia Professorships (1824)*
  - School of Natural Philosophy (including applied mathematics, engineering, physics, and astronomy)
  - School of Natural History
    - Botany
    - Zoology
    - Mineralogy
    - Chemistry
    - Geology
    - Rural Economy

prospects, cultural variation, and human potential, and would have been a better orientation towards the slavery of African Americans and the treatment of Native Americans.[12]

His sense about the Enlightenment was abstract, perhaps too abstract, and not grounded in the literature of social belonging, a democracy linked to both individual initiative and a measure of social progress by the inclusion and social harmony sought amongst individuals: he had a major focus on science and nature. Jefferson's conception of nature and of what counts as social progress and harmony may not resonate with many naturalists and democrats today; however, what is important about Jefferson's views here is that they, like the man himself, were open to revision in the face of the limits of the situation. From this, a critical realism with an evolving sense of nature is possible.

Nature's beauty became a normative part of the parcel of even the most visionary Enlightenment thinkers; to take for granted the grandeur and elegance of nature and its endless beauty was a common currency in the 17th and 18th centuries.[14] On this continent, nature was part of the surveying of the land for possibilities to build and remake. Of course, it interacted with endless expansionist predilections.

The pragmatist journey to demythologize nature to sane proportions is also a vision of ourselves as a culture-producing species rich in duties to nature, and a realistic sense of our limits and possibilities. This vision entails Jefferson's, but goes beyond his as well. Nature and culture are not conceived as separate; we are cultural animals. But omnipresent is our evolution and our devolution, a predilection for the devolution and expression of gluttonous stupidity ripe with short-term modes of satisfaction, an outlook that is just as big as our noses, amidst longer-term predilections emboldened by our cultural evolution.

Kant and Peirce

If we naturalize Immanuel Kant, we can then ask: what are the cognitive conditions for the possibility of knowledge? What figures in the determination of causation? What are the factors in time, the succession of objects, the permanence of objects, and their representation in space? The science of space and time in cognitive systems was addressed in an empirical endeavor to determine possible objects of possible experience.[15] Space, time, causation; what is the conceptual apparatus required for cephalic adaptation, coherent action, and the epistemological tools for investigation and inquiry? Which concepts underlie human judgment? In these questions, Kant was a forerunner of the modern cognitive revolution; demythologize Kant and naturalize his orientation,[16] and he can look quite modern. Rules provide coherence and are empirically embodied in human experience;[17] they represent a logical syntax of cognitive order.[18]

C. S. Peirce labored endlessly with Kant, and a close reading of Kant forges links between Kant and Peirce. Both inquired into the conditions of the possibility of knowledge. Kant had the transcendental edge, linking the transcendental to the pragmatic. For Kant, conceptual pragmatism is a natural predilection.[19] Peirce, on the other hand, focused on the range of categories, and our capacity to induce events on a natural predilection. For him, ideas are constrained by the contours of nature. The understanding mind requires thinking about our evolution, our problem-solving capabilities, our ability to predict events, and our sense of probable events in space and time as tied to habits of intelligence and problem solving. These are some of the essential features in the architecture of Peirce's thinking.

Cataloging objects in nature, then kinds of objects in nature and their relationships, is at the very heart of keeping track of kinds of things, organs of physiology, variation in expression, and functional consequences of that variation. Keeping track of objects is a basic predilection built into cephalic sensibility and reaching into and part of everything we do; there is no separation of the activity of cephalic predilection and exploration of the natural terrain. Imagined activity is a core feature of the cephalic capability; movement is not necessary, but cephalic exploration, categories and touch, action and reflection, modification or adjustment, are core features of being born to categorize and to know.

German and American romanticism

Like Kant or Coleridge, Ralph Waldo Emerson was awestruck and saw in the experience of nature a sense "which yields the activity of man an infinite scope."[20] A romanticism set forth about nature, joined to an enlightenment sense about the capturing of knowledge "amidst a chain of being," initially frozen in time and fixed for eternity, but now emerging into an evolving sense of nature diversifying: these are the signs that the 19th century had emerged.[21] A romantic conception of the beauty and sustenance of nature had converged with the expansion of its use, devolution, and extinction.

In Romanticism, as in *Wanderer above the Sea of Fog*, see Figure 1.1, the glory of nature found sustenance in a broad area fraught with human possibilities in an expanding environment with what seemed like limitless possibilities; a culmination of an historical idea.[22]

A conception of nature brimming with possibilities was in the effervescent pens of Romantic expression; consider Kant, Rousseau, Heine, Goethe, Schiller, and Schelling. It was a sense of nature in which aesthetics was a fundamental feature, in which education was to be knotted to an "aesthetic condition,"[23] and where "spirit is natural philosophy."[24]

Perhaps it is not surprising that Barbara Novak,[25] in describing American landscape painting in her beautiful book *Nature and Culture*, would describe a sense of nature and wonder as so pervasive.[26] A conception of biology and naturalism was tied to this sense of natural wonder embedded in the romantic sensibility. Natural philosophy,[27] the exploration and discovery of nature, was part of human self-reflection

Figure 1.1 *Wanderer above the Sea of Fog* by Caspar David Friedrich, ca.1818

and development, was rich in organic self-regulation, sustenance, and rejuvenation, and was ripe with sublime natural and poetic beauty.[28]

Nature was and is something to be understood by a knowing subject, who realized that, from the beginning, the human species was part of that same nature. Johann Wolfgang von Goethe combined the sense of the wonders of nature with the science of vision and is an exemplar of the "knowing spirit."[29] Goethe, a lawyer by training like his rationalist precursor Leibniz (who had a vision of universal logic before Peirce and Frege), combined a theory of vision (modern in form) as a precursor to what would eventually become an opponent process theory of color vision, a conception of change through metaphors and an eye for details of plants and metamorphosis.[30] He was also a romantic teller of tales of loss and the threat of the myth of knowledge and the fall, and, like Friedrich Nietzsche, was a man dependent on trips to Italy for vitality and spirit.[31] For Goethe, the broad view of science was embedded in a spirit of inquiry linked to language and translation, history, and natural history.

Wonder, nature, and discovery are prevailing themes for all these Romantics. Possibilities seemed endless and awe-inspiring, amidst an acknowledgment of the constraints that make cognitive capability possible, and thereby invention and discovery. They placed our capacity to reason into a more scaled down and continuous context with science.[32] They demythologized expectations about reason. They preserved it but sized it down in the modern era to more sanguine and learned possibilities.

Nature was also tied to a burgeoning notion of experience with respect for, rather than the deification of nature. Emerson particularly had an expanded sense of experience, which would become a common theme in what would later be pragmatism. This experience was an exploration as reflection on oneself and the exploration of nature. The experience counts, and that expanded sensibility is at the heart of our cultural revolution.

Fields, forests and fences pervaded an ecological landscape that traversed a vast ground, a field animate with brewing possibilities,[33] and a place in which there were few inhabitants, and what seemed like endless cultivating space. Of course, there was also an endless curse of loss, distrust, and disrespect. Perhaps not surprisingly, at the other end of this continent from Emerson's New England, John Muir came to represent a sense of the wilderness[34] – a sensibility that is transparent and perhaps, in some stances, transformative to the civilizing factors that surround, a sensibility that can embrace nature, demythologized and placed in a context of conservation and responsibility – no small order. A Romantic wonder would provide an aura of the civilized amidst the wilderness.[35]

A primal sensibility was pervasive in a continent brewing with possibilities and emboldened by a rich sense of nature; it was cataloged and noted.[36] Romanticism fueled an appreciation of the natural landscape, understood by Thoreau, mystically infused by Emerson and demythologized by Dewey. Technology was not necessarily an enemy, but could be when it became identified with knowledge and the "scholar disappears."[37] Most human inventions, after all, are ripe with possible dangers; remember, for instance, the dogged warnings and radiant appreciation of nature by Rachel Carson,[38] as pictured in Figure 1.2, a mid-20th century proponent of our dependence on nature and our embodiment within the cultural milieu.

Human expression is diverse; amidst the respect for and appreciation of the elegance of nature was the slaughter of the bison, the degradation

Figure 1.2 Rachel Carson

of nature, the devolution of our surroundings. Of course, human ingenuity is the wonder embodied in cephalic capacity. As Dewey always noted,[39] we invent, we change, and orient to future outcomes. We build, and we can degrade the damage we do, the devolution we impose at least for now, and certainly in the past.

A pragmatism embedded in a respect for nature, something reachable by pragmatist sensibilities, is a formidable perspective, albeit difficult to realize.[40] It is a natural piety that reaches back to Emerson and can be found in Dewey.[41] It produces an ethics of nature that emboldens a respect for nature and one another,[42] a recognition of the value of resources, where valuation is a highly embedded cognitive predilection in our valuational architecture, where the diverse valuational features pervade human experience.[43]

My view of the virtues of pragmatism, though frail and fraught with limitations, is also linked to a conception of progress, a respect for science, and a conception of the continuity of philosophy and science.[44] Progress is linked to a conservation ethic of the resources, natural piety, and wonder.

In fact, the new continent for Europeans was shaped by the people they found here: American Indians. The aboriginal inhabitants of this continent fermented the evolution to a sense of place, a place within nature , which impacted the naturalism that evolved as an important element of pragmatism. Cultural richness was pervasive across Indian populations; sentiments about nature, about birds and human well-being resonate profoundly across our continent, fragile, but emboldened with a "native pragmatism."[45]

Henry David Thoreau, a writer of nature much the way James could write about human psychology, had a flair for expressing human experience. For Thoreau, nature is all-pervasive: the forests in New England, the walks, the cold, the trees, and what he called "the wonderful poetry of nature."[46] "Nature alive," a phrase of Whitehead, lines the pages of Thoreau's prose so that he asserts, like many others across a wide area of human investigation (e.g. Taoism, Buddhism), "no dominion is closed to man."[47] But there is, as we all know, an "untamable" feature to our experience of nature. It is humbling, awe-inspiring, expanding of our sensibility; it is rich in aesthetics as we embrace what we know and what we do not; as we recognize what we have some power over and what we don't. This is a sense or feature dominant in many cultures of being part of "heaven and earth" – very much a part of nature in the sense that *heaven* signifies an ideal to be effected in this living natural world, not something supernatural or post-nature.

Thoreau had a social conscience. He would chastise a friend for being out of jail while he was setting a condition of civil disobedience, a concept important later for others (Gandhi, King). He selected social topics to protest against (slavery) while he remained sequestered in his solitariness, his want for isolation, and his walks away from the clutter of the little Massachusetts crowds. In his wake, a "flowering" of expression would emanate from New England.[48]

Thoreau's solitary walks remind us of Rousseau, another lover of nature, but much more the social provocateur, not the suggested expositor of rebellion from afar. Walden is far from Paris, let alone Boston. Thoreau successfully captured the "wildness in nature," a virtuous sensibility of the "animal spirit,"[49] a primitive sense of belonging to the class of living things emanating from nature in his work *Walden*, pictured in Figure 1.3.[50]

For me, Thoreau's nature is a bit mythologized as a harbinger of virtue, but nature should be demythologized, not deified as Thoreau tended to do. However, our endless expression of shortsightedness is omnipresent, and our exploitative and short-term consumer-based culture continues,

WALDEN;

OR,

LIFE IN THE WOODS.

BY HENRY D. THOREAU,

AUTHOR OF "A WEEK ON THE CONCORD AND MERRIMACK RIVERS."

I do not propose to write an ode to dejection, but to brag as lustily as chanticleer in the morning, standing on his roost, if only to wake my neighbors up. — Page 92.

BOSTON:

TICKNOR AND FIELDS.

M DCCC LIV.

Figure 1.3 Frontispiece to Thoreau's *Walden*

particularly in our era. So many of the findings in the decision sciences highlight the diverse ways in which we minimize future concerns. It is not simply that life is brutal and short, paraphrasing Hobbes. Certainly life can be that and has been, but it is just that we are also prone to terrific adaptations amidst a reminder of our endless vulnerabilities; vulnerabilities to over-calculate some things and to underestimate others, to skew everything by imposing a mistaken framework.[51]

It is not a foundational framework that is the goal, but workable, well-justified frameworks that satisfy our concerns.[52] Evolution provided us with the ability to be a sort of problem solving, generalist, expanding species, combined with the curiosity and eventually, the leisure to realize diverse cognitive goals. Vulnerability to diverse forms of cognitive error, however, is a feature of devolution, a fall downward and a chronic reminder of where we might end up, and perhaps a modest prophylactic against arrogance.[53]

Evolution and naturalism: Charles Darwin and the Metaphysical Club

Evolution impacted the wide area of all thought. We have recently celebrated Darwin's 150th birthday.[54] The fight to establish an evolutionary perspective, a feature of the 19th and 20th centuries, amazingly still lingers in many parts of the U.S. But Darwin changed everything, and the impact of the cumulative view of biological evolution became a common parlance of many thinkers. One in particular is Chauncey Wright, one of the least-known of the 19th century thinkers in what was called the "Metaphysical Club."[55]

The Metaphysical Club was an interesting group of thinkers that included Oliver Wendell Holmes Jr., William James, Charles Sanders Peirce, and Chauncey Wright. The group was set on discussing the contours of thought in the second half of the 19th century, and was steeped in the elite of New England. Holmes's father, for instance, was a leading physician. Henry James Sr., father of William, was a theologian and literary figure. Benjamin Peirce, the influential mathematician, was the father of Charles Peirce.

This was no ordinary group of men, and the ideas they expounded, as Menand has so gracefully shown in his book *The Metaphysical Club*, impacted profoundly this country from the common law, to the sense of inquiry, to the embrace of an informed psychology, to the engagement of biology. The Metaphysical Club had close ties to Emerson and Thoreau, but also to Unitarian abolitionist elites, and even contained a pro-Southerner (Peirce) amongst them; but they were intellectual to the core and tied to a variant of naturalism and pragmatism grounded in community, experiment, and sustained action.

One very important figure dominated the intellectual landscape for this club and many others: Charles Darwin. A shy, diminutive man, Darwin was one of a line of thinkers for whom evolution was a part of the intellectual milieu. His grandfather, Erasmus, would write poetry with a conception of evolution; Darwin himself would later correspond and interact with the dominant intellectual expositor and defender of evolution, Huxley.[56] A voyage around the world with a cranky melancholic captain would result in diary about the voyage on the *Beagle* and lead up to the great books on the origins of humankind. The marvelously diverse creatures in the Galapagos would fill the pages of this diary and then the great books on evolution some 25 years later, after some pressure of being scooped by Wallace, another eccentric Englishman.[57]

Journeys for riches and discovery were a common currency for many ages, but, in the 18th and 19th centuries, they reached a particular prominence as a glorification of nature, a combination of Enlightenment vision, humanism, and romanticism, linked to experimentalism, in which aesthetics was integrated into science and the culture of inquiry was broad, not narrow.[58] The great journey was about biological discovery, embracing biological diversity, coming to understand something about biological possibilities and their evolution. In Alexander Humboldt's case, it also involved the conception of the diversity of human expression to be appreciated as part of a large family of species.[59]

Humboldt, much appreciated by the likes of Emerson and Thoreau, and with very strong affinities for Thoreau, contributed to their narrative of the human condition, their sense of nature, and our evolving culture of human expression. Humboldt was a colleague of Jefferson. He was both an artist and a naturalist. His friend, Charles Wilson Peale, constructed the first museum in the U.S. He was also linked to a sense of conservation and its movement in this continent through the eyes of Muir.[60] Humboldt was a seminal figure. For him, this continent was a thing of beauty, nature's beauty.[61]

Like many other investigators, Humboldt's depictions are delightfully aesthetic, such as his description of the land. Also, consider another work by Caspar David Friedrich, *Two Men Contemplating the Moon,* pictured in Figure 1.4. Humboldt appealed to a nature that is liberating and accepting from a narrow purview, emboldened into an aesthetic of language and poetry that broadens the sciences.[62] It is a naturalism easily tied to "critical realism" of the sort offered by Roy Wood Sellars,[63] in which nature is not the same as what we can encompass simply in our minds. This is a move beyond Emerson and towards nature and adaptation,[64] something endemic to a broader view of naturalism and human inquiry.[65]

Humboldt's book *Cosmos*, romantic to the core about knowledge and nature,[66] was a key book for Darwin, as it was for the Transcendentalists' pontificating about the unity and wonder of nature. Laura Dassow Walls, a professor of Letters, who captures the sense of nature in her prose and interests, quotes an early Peirce in her book on Humboldt: "It is true that the progress of science may die away, but then its essence will have been extracted. This cessation itself will give us time to see that *Cosmos*, that esthetic view of science which Humboldt prematurely conceived." Humboldt's imprint on our perception of nature and realism is pervasive. But he was part of a much larger intellectual brew: a romantic conception of the individual and of fraternity.[67]

No less a figure, William James, with his teacher at Harvard, Louis Agassiz, was a student of Humboldt. But James had a very different attitude towards nature and humanity,[68] and would himself make

Figure 1.4 Two Men Contemplating the Moon by Caspar David Friedrich, ca. 1824

two special journeys: one to Brazil and the other to Germany. Both could be characterized as old and new worlds, and both are. What the trips represent is an exploration of nature via the wildlife of Brazil and the rich intellectual scientific culture of Germany in the 19th century.[69]

Evolutionary considerations of the Lamarckian sort, where habits influence form by desire (something long disputed but still living and indeed somewhat in vogue again[70]) concerned many pragmatist thinkers. Volition and adaptation underlies the Lamarckian perspective of the first part of the 19th century, a feature that could be found in both Erasmus and Charles Darwin. The important point for pragmatists was that purpose, or rather function, underlies adaptation. The Metaphysical Club would engage those issues with a predilection towards Lamarck, except for one: Chauncey Wright.[71]

A sense of the struggle derived from Malthus and Ricardo, speciation from Lyell, and a conception of change would line the pages, along with a 19th-century prose, but with punctuated moments of astute clarity and insight: from plants to platypus, from bird form to iguana form, pages lit on fire with wondrous nature, including exotic animals such as that pictured in Figure 1.5. Wright understood, as did others, the deep meaning of what was occurring.

Figure 1.5 Brazilian wildlife

Chauncey Wright, America's first philosopher of biology, and some consequences of an evolutionary perspective

Wright, who died young, was thought of by many as the most brilliant of a group that included Holmes, Peirce, and James. Indeed, Holmes also thought Peirce's genius was overrated, though he is now seen by most, including myself, as the most original thinker of the group. A product of South Hadley, Massachusetts, Wright was never able to get his intellectual feet on the ground. Wright was shy and self-destructive, vulnerable to depression, alcohol consumption, and withdrawal, yet nevertheless he was well-known in a small circle. The future president of Harvard, Eliot, and of course Charles Darwin, with whom he both met and corresponded,[72] had great respect for his thought.

Darwin was essential for Wright. He was the most thoroughgoing evolutionist of the Metaphysical Club. Wright embraced the new insights emerging from evolutionary theory; it is largely due to Wright and his efforts that evolution is a key cornerstone in pragmatism.

Chauncey Wright, who understood the importance of Darwin and the concepts of evolution, would write about the logic of evolutionary theory (in 1870), and would help introduce American thinkers to this paradigm shift in thought. This would dovetail more generally with his critique of science, and more generally on the continuity of animal species towards self-consciousness, or self-reflection. Wright argued for the continuity of animal and human adaptation and problem solving.[73] And of all the 19th-century thinkers that read Darwin in the U.S., Wright was the closest to Darwin in his scientific orientation towards biology, while Lamarck hovers, to various degrees, over all the classical pragmatists:

> The inquiry as to which of several real uses is the *one* through which natural selection has acted for the development of any faculty or organ, or stands and has stood in the first rank of essential importance to an animal's welfare in the struggle for life, has for several years seemed to me a somewhat less important question than it seemed formerly and still appears to most thinkers on the subject. The reasons you give why sexual selection should have had much to do with several of the features, of which I have spoken, are still perfectly valid. The uses of the rattling of the rattlesnake, as a protection, by warning its enemies, and as a sexual call, are not rival uses; neither are the high-reaching and the fore-seeing uses of the giraffe's neck rivals, but are in the most intimate conspiracy to the

> same effects. Furthermore, it seems to me presumable that in a long course of development, even in cases of highly specialized faculties, existing uses have risen in succession or alternately to the place of first importance, as in various uses of the hand.
>
> This principle of a plurality of existing uses involves a very important influence on secondary uses, whether these are incidental or correlative acquisitions, or are the more or less surpassed and superseded ones. They seem to connect in some cases the action of natural selection with the inherited effects of habit and exercise. An animal may, for comfort or convenience, which bears but little reference to its essential welfare, be indirectly furthering, through exercise, certain faculties which, though rarely called into exercise in functions of prime importance, may nevertheless have, or may come to have, such functions.[74]

Wright tended less towards saying important things about the internal milieu, something the literary genius of William James in his great book, *The Principles of Psychology*, would do. Wright, on the other hand, faced squarely and unapologetically the importance of embracing an evolutionary conception, and tied it to the logic of science and a conception of "real kinds," and with the consequences of our actions and ideas in real palpable change:[75] "Thus, man is a geological agent. He affects and alters unintentionally the physical forces and conditions of the globe. He changes climate even, and its consequences, by actions designed for other effects."[76]

Chauncey Wright was keen on drawing attention to the expansion of attentional abilities that underlie the evolution of self-consciousness, and noted that "natural realists" ought to be "evolutionists."[77] He corresponded with Darwin over a period in the 1870s (1871–1875) until Wright's death in 1875. The topics were varied, from the origins of species, to cause and effect in science, to agency. Flattered by Darwin's attention and pleased by "kind expressions respecting his thoughts" (August 1, 1871), Wright aligned, more than others in The Metaphysical Club or any other group on this continent, with Darwin regarding the new biology and its implications for a functionalist conception of cognitive function.

Wright would also be critical of what would become "Social Darwinism" and its variants. Social Darwinism built upon the Spenserian emphasis on social struggle, survival of the fittest, and the worrisomeness of the unfit, which came to later have a disproportionate expression in North America and Europe, coinciding with a world-view that placed the pinnacle of thought in a white straight male. This variant of racism and sexism combined had percolated over many centuries, but in the 19th century permeated an outlook embedded in xenophobic nativistic exclusionary practices simmering across the U.S. and Europe then and, indeed, still.[78] An historical sense of progressivism had appeared,[79] yes, but it was coupled with the development of eugenics and the beginning of a construction of the "superhuman," and "manifest destiny."[80] Several pragmatists were quite favorable towards eugenics.[81]

For example, a future jurist of the Supreme Court, who would argue in his book on common law that "the life of the law has not been logic, it has been experience."[82] Also, the son of a Boston icon in medicine would be a proponent of eugenics, as would Margaret Sanger and many others: perhaps he should also have emphasized more logic and experiment. Oliver Wendell Holmes Jr., part of the core group of early pragmatists,[83] would vote on the Supreme Court in ways to undermine African American voting security in the devolution of the 14th and 15th Amendments (voting rights) that were in vogue in Southern States.[84] "Survival of the fittest," a Spenserian embedded expression, is central to the social Darwinism that dominated a portion of the intellectual landscape, and which became associated with a variant of scientific naturalism, a naturalism linked to all sorts of group xenophobic, elitist sensibilities whose justification was hidden by the tapestry of the new biology.

The fall downward, the devolution, is always easy, and just a half a neuron away. Of course, on the other side of Holmes, the great impartial jurist, was a love of science, a liberal sentiment that allowed for the inclusion of other ethnic groups with little sentimentality. He was part of the changing landscape of American universities.[85]

Social Darwinism came to dominate and to justify a variety of social orientations, most of which were exclusionary towards others. While pragmatists like James and Dewey objected, others like Peirce and his father (indeed like most other people) looked at the new biology as a legitimating factor for diverse forms of racial theories that stretched back to many thinkers, from Jefferson to Lincoln. It was the lure of the day, embedded in Enlightenment sensibility under the cloak of an elitism that was exclusionary. The hierarchy of empirical chauvinism, western progressivism, was a bastardization of the new realism, the new biology.[86]

The impact of Darwin on philosophy was profound.[87] What appeared in his wake was a critical realism and naturalism grounded in object knowledge and linked to organic habits, natural inference, and grounded reason.[88] At times, the grounded reason was Aristotelian and oriented to object essences.[89]

Dewey rightly would suggest that with regard to the mind and adaptation, questions are rooted in terms of change and function, context and resolution. As Dewey put it,

> Interest shifts from the wholesale essence back of special changes to the question of how special changes serve and defeat concrete purposes: shifts from an intelligence that shaped things once and for all to the particular intelligences which things are even now shaping; shifts from an ultimate goal of good to the direct increments of justice and happiness that intelligent administration of existent conditions may beget and that present carelessness of stupidity will destroy and forego.[90]

These philosophical inquiries were forged by interaction with kinds and events, linked to what Peirce called "Scottish common-sense realism," in which interactions are with objects, not sensations,[91] or what Whitehead called "misplaced concreteness,"[92] and fermented to a general form of reason linked to objects. Thomas Reid,[93] beacon of common-sense realism, suggested (and Peirce concurred) that our orientation to objects is anchored to inductive mechanisms that provide coherence for an active mind foraging and making sense of its surroundings.[94] Peirce was rooted in real kinds, in objects encountered, not simply invented. The active mind is prepared to respond to diverse forms of objects.

For Reid and then for Peirce, James, Dewey, and classical pragmatists in general, being anchored to objects is of primary epistemic importance. Looking at a particular object is noting a general feature, or something quite general about a tomato, face, heart, etc., despite the variation; an orientation toward the generic properties of an object.

Cooperative inquiry, critical realism, social hope

Early critical realists displayed a strong cooperative sensibility. Indeed, they published a book entitled *The New Realism: Cooperative Studies in Philosophy.*[95] One of its aims was to elicit small groups of individuals (e.g. Edwin Holt, William Pepperell Montague, and Ralph Barton Perry, and several others suggested) to engage in cooperative inquiry towards a common end. It was a rather unique thing in philosophy, but then this was something quite different; a philosophy that emphasized community, purpose and expansion of human experience, not isolated, but socially configured, responsible to one another and working together in a joint enterprise.

Cooperative behaviors were linked to a sense of human progress, grounded in a new sense of biology and human possibilities. Indeed we know that our evolutionary success is linked to social cooperative behaviors, something emphasized by both classical pragmatists and critical realists.[96] One pragmatist, who continually wrote about cooperative behaviors, an essential feature of our evolution and our intelligence, was John Dewey. Social behavior is rooted in our psychobiology, anchored to an individual who is social in nature. Moral experience is pervasive in our social nature. From it develops an ethics that sets the condition for a participatory democracy, which emerges with our capacity to work through endless conflicts.[97]

Dewey, as most of us pragmatists want to do, aimed to anchor a rich sense of human experience to the social context of civilized action, and also to anchor it to cephalic (i.e. whole-brained) propensity embedded in a sense of objects. We are rooted in objects, as he often put it.[98] We are forged in communicative social contexts of adaptation while coping with diverse forms of precarious experience. One key is to anchor the prosocial sensibility into adaptive, culturally, and socially bonded individuals. Diluting differences, encouraging engagement, and as Dewey often noted, "all human experience" has a social component.[99]

Perhaps this was made clear in social charitable behaviors that emanated from the Chicago Pragmatists, of whom George Mead is one, and Jane Addams another.[100] The Chicago School of Pragmatism expressed itself as a moral philosophy concerned with social change.[101]

Jane Addams,[102] a core member of the Chicago pragmatists and pictured in Figure 1.6, was a founding member of the NAACP (with Dewey). Hull House, a magnet for embracing others, a place to nurture youth and their development towards citizenship, was a practical expression of the Chicago school anchored in civic action and a broadening of the consideration of other experiences; perhaps the broadening of American pragmatism and an orientation towards the acceptance of others in the social milieu is the broadening of the social space of gender and gender roles.[103]

Jane Addams is in part an expression of the important link between pragmatism and feminism, and a philosophy that embraces the larger cultural milieu, linked to a "radical pragmatism,"[104] and a sense of human progress and liberation,[105] for which, moral reform and social virtue are prime motivators.[106] Indeed, Jane Addams, a Nobel Peace Prize winner in 1931, first provided a philosophy of social progress.

Figure 1.6 Jane Addams, on the right, protesting

She grew up with parents in Illinois thinking of Abraham Lincoln as mythic in the first part of the 20th century, and she created a laboratory school in the midst of change, steering towards a link between ethics, righteousness, and social democracy, with a vigilance that was a moral equivalent of war.[107]

Underlying this broadening of pragmatism was an inclusiveness of the human experience tied to "cooperative intelligence,"[108] a pluralism of human experience to be embraced and part of the larger social fabric of pragmatism. This is the sense of naturalism consistent with "social hope." A prophylactic predilection against the narrow scientism often associated with pragmatism is this important broadening of pragmatism away from a narrow one-gender and Eurocentric understanding of everything. Instead, a pluralistic universe of individuals is knotted to a larger vision of social cooperative behaviors as an important normative feature of pragmatic naturalism and the larger sense of social progress.

Inferences about kinds

There is more than one sense of critical reason, just as there is more than one sense of naturalism or pragmatism.[109] One important point is being anchored towards objects, to kinds of objects for which we are forging coherence in action.[110]

An "animal faith," as Santayana put it,[111] is a grounding metaphor of our rootedness in nature, in an evolutionary past. For a visual species such as ourselves, in which the visual cortex has expanded to such a great degree, a statement by one of the proponents of critical realism is more than warranted: "Nature is at once recognized as preeminently the visible world."[112] Capturing a sense of nature became a larger philosophical endeavor more generally aimed at identifying kinds of objects amidst a sense of nature rich and varied with endless possibilities of evolutionary selection and human invention and ingenuity, with a sense of adventure and play of ideas, about nature and the lure of possibilities of human understanding.[113] Marjorie Grene,[114] a philosopher rooted in biology, heralded in her philosophical testament "the primacy of the real" that was rooted in bodily adaptation.[115]

We are anchored to kinds of objects. They are not simply given, but require sustained investigations. But this is no panacea, and it could be misleading and complicated in terms of determining whether something is revealed by investigations. And most events should not be simply taken for granted as real, perhaps by degrees of stimulus sensitivity, or what Quine would later call "degrees of theoriticity."[116] As Peirce noted, there are no precognitive states,[117] and understanding is a core feature of pragmatic naturalism.

The cognitive apparatus is oriented towards objects carving out objects, responding to them, drawing reliable inferences, and the assessing of "cognitive significance,"[118] in the context of grounded reason.[119]

Reliable inferences take the place of certain ones in a context of what are commitments.[120] Projectable predicates and cephalic expressions that are embedded in adaptation cast the problems of induction in a new light, in a context of expectations that matter,[121] and in a broad view of inquiry that is not knotted to sensations and diverse dualism which undercuts inquiry and nearly all pragmatists eschew (e.g. fact/value, theory/practice, etc.). It is, nevertheless, tied to others and a sense of community, with social cooperative experiences as a normative goal.[122]

The grounding of induction in human cognitive expectations is put into a context of action.[123] Epistemic concerns are grounded in human decision making and action,[124] and in naturalized epistemology.[125] Knowledge is rooted in our evolving place in nature, in which inferences to a best explanation, or at least a good and satisfactory explanation, are made given time, circumstance, facts, and grounded possibilities. Mythologies of grandiosity are replaced with humility. At times, especially amidst the toil of experimental and theoretical labor, an unexpected satisfaction arises that an idea and the weight of the evidence might be so positively linked that a hypothesis must be brought to test and fruition.

What we have discovered is that context and background, along with the scaffolding to practice that expanded memory and capacity,[126] pervade human competence and performance. Worlds of cognitive predilection and ecological and social viability converge in everyday expectations and practice.[127] Memory reaches the boundaries of the literal brain both central and peripheral to the social historical milieu. Our cognitive capacity is enriched by the contours of our broader historical and ecological context; meaning and memory are strictly beyond the central nervous system into the outer world in which one is adapting, constantly forging new and meaningful connections.[128]

Dewey, certainty, and community

Myths of certainty, or, as John Dewey, icon of American philosophy, with sanguine progressive sensibilities, understood them, "quests for certainty," are replaced with frameworks that can be justified and inferences that seemed warranted.[129] Dewey felt that we often live "in a world of hazards" and feel "compelled to seek security";[130] the perilous pervades amidst moments of reprieve. Amidst the ontologically pervasive, the natural desire is to seek comfort in certainty, as a blanket cover against the endless precarious sensibility in which we find ourselves. The forms for this are as diverse as humans are as a species.

With no Archimedean starting point, just good heuristics of problem solving embedded in cephalic capabilities, we move forward. Edifices of stone are replaced with more labile edifices, changing frameworks, workable, usable, reliable, predictive, and anchored to self-correction, itself bound to inquiry and correction. The endless quest to escape the predicament, to quell the need for certainty, and to eradicate the choices that we must make, is the omnipresent urge to forge and live in an edifice of comfort.[131]

Dewey embraced this desire and coupled it with intelligent action and sustained humility. He may have exaggerated our conception of ameliorating the human existential condition of endless angst, as many of his critics griped.[132] But he sought an acknowledgment of our vulnerabilities amidst a steadfast conception of how ends blur into means of action – action replete with a thoroughgoing experimentalism, aimed towards controlled inquiry. He made no separation from other parts of our life; for him, it is just that permutations are a dominant experience, and they put the sense of control in perspective. It is no illusion. It is just never grandiose, and it need not be.

A thoroughgoing consequentialism amidst the test of ideas pervades Dewey's philosophical sensibility, where ends merge into means, and where action is guided by ideas knotted to a thorough sense of the experimental and the empirical. The testing of ideas is a normative goal in Dewey's experimentalism, tied to the reduction of the quest for certainty. As Dewey nicely expressed it: "The road from perceptible experience which is blind, obscure, fragmentary, meager in meaning to objects of sense which are also objects which satisfy reward and feed intelligence is through ideas that are experimental and operative."[133] Problem solving or intelligence are demythologized and naturalized in this intellectual landscape. The naturalizing of intelligence, viz., problem solving as anchored to human adaptation and human ideals of worth, is the normative goal of a thoroughgoing naturalism ripe for the human condition.

The "promise of pragmatism"[134] was undermining excessive authority and opening sensibility for endless inquiry, an affirmation of beliefs and action and judging the consequences. This afforded continuity for meaningful experiences, rich in possibilities and social sensibilities.[135]

What makes Dewey's naturalism relevant today is that it is not anchored to an overzealous reductionism in which one explanation dominates at the expense of all else.[136] The view of this pragmatic naturalist is that levels of explanation are a dominant way in which to understand our orientation to nature. We move easily to different levels of explanations, from kinds of objects to abstract entities one never sees but posits. Respect for nature entails a respect for levels of explanation, from the molecular to the molar. This is a very modern perspective.

Dewey's naturalism is also emboldened by his sense of the action of ideas and the continuity of philosophical perspective and science. His naturalism is embedded in his realism, which is bound to ideas in action: transforming and yet responsive to changing landscape, and to an expanding sense of social hope and social well-being, as well as an expanding sense of participatory democracy. It is a philosophical naturalism and realism embedded in developing predictable coherent forms of action.

Indeed, as one pragmatist philosopher of science notes with regard to Dewey:[137] Dewey understood cognitive systems as adaptations to environmental events, especially to the uncertainty that we face. Moreover, the vast instrumental capabilities of diverse cognitive systems that have evolved are linked to aesthetics (see Chapter 4). Cognitive systems evolved to cope with diverse forms of complexity.[138] Dewey's view covering the last 40 years of his life are quite consistent with modern views of philosophical naturalism, which are less about essences to be discovered and more about coherent action.[139]

Naturalism is demythologized in terms of use and human duties, human interests, and wondrous awe; gluttony and abuse are as readily available to our cephalic machinations as our control, inhibition, and considerations of future needs (despite the limitations). Clusters of properties hover around diverse forms of explanation,[140] which embolden our inferences – properties tied to a real sense of the growth of knowledge, and a respect for our limitations.[141] For all of which, a community of inquirers is the foundation,[142] international in perspective.

Core themes of pragmatism

The classical pragmatists were rooted in common core themes. Indeed, I believe these principles are woven into a well-warranted hypothesis.

- Undermining dualisms of diverse sorts, e.g., facts and values, cognition and action, perception and cognition, analytic and synthetic separation.
- Elaborating a sense of experience beyond classical empiricism, with its emphasis on sense data or modern positivism with the validity of statement being lodged in sense datum; James's radical empiricism, with its emphasis on relations, is rich in action, with self-control with cognitive systems embodied in human performance.

- Continuity in action, replete with human meaning and purpose, intelligence continuous with adaptation derived from biological consideration, philosophy continuous with but not reduced to science, an appreciation of nature and the continuity of animal and human cognitive systems, human meaning bound with human community, scientific and otherwise.
- A sense of the common sense of critical realism; abstractions were not more real than the objects encountered, in fact staying anchored to objects was a core theme. Experiments and experience are at the heart of human investigation. They cut across all avenues of human endeavor. The cultivation of intelligence is a lifelong endeavor, instantiated early on in educational settings in which the life of the mind can take place along vectors of possible and long-sustained value, in order to maintain democratic inclinations of value in a liberal democracy.
- Aesthetics is a fundamental feature of the human endeavor. It is not separate from human cognitive purpose or human activity, rather it is pervasive: the enrichment of human experience is linked to aesthetics and tied to our representative capabilities.
- Pragmatism is linked to other people, a cultural proclivity in which participation is at a premium. Participatory democracy is a key variable, a prized way of being, a fundamental part of a way of life.
- Hypothesis formation and correction are at the heart of inquiry and our evolution.
- Social cooperative behaviors are the cornerstone of our cultural productivity.
- Pragmatism is linked to an expanding sense of human capability and rights, tied to the deepening and depth of human experience, human empathy, and compassion.
- Pragmatism is about science and not scientism.
- While pragmatism is open-ended, the limits of human function put our advances into a perspective in which evolution of function is just as apparent as devolution of function.

Pragmatism was the alternative to positivism, a reunion of sorts of the big questions without the lumping into sense data and needless scientism: a breakdown of dualism of diverse sorts that ran the gamut of philosophical positions.[143]

Rejecting dualism, pragmatism requires lots of engagement of ideas expanded in use and continuous with the contours of science; but it is far more than a handmaiden to science. Pragmatism values clarity of positions. It is not just a way of tidying up the messes, but an integral part of the ongoing practice of inquiry that is open-ended, replete with signs to be interpreted in diverse symbols rich in metaphoric expression and linked to forms of grounding in which inferences can take place.

Value or appraisal is a richly cognitive activity embedded in a social network of meaning:[144] an axiological framework is at the heart of the knowing process. Growth, as Dewey and other pragmatists understand it, underlies the knowledge process. The pervasiveness of community and moral progress, linked to inquiry and social justice, are an ongoing process embedded in the experience of moral agents. With a moral imagination that is not detached but situated in context and community with a normative goal of participating individuals, the human inquirer is embedded in a sense of inquiry rich in valuation. Of course, what Dewey commented less upon was devolution of function. Not just evolution: devolution of function is also a formidable fact in our biology and culture.

Conclusion

What makes classical pragmatism so useful is its tie to inquiry, and the appreciation of science without reducing all knowledge or, simply, experience to a narrow one-dimensional view of science. A civic social concern for the good amidst a conception of the prominence of evolution and devolution immersed in our experiences is a common thread, a thread rich in the experience of philosophical concerns, socially emboldened by concerns that matter for an evolving culture with high expectations amidst the frailty of the outcome, their realization a faint glimmer amidst social hope.

Philosophical critical realism anchored to endless inquiry, with reprieves and rejuvenation, is a continuous calling for the classical pragmatist. Civic conscience, and an appreciation of the plural without squandering the individual search for human meaning, is the bedrock of pragmatic philosophical meaning. Evolution and devolution are recurrent features amidst attempts to integrate a philosophical critical realism with the many meanings of inquiry, social solidarity, and aesthetic sensibilities.

What motivated classical pragmatism was a naturalism without excessive scientism, which evolved into levels of analysis within inquiry. Respect for phenomenology, including the experience of individuals, is a cardinal feature of this view. Much in classical pragmatism always had a naturalistic bent tied to an expanding sense of human experience and the reconstruction of philosophical orientation, including devolved barriers towards this end.

As I have indicated, there are a number of strains of pragmatism.[145] The one that I am most comfortable with and find most useful is the more classical version, a version rooted in naturalism and critical realism, never completely naïve, though naïve enough to see intelligence grounded in problem solving, to root problem solving to endless human problematics.

Pragmatism is interested less in the mythology of grandiosity than in slugging it out in the trenches of problem solving. Pragmatism is understood as engagement and discovery through ideas in action as no mere abstraction.[146] The notion of the diverse sense of real events, etc., is the recurrence of pragmatism into the basic fabric of meaning and social connectivity, the naturalization of intelligence in morals and practice,[147] with practical consequences.[148]

2
C.S. Peirce: A Warranted Perspective on the Culture of Inquiry

Introduction

Charles Sanders Peirce is not only among, but *is* the greatest American philosopher. He glorified the community of inquirers, the termination of inquiry by labor and discovery, invention and application by the toils of the many. But he never could fit into any community; he was a community of one. Yet the margin and validity of science was the inter-community verification of the findings of others, the settled data that comes to be accepted by the intellectual lens of the community of inquirers.

Peirce is a fistful of contradictions, except one: he was, unequivocally, an exceptional mind. He was born into pedigree and privilege, a heritage that placed him in the intellectual womb of 19th century America – New England, more specifically Cambridge, Massachusetts. Around him were many great minds, including his father, Benjamin Peirce, the first great American mathematician, who worked on a new attempt to calculate quaternions, did original work in linear algebra, and was regarded as a mystic.[1] For Benjamin Peirce, all things were in mathematical space; as far as his mystical side, as his son, Charles noted, "he had a superstitious reverence for the square root of minus one."[2]

Benjamin Peirce was the co-founder of the National Academy of Sciences in the United States. Below, in Figure 2.1, with Abraham Lincoln in the middle and Benjamin Peirce on the left, is a mural of its founding.

Charles Sanders Peirce was introduced to privilege on all fronts.[3] After the American Philosophical Society, which, as noted in Chapter 1, was founded by Benjamin Franklin (an endless inventor on the practical side of the theoretical, Franklin was a prominent figure in the

Figure 2.1 Group portrait of Benjamin Peirce (far left) with fellow founders, including Abraham Lincoln, of the National Academy of Sciences of the United States of America, on 3 March 1863 by Albert Herter, ca. 1924

Enlightenment perspective in the United States and a pragmatist to the core[4]), the second major organization for the sciences on the grand scale was the National Academy of Sciences. Benjamin Peirce's son, Charles Sanders, was young when inducted into the National Academy in 1876 – just a few months shy of his 37th birthday.[5] He was elected for his work in mathematics. Such are the opportunities of privilege.

In this chapter, I begin first with a discussion of calculus or calculating objects in space. Peirce, a very modern inquirer, who set a standard for both doing and understanding inquiry in pragmatism, understood something integral about the genesis of ideas and what he called "abduction." Abduction is placed in a social context with social fields of meaning. It is ironic that an isolated individual with his personal plights (which we discuss in this chapter) would anchor human existence to social meaning. Thus, Peirce is placed in a context in which to understand something about him, before anchoring his story into the main narrative themes of this book, namely pragmatism, naturalism, and social cooperation. After all, Peirce's core notion of inquiry is social; its aim is what the community of inquirers, through rigorous tests, comes to accept. I then extend a perspective of my own, derived from Peirce, with regard to hypothesis formation, experiment, and semiotics, into the contemporary contexts (embodied numerical reasoning and enactive cognition) amidst a laboratory frame of mind, and social history of discovery and statistical inference, themes and issues dear to the heart of Peirce.

Calculus and logic

Peirce spanned the bulk of the 19th century (b. 1839) and died in the early part of the 20th century (d. 1914). He was a synthesizer of science. Broadly educated, he was as much at home discussing Platonic scholastics (he was a big fan of Duns Scotus, the scholastic realist) as Kant. He had read Kant quite carefully for years and years. He was particularly at home with mathematics and would remain so, especially with a new version of mathematics to which he made a significant contribution: symbolic logic, or the logic of relatives.

Leibniz, the inventor of calculus – something Newton called "fluctions" to predict objects' whereabouts and trajectories in space – also had a grand vision of reducing mathematics to a symbolic logic, as a more formal general typology to link all of the formal sciences. This idea led, in the 19th century and early 20th century, to diverse attempts to take all known mathematics and reduce them to a symbolic logic. Frege, Boole, DeMorgan, Schroder, and Cantor are well-known individuals in this field, and others include Bertrand Russell and his teacher and senior colleague Alfred North Whitehead. Russell was steeped in Leibniz's sense of the philosophy of mathematics. Russell, Whitehead, and others (e.g. Dedekind, Frege, and Hilbert)[6] attempted to formalize and reduce mathematics to symbolic logical terms in their great *Principia Mathematica*, an extraordinary three-volume work. Peirce, in this same tradition, thought of himself as a chemist and logician.

Peirce, a peruser of books, an iconoclast solitary irritant to the establishment, was a deep historian of the history of logic. He was well steeped in the Greek exposition of logic and mathematics through the scholastics and the modern era, and well versed in the works of mathematical logic from DeMorgan, Boole, and Whitehead. In fact, there was little in science with which he was not acquainted. While he fancied himself a chemist, he was in fact a mathematician, worked as a psychophysicist, and did psychology experiments.

Peirce's contributions to logic cover a broad array, including discussion of the leading mathematicians and logic systems, particularly what he called the Calculus of Logic, or the Algebra of Logic: exact logic, numbers, and the logic of relatives, the systematic connectives of events.[7] In fact, Peirce was always trying to show the diverse ways in which events are linked to the essence of reasoning.[8]

Peirce was prescient and well placed to overview what became dominant themes in the philosophy of science: the status of propositions, reasoning about inferences, a philosophy of induction, a formal representation of inferences to good adaptive explanations given the data, an orientation to hypothesis generation, something he called, variously, abduction or retroduction. He provided a logical place for the relationships between induction, deduction, and abduction. This concept can be found in all his works from 1860 to the end of his life in the first part of the 20th century. In fact, the kernel of his whole thought was abduction, how ideas emerge, their explanatory range, and their long-term consequences.

Pragmatism, or pragmaticism, Peirce regarded as the method to determine consequences writ larger, a move away from nominalism to diverse forms of realism, towards the realism of Scotus and Reid about kinds of objects and our knowledge of them. For Peirce, pragmatism is knotted to *abduction*.[9] Abductive events are at the heart of inquiry; abduction is about hypothesis creation. Abduction, in other words, is about hypothesis formation and testing. It is grounded in objects and action, as well as reference and a natural sense of objects.

The richer the hypothesis is, the larger the implication: the deeper the hypothesis, the better the understanding. Alogic of discovery was the driving force behind Peirce's pragmatism and the emphasis on abduction and hypothesis formation in his notion of inquiry.[10] For Peirce, abduction is "preparatory"; and induction is a "concluding step" grounded in natural outcomes.[11]

Peirce understood a lot about the science of discovery.[12] He was really a creative genius, who appreciated and demythologized the discovery process, and in doing so pointed towards heuristics, a toolbox of problem solving devices.[13] Peirce, who read Aristotle and Kant more than any other philosopher, offered an alternative to rationalism and deduction, as well as endless empiricism and induction, by introducing a term for hypothesis generation, abduction.[14] In more modern terms, search engines underlie inquiry.[15] They can be well grounded or not, but a lens serves as an orientation.[16]

Peirce was searching to understand "thinking machines" in his original work in logic and mathematics. His idea of "cerebration and the brain" and his tie to simple thinking machines is reminiscent of the first Turing machine. To move from Alan Turing, the great 20th century British logician, and his envisionment of machine knowledge, in reality very low-level computational devices, to the tiny microchip-driven devices that we carry in our pockets in such a short time period is astonishing. But we are in some ways no wiser than either Peirce or Alan Turing.[17] The pulse of the information age is our ever-widening fear of the same human nature with greater tools of destruction in the hands of more nations.

These machines needed to be embedded in the larger world of semiotics. Inferences to better explanations must follow something like Peirce's deduction, induction, and hypothesis. Peirce held that "settlement of opinion is the sole object of inquiry."[18] Of course, that is misleading since inquiry is broader, but ideally, as a normative goal, finding a level of agreement about the investigation holds for most forms of human activity (e.g. law). Peirce was in a perpetual state of modifying his categories and was, like Kant, continually asking questions. These are the cognitive conditions for knowledge acquisition that have been at the basis of modern philosophy.[19]

This otherwise socially inadequate person, unable to secure an academic position for most of his life, became esteemed by the philosophical community of his local institution, Harvard University. This community included not only William James but also, a bit later, Josiah Royce and Morris Cohen. Cohen, a Ph.D. student at Harvard and icon of City College of New York, eventually managed to present Peirce's work to the world, followed later by two wunderkinds, Paul Weiss and Charles Hartshorne, graduate students of Alfred North Whitehead at Harvard. Their more comprehensive job of rescuing Peirce's work, starting in the late 1920s and emerging in the 1930s,[20] finally resulted in the Harvard University Editions of Peirce's corpus. But Morris Cohen, an Aristotelian rationalist with a penchant for naturalism and history as well as logic,[21] helped educate generations of Peirce followers at the City College, many of whom would transform the cultures of leading higher education in the United States by embracing the larger scientific goal of inquiry.[22]

Logic was always the cornerstone of Peirce's treatises: how to improve thought and the rationalization of categories. "Firsts to thirds" was one of his central methodologies. Firstness was the quality of something. Secondness stood for some sort of resistance or struggle, and thirdness was an idea, a connection. Thirdness was the generality that we search for and find. Peirce had a lifelong fascination with the categories that render knowledge possible, echoing Kant and, in part, Hegel. Triads were a core common theme, with a predilection towards objects. This relationship to kinds of objects was a direct cephalic adaptation and a form of rooting the cognitive predilection and the objects encountered.[23] One list of his broad categories is:

Firstness: Quality
Secondness: Reaction
Thirdness: Representation

"The triads," as Peirce liked to call them, were for Peirce "the first of all the conceptions of philosophy are that of a primal matter out of which the world is made."[24] It is worth emphasizing that they are not separate. In fact, at the level of quality, appraisal is embedded in theory and purpose, and reaction is an ongoing affair with cephalic adaptations and capability and devolution and evolution of function in individual lives.

Reason and conduct were intimate; exact thought was paradigmatic of the new logic, symbolic and calculative. There was, for Peirce, clarity without subjective foundations, no Cartesian edifice of certainty, just real, live adaptations. Moreover, there was no detached starting point for the eradication of all doubt, and doubt for Peirce is contextualized as the prolegomena for beginning thought.[25] The naturalism and pragmatism of Peirce are the cumulative results of inquiry, logic, and experimentation, within something akin to an instinct hitting on the right idea: abductive moments[26] – an instinctive adaptation.

Hypothesis formation lies at the heart of inquiry in the drawing of inferences in the context of deducing consequences and inducing noted relationships. The causes and the consequences are embedded in larger based frameworks,[27] near what Whitehead called "misplaced concreteness" when referring to sensation in the context of legitimating scientific claims, as it would later be argued.[28] For Peirce and for many others that followed in the 20th century, the issue is not sensations for legitimacy, but the entrenchment in perspective in the knowing process tied to prediction and coherence. The degree of dependence is not absolute separation between analytic and synthetic statements,[29] but it is found in the integration of the sensory pull amidst the cephalic determination and lens for seeing relationships.

This was the power of *retroduction* or *abduction*. (Peirce should have simplified his categories instead of continuing to use different terms; he needed the edge of Ockham's razor. I will predominantly use the term abduction herein.) The making and the doing are not separate, as is the theory and action, sense data and frameworks; all cognitive thoughts have some fundamental link to sensory properties, as many of us would argue. What we call intuition Peirce understood as "immediate cognition."[30]

Peirce was anchored to "events," to a critical common sense fallibilism, and to many other terms that he used to describe himself. Scottish common-sense realism, fallibilism, and humility, and social replication, all these terms were used, over the long run, to denote whatever turns out to be a normative measure of truth. For Peirce, truth is social by definition, and we can, under suitable conditions, hit on the right hypothesis about an event because of the way cephalic capabilities are suited to discern and understand the events that we might be embedded in.

But within his essays, he also described the other side of his approach as that of "laboratory philosophers." Fallibilism, a term dear to Peirce and often used by him, was paramount to this approach. This is not surprising, because he also understood something about the new tools that were emerging in the statistical sciences, such as the properties of chance and probability that involved looking for long-run frequencies, determining statistical support, and flushing out the controls and the control groups from experimental design.[31]

One orientation was statistical inference, the flip side of which is probabilistic errors. Peirce, like all formal scientists, was searching for law-like relationships, except now the laws were couched in terms of likelihood, not necessity. Empiricism thrived amidst a rationalist predilection in the sense of inquiry that Peirce proposed. At his worst he had an inordinate fondness for making long lists of categories. At his best he anchored all thought to the actual practice of science (theory and experiment with invention and advance). But at the intellectual heart of his ideas was logic and inference based on probabilistic relationships. Here, Peirce was in the intellectual vanguard, well positioned, and part of the advances in logic, in statistical inferences, and diverse sciences.

In fact, he spent years measuring events for the government (a job his father helped secure for him). Protected by some of his bosses, reviled by others who found him impossible to work with, he had his hand in measurement and the practical side of applied science for the great bulk of his life. He held a job at the United States Coast and Geodetic Survey (again through his father, who at one point was the head of the U.S. Coast Survey as well as a professor at Harvard).[32] This was Peirce's main, secure occupation and source of income for some 30 years.

He measured a wide variety of "swinging pendulums" (see Figure 2.2)[33] for measurements in astronomy, and to measure surfaces of the earth and changes in the earth's gravitational pull. Peirce was lucky enough to witness a solar eclipse with his father and brother in Sicily. Benjamin Peirce's help and influence were omnipresent in the first part of his career, and allowed him into circles that he perhaps would not have gotten into otherwise.

The "Coast Guard Reports" about the solar eclipse in 1869, along with his other work with this government agency, kept him both employed and on the practical side of science. It is not surprising that he could

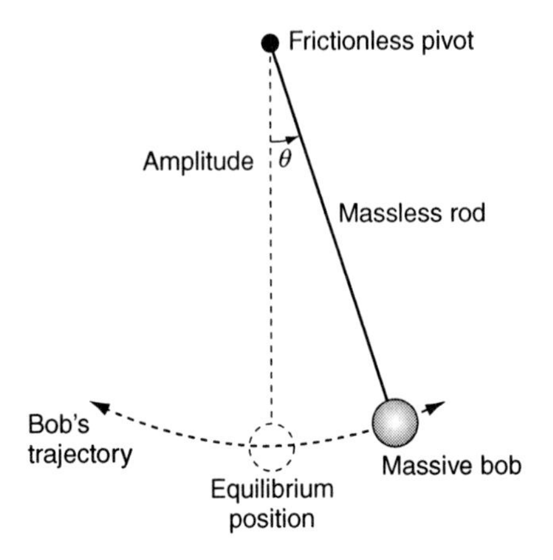

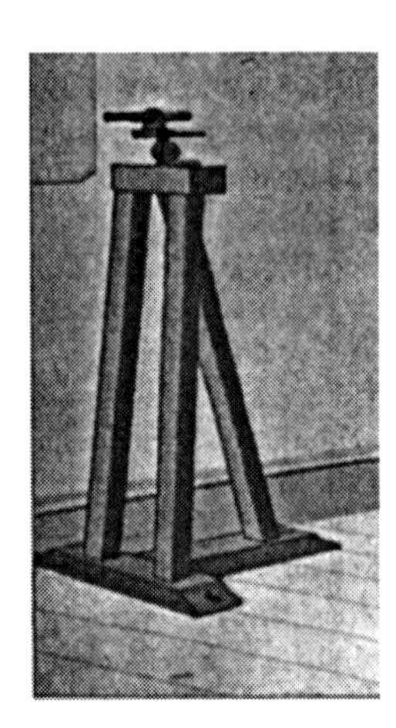

Figure 2.2 Examples of a pendulum

write for *Popular Science*, or that his philosophy would be grounded in action and in experiment. In fact, he was never very far from the experimental world. His reclusiveness and the sentiments of disconnectedness and discord which so alienated him from others were set in contexts where he gave presentations about observations of the solar system, gravitational pull, the practical side of logic, and the more formal side of the logic and structure of mathematics.

Depicted in Table 2.1 is a typical set of activities in Peirce's fieldwork exploring and measuring the earth with the use of the pendulum.[34]

But through most of Peirce's tragic personal life, he continually alienated those around him, and fell out of favor when new management moved into the bureau that paid his salary and supported his research. He had always been well-known for his cantankerous genius, but the genius part was now suspect.[35] He let fools know they were fools in no uncertain terms. His rationalism about his research findings on gravity was absolute. Peirce had the certainty of a mathematician and the theorizing of an experimentalist; he was quite certain, yet his philosophy was that of fallibilism. Like most of us, he was ripe with contradictory sentiments. But his sense of precision was quite beautiful, the insight of the curvature of space, the possibilities of diverse forms of geometrical representations quite modern, as they emerged from some his best work during the thirty-year period in which he received a salary from the government. An article in *Science* in January 1892 announced the separation of Peirce from the United States Coast and Geodetic Survey, mentioning in particular Peirce's work on gravity.[36]

Table 2.1 Typical log of Peirce's pendulum measurements

August	14–21	Measurements of length.
September	5	Swinging, heavy end down; knife, 3–4. Swinging, heavy end up; knife, 7–8.
September	6	Swinging, heavy end up; knife, 7–8. Swinging, heavy end down; knife, 3–4. Centre of mass determined. Interchange of knives. Centre of mass determined.
September	7	Swinging, heavy end down; knife, 7–8. Swinging, heavy end up; knife, 3–4.
September	8	Swinging, heavy end up; knife, 3–4. Swinging, heavy end down, knife, 7–8.
September	10–13	Measurements of length.
September	14	Swinging, heavy end down; knife, 7–8. Swinging, heavy end up; knife, 3–4.
September	15	Swinging, heavy end up; knife, 3–4. Swinging, heavy end down; knife, 7–8.
September	16	Determination of centre of mass. Interchange of knives. Determination of centre of mass. Swinging, heavy end down; knife, 3–4. Swinging, heavy end up; knife, 7–8.
September	17	Swinging, heavy end up; knife, 7–8. Swinging, heavy end down; knife, 3–4.
September	18–25	Measurements of length.

Hopkins experiments: origins of a laboratory in psychophysics

Peirce, it has been suggested, established the first experimental laboratory of psychology in America. It was at Johns Hopkins, where his students (e.g. Joseph Jasnow, Christina Ladd-Franklin)[37] worked with Peirce in psychophysics experiments (noticing small differences in sensation recognition), in the tradition of Wundt, which Peirce acknowledged. Later, Peirce followed Fechner and Helmholtz and the German experimental traditions that had emerged in the psychophysics of perception. He embarked on some simple experiments based on noticing small differences in tactile sensation and weight determination. James was quite vocal in promoting Peirce for the position at Hopkins.[38]

One of the key contributions that emerged from psychological discovery is experimental design.[39] In fact, the logic of experiment combines rational expectations with hypothesis testing. Peirce, much more than James, understood something about this from his work on the logic of psychophysics; but he was on to what would be the cornerstone of what later became experimental psychology. Both Joseph Jasnow and Christine Ladd-Franklin (who was a member of the Hopkins graduate group despite the fact that it was supposed to be an all-male establishment) were students of Peirce, who went on to become well-known psychologists. Ladd-Franklin did not formally receive her Ph.D. from Hopkins for nearly 35 years, despite publishing papers in logic (on the algebra of logic before she left, and eventually important papers in the theory of color vision). Had she been a man during that period, and so inclined, she would have achieved great academic success.[40] Moreover, she was well suited to interact with Peirce. She was a first-rate mathematician/logician and an experimentalist,[41] whose discoveries for an opponent process theory of color vision became the received wisdom in psychophysics. She would later suggest that Peirce, by the early 1890s, was beginning to "lose his mind."[42]

Peirce's experiments on tactile sensation and weight determination were well controlled for his day. Experimenters were not able to see which objects were heavier than others.[43] The design was, in part, randomized, with the weights varied in the order in which they were presented. The subjects assessed the weighted objects with blindfolds on, as measured by the pressure to the finger. Later, with his student Jasnow, Peirce started to look at probability of error and degree of confidence.

They provided the following formula:

$$m = c \log (p/1-p)$$

where m is the degree of confidence; c is the index of confidence; and p denotes the chance of being right. They used the equation in the evaluation of probability judgments.

The experiment did not actually work very well in the long run, but Peirce clearly thought as an experimentalist in its theoretical design.

Johns Hopkins modeled the German tradition in scientific research in the 19th century. Peirce was right at home there (see Figure 2.3). Close to Helmholtz and Herring in rigor of experimental mind and in the conception of unconscious inference, Peirce was nevertheless an endless irritant for College Presidents Gillman at Hopkins and Eliot at Harvard, while both, perhaps to their chagrin, acknowledged his talents.

Figure 2.3 Charles S. Peirce (second from the left) depicted in a mural located at Johns Hopkins University campus, which illustrates the university's beginnings in the late 1870s

Gillman, in a note to his board of trustees, would characterize Peirce as "proficient in more than one department of science" and offering the Hopkins students "the principles that underlie scientific truth."[44]

A work group consisting of a number of Peirce's students contributed to a publication about logic. All the major students contributed to the book, including Christine Ladd-Franklin.[45]

Peirce understood that "thinking is a species of the brain and cerebration is a species of the nervous action."[46] Peirce always noted that there are no precognitive events; for Peirce, degrees of cognitive systems underlie perception, attention, and action.

The theory of inquiry, of hypothesis testing, is rooted in this cognitive perspective. The fixation of belief is rooted in the organization of action.[47] The orientation is not simply reactive, but anticipatory as well as responsive to discrepancy with expectations. German cognitivism was part of Peirce's depiction of what he called "Scottish common sense realism." Indeed, Max Fisch, a scholar of Peirce, has noted that Kant and Alexander Bain were particularly influential with regard to the origins of pragmatism. Alexander Bain, less well-known, had written an influential book in the early 19th century on the will, linking beliefs to a tendency and strength of action.[48]

Indeed, we know that human organization is replete with anticipatory cognitive systems, most of which encompass the vast cognitive unconscious.[49] Action sequences are well orchestrated and are embedded in successful survival for both short- and longer-term expression.[50] When Peirce noted correctly that "it was impossible to know intuitively that a given cognition is not determined by a previous one,"[51] he was close to recognizing that cognitive resources figure in the organization of action, something John Dewey would, some 30 years later, note in his critique on the "reflex arc."[52]

Of course, action is often habitual routine. Memory, attention, and other cognitive resources are minimized during diverse routines. Cognitive capabilities are then recruited elsewhere in the ongoing action. It is the breakdown that helps generate further action and cognitive resources to learn, attend, and construe new resolutions and new forms of adaptations. It also effects part of the clarity required to determine the diverse consequences of the new set of hypotheses determinant in human action, especially in the long-term (e.g. "How to Make Our Ideas Clear").

Epistemology is rooted in nature. Since the categories are constrained by nature, the range of hypotheses has an instinctual component: they are not infinite but finite. Peirce pulled up reason by linking it to action for consequences and instinct for some basic responses. It is interesting, again, that Peirce mixed writing these philosophical articles with items for wider-audience journals. He was rooted equally in the practical world and in the rarified world of his father, with his roots in the Metaphysical Club at Cambridge.[53]

Part of Peirce's purpose was to establish warranted beliefs and legitimated habits of action, routines towards goals that serve diverse purposes. Science, like conduct, is established by expectations. These expectations are not reduced to simple sensations that legitimate action. Detecting noticeable difference in sensation is one thing; grounding all of epistemological legitimacy is quite another. Action is rich in cognitive resources; the automatic perception of events, the orchestration of action, vital for diverse social behaviors, has long been noted. One exception is that we have re-envisioned the motor regions below the neocortex with regard to the codification of action as we have expanded our notion of cognitive systems; there are many diverse cognitive systems that underlie the organization of action.

Discussions of classifications are exhaustive, exhausting, and exhilarating in Peirce's iconoclastic lexicon. What draws many of us to him are his deep insights into the organization of thought, amidst the endless invention of new terms. While he himself needed some intellectual pruning of his mental arsenal, he could be highly critical of the same trait in others.[54]

William James, given credit for shaking loose psychology in America, actually did no experiments. James was never an experimentalist, and that was part of what made Peirce turn away from James and drive his need to distinguish himself and pragmatism from Jamesian pragmatism. Pragmatism can be highly theoretical as well as experimental; consequences with regard to what works in the short-term can take the turn of cheap pragmatism, where what works is confused with what becomes true by rigorous hard-earned experimental failures and successes.

To the end (he died in 1910), James tried to help his friend Peirce secure an academic position (for instance, appealing to Eliot at Harvard, and Gilman at Hopkins). He also sent him money. Peirce, the eternal self-imposed outcast, tended to enrage everyone around him, and usually castigated those who tried to help him. Peirce thought of James as soft-minded and confused, and changed pragmatism to pragmaticism to keep the concept away from the foolish James. James, on the other hand, was supportive, trying to get Peirce work.[55] At Hopkins, Gilman eventually got rid of Peirce and would not engage him later, even after Peirce wrote to tell him he would dedicate a book to him.[56]

Nevertheless, his time at Hopkins was a creative period for Peirce, and he had a lifetime effect on his students there. His work on logic during this period was published; his experiments were some of the first to really use probabilistic reasoning. He also reasoned historically about great thinkers and their origins, and he established courses in logic and reasoning, in addition to core courses in history. However, he had, as I indicated above, a secure job as a national surveyor of land and gravitational influences until he was pushed out.[57]

While not a close student of Peirce's, Dewey was a graduate student at Johns Hopkins and under the aura of Peirce (Dewey did take a logic course with Peirce, but the two did not develop a tighter relationship beyond the classroom). It would be Dewey who would realize the social context for Peirce's ideas.

Semiotics and cephalic capability

Many of Peirce's works are a consideration of signs and symbols, community-packed social discourse and meaning, viz., human understanding. This is nicely captured in an exchange of appreciative letters between Peirce and Lady Welby (depicted in Figure 2.4), a British aristocrat interested in the problem of meaning. In the letters, core themes such as triadic relationships, quality, effort, and generality (or first, second,

Figure 2.4 Victoria, Lady Welby

and thirdness, respectively) are discussed,[58] as well as the problems of meaning, understanding and hypothesis testing.

She writes,

> My dear Sir,
>
> I have ventured to request Messrs. Macmillan to send you a copy of my book 'What is Meaning?', because if you do me the honour to read it any comments from you will be of special value in my eyes. I do not pretend to be able to follow the course of your technical arguments, being quite untrained in that direction; but I have constantly come upon points in your writings which have for me a keen interest from my special point of view. This is markedly the case in your contributions to the Philosophical Dictionary.[59]

> Your term Retroduction seems to me much needed. In my humble way I claim to be a Muser, though I see that entrance into the world of Musement needs – for [me?] at least – an ungrudging study of the conditions of a health exploration.... Of course,[60] "all thinking is performed in signs" and "a concept is intentional, – has meaning."[61]

Most of what they wrote to each other centered on the wide range of communicative functions in everyday life – semiotics, along with logic. For a period, they were intellectual soulmates, both outsiders to the academic intellectual life: he because he was endlessly obnoxious, she because she was born a woman.

But both were also drawn to eugenics and instincts. She writes to him:

> I venture to enclose a brief and closely compressed Paper which I was asked to contribute to the discussion on Mr. Galton's momentous Lecture on Eugenics (well reviewed I am glad to see in the 'Nation'). You will there see my interpretation of your statement that "reason blunders so very frequently that in practical matters we must rely on instinct and subconscious operations of the mind as much as possible, in order to succeed." But in *my* logic (if you will allow me any!) I see no great gulf, but only a useful distinction between methods proper to practical and theoretical questions. So then 'Never confound, and never divide' is in these matters my motto.[62]

Semiotics, Peirce thought, are never as precise as one might want; vagueness and generality are ingredients in the perceptual availability of everyday transactions. What permeate the cephalic social space are signs that serve as icons, indices, and symbols, as rich bearers of information – an information in which the knower and the known are in contact. There is no Cartesian space of separation in a fabricated rationalistic space. Rather, there is engaged activity, the signs rich in scaffolding towards conceptual sanity and expansion, an "extended mind."[63] One essential feature of cognitive events is forging links between events. Semiotics was the connective glue.

These ideas can be traced to the 1860s, along with fields of inquiry around semiotic meaning, behavioral expression, and human and animal understanding.[64] For Peirce, animals come prepared to discern prelinguistic meaning, which is pervasive and rich in signs. The range of cephalic capabilities is reflected in the richness and flexibility of semiotic expression. Evolution favored both specific local adaptations and broader, endless semiotic expression (as in our species). This, however, is something Peirce adumbrated but did not fully engage.

Peirce, unlike Dewey for instance, was no naturalist with regard to other species, though many students of animal behavior and ethologists recognize and make use of Peirce and his work in semiotics.[65] He was anchored to social discourse and logic, chemistry and exactness. His philosophy was tied to objects and consequences, broad considerations, but not a sense of nature. His thought still reverberates within our considerations of the natural world. His work, both experimental and theoretical, is always within the sciences. Not surprisingly, many ethologists of animal communication acknowledge his contributions to the natural understanding of semiotics.[66]

The different social calls, for instance, serve diverse roles in different species. Almost always, they are rooted in social functions. Like language and other forms of animal communication, the calls are semiotic, rich in information about territory, social groups, alliances, predation, danger, and resources, etc. The life-blood is embedded in the diverse cognitive systems for retrieving and signaling information. Our brains are prepared at each level of the neural axis to participate in the orchestration of, say, facial expression to the appropriate context, or, for example, the hooting sounds to the social milieu in which the chimpanzee finds itself.

Peirce did not glorify the language organ, a popular trope since the 17th century, but gravitated towards a toolbox full of problem-solving capabilities rich in a semiotic of meaningful space, natural and cultural.[67] It is ironic that a person who so emphasized the social milieu would himself be an endless outcast in the community of inquirers, a community in which social health, as maybe Peirce might have realized, is manifest and emboldened.

Peirce, the reviewer of books, aggravator of people, lover of inquiry

Peirce was a lifelong reviewer of books for *The Nation* and other periodicals. This probably served as one of the ways he acquired books, as he had little money for the last part of his life. His reviews, like his personality, are direct, clear, and, at times, brutally honest. Arrogant and self-destructive (he self-medicated with opiates and other pain killers, and like other individuals in his time such as Freud and the great Hopkins surgeon Halstead, he was addicted to morphine),[68] Peirce was as unlikable personally as his work was elevated. His deep knowledge of the conduct of inquiry, the development of good thinking, the laboratory arsenal to test out the idea, the invention of instruments to measure and expand our horizons of discovery,[69] all continue to demonstrate his extraordinary achievements as well as his own inability to use it for his own social good.

Peirce made anti-Semitic comments about Spinoza. He was a supporter of slavery and the South during the Civil War.[70] He was obnoxious, arrogant, and ungrateful. Penniless, with few friends, and excluded from the rich environment in which he grew up, Peirce looks pathetic from this point of view. The apparent heir to his father, he still thought of and advertised himself as the teacher of right thinking, and advertised correspondence courses for money on "the art of reasoning."

Given his noxious personality, why should Peirce matter? Because he outlined the contours of modern inquiry; he broadened scientific thought beyond narrow positivism, and he was prescient about this and much else that matters. There is simply no one else like him, with an approach simultaneously philosophical, technical, formal, and experimental. His elitism and his 19th century prejudices were misguided. While I don't believe in original sin as a theological concept, our social original sin is, in him, omnipresent. He was not "glassy eyed," though the metaphor of the glassy essence fits him.[71]

Peirce was bold. He chose to go after the greatest riddle: what are the key categories in knowledge acquisition? He speculated about the "law of mind."[72] Simply stated, ideas have an essential tendency to connectivity and connectedness – what he called "synechism." Peirce made sweeping non-empirical judgments about the mind in his later period,[73] and remained steadfast about a critique of determinism by invoking the possibility of chance, in an absolute sense in which chance is pervasive in the universe.

Peirce always understood science without pernicious scientism. He provided for experience but recognized the value of pure theory. Even if his elitism was repugnant and his lack of social grace was a repellent that deprived him of a stature that all knew he deserved, his significance for us today must not be underestimated. He was his own worst enemy, yet the rest of us benefited. The 20th century built on his insights.

Peirce, categories, and an historical sense of science

Peirce was of one of the few philosophers of science who was not only steeped in the laboratory and was quantitative and inventive in logic and mathematics, but he was also historical about the origins of science, its advancement, and the role of the individual scientist. He wrote extensively about the history of science. In fact, the logical foundations of science crumble if they are not couched in historical terms, as Peirce understood. Everything is to be anchored in historical terms with contextual insight.[74] For example, Peirce's curriculum for his Hopkins students included the history of the great thinkers, their temperament, their discoveries, and their tools and mode of discovery.

Peirce was enough of a rationalist, however, to separate logic from psychology or psychologism, a core feature of rationalism that underlies his orientation. But ideas are tied to laboratory skills and epidemiological concerns. A Peirceian orientation is historical, even about objects. It rests on edifices of cognitive predilections, while scaffolding on expanding practical and memory systems,[75] all while our universe of interactions expands through our instruments and ways of seeing objects and constructing events.[76]

Modern forms of rationalism, in which the process of knowledge rests on primitives with little discussion of the interaction of individuals within communities, are just remnants of the past in comparison.[77] Core categories about objects – e.g., whether they are animate or not, the predicative capacity of events, the spatial and temporal events, the belongingness of events[78] – are footholds into the world for interaction and coherence. Pragmatism does not mythologize events.

Moving beyond Peirce: numbers, natural frequencies, and cephalic capacity

Peirce rejected psychologism, but, ironically, he set up one of the first experimental psychological laboratories in America. I do believe that we discover real properties and are emboldened by a coherent framework of categories and social practices. What are categories and the basic practices that orchestrate, sustain, and nurture inquiry? These are the fundamental questions with which Peirce was in tune and that still make him relevant for us. Towards the end, Peirce's general courses were offered to a general audience of the educated and the eager. One wonders what the attendance rate was for the lecture advertised in Figure 2.5.

What those who attended would have learned is that evolution did tag, as Peirce suggested, cognitive adaptations towards objects, which makes their ready use and understanding available: a veritable set of cephalic heuristics.[79] We, in short, arrive at some ideas rather quickly.

Natural frequencies are more readily understood in diverse contexts. This goes for individuals with low numerosity capabilities, as well as

Cambridge *Conferences*

REVISED ANNOUNCEMENT

Mr. CHARLES SANDERS PEIRCE

of MILFORD, *Pennsylvania*

ILL give a course of Eight Class Lectures on REASONING AND THE LOGIC OF THINGS, at the rooms of the CAMBRIDGE CONFERENCES, Studio House, 168 *Brattle* Street, on Monday and Thursday Evenings in February and March, 1898, at eight o'clock.

Figure 2.5 An advertisement for Peirce's lecture on *Reasoning and the Logic of Things*

those who are number oriented. For example, physicians and patients are better able to grasp decisions about risk when the facts are presented as natural frequencies vs. probabilities. Subsequently, the depiction of distributions of events is more transparent, more figurative, less abstract, and perhaps easier to understand. Ecological validity or primacy links cephalic capabilities towards enhanced problem solving.

Anchoring our decision making to useful heuristics with a cephalic orientation about number and frequencies is very useful. The way problems are framed influences judgment and decision making. Keeping track is demythologizing the decision making by making it transparent and readily understandable for patient and physician. No easy chore, but knowing something about relationships between events – the origins of statistics being linked to keeping track of events[80] – it makes sense that health statistics can be used in this way, as a fundamental system for growing old, and for managing health.[81] One way we do this is by representing statistical relationships and the quantification of data in terms of probabilities and natural frequencies (see Table 2.2). Keeping track of events is expanded by the scaffolding to diverse forms of information.[82]

Table 2.2 Bayesian inference and information representation – natural sampling of frequencies and standard probability format

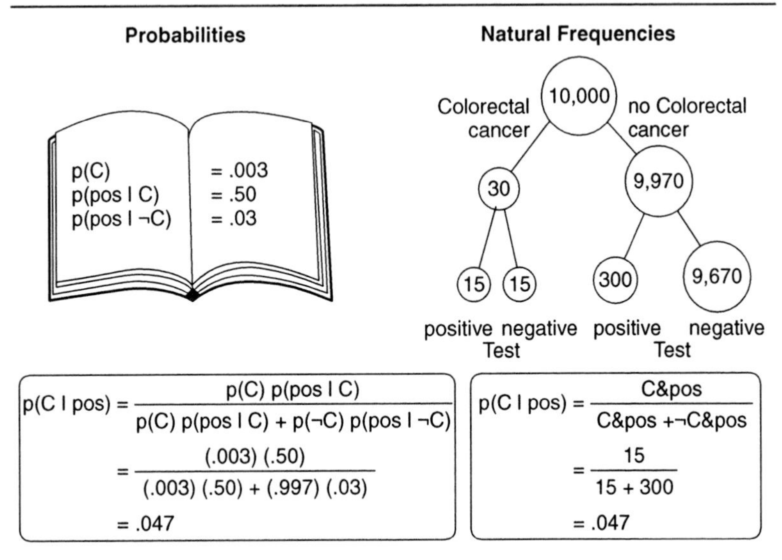

Source: Gigerenzer (2000).

Abduction in context, constrained by context and ecology, was something Peirce appreciated. Problem solving in context drove his whole thought process. Our sensory capacity is keen to detect objects that afford sustenance or harm.[83] Fast forms for detecting information can change the structure. Diverse forms of heuristics, fast ways to solve problems both specific and general, evolved along with the evolution of our brain.

We come prepared to associate a number of events linked by causal building blocks in cephalic structures by worldly events. *Ecological rationality* – viz., readily available heuristics well grounded in successful decision making,[84] as the classical pragmatists understood – places decision making and the use of statistical features within cephalic predilections about numbers and representations of frequencies in real contexts.

One feature is a sense of numbers as a general feature of the cephalic capability. Peirce spent a lifetime thinking about numbers, their application in inquiry, their status in our cognitive architecture. Numerosity is easily expressed, and easily triggered by exposure, as Plato demonstrated with regard to other mathematical abilities (geometry). But a predilection, as Plato noted in his depiction of the slave, never exposed to geometry, who nevertheless demonstrates some capacity for it, is just that. We are prepared quite readily to express degrees of mathematical competence. Unlike language, only a few are really good mathematicians, whereas most of us are good language users. The native gift of hardware endemic to our cephalic capabilities varies in kind and degree.

Indeed, reasoning about numbers is expressed early in ontogeny. It is a generative cephalic capability that serves many problem functions. An essential one is tied to numerosity.[85] One suggestion is that cephalic systems embody a "number sense."[86] Number sensibility is an essential of human cognitive capabilities, part of the brain. Moreover, mathematics more generally is linked to our systematic instincts to action. It is continuous with cognitive exploration and problem solving. This is certainly consistent with Peirce's view of numbers.

As Lakoff and Núñez suggest, "mathematics is a systematic extension of the mechanism of everyday cognition,"[87] and thus mathematics may be the most general form of abstraction of human cephalic capabilities (see Table 2.3).[88] It resides in commonplace features, is rooted in objects

Table 2.3 Some properties of external objects that are characteristics of mathematics

Universality	Just as external objects tend to be the same for everyone, so basic mathematics is, by and large, the same across cultures. Two plus two is always four, regardless of culture.
Precision	In the world of physical subitizable objects, two objects are two objects, not three or one. As an extension of this, given a sack of gold coins there is a precise answer to the question of how many there are in the sack.
Consistency	For any given subject matter, the physical world as we normally experience it is consistent. A given book is not both on the desk and not on the desk at a given time.
Stability	Basic physical facts – that is, particular occurrences at a given time and place – don't change. They are stable over time. If there was a book on your desk at 10 a.m. this morning, it will always be the case that on this day in history there was a book on your desk at 10 a.m.
Generalizability	There are basic properties of trees that generalize to new trees we have never encountered, properties of birds that generalize to birds yet unborn, and so on.
Discoverability	Facts about objects in the world can be discovered. If there is an apple on the tree in the backyard, you can discover that the apple is there.

Note: See Lakoff and Núñez (2000).

and the physical world of transaction, and is embodied with enactive cognitive adaptation.[89]

Mathematics pervades everyday forms of reasoning. It is inherent in the way we structure our world, from space and time to predictions about dangerous or opportunistic events. It is taken for granted because it does indeed pervade so much, as the Socratic self-disclosure of the slave suggested.[90] As Lakoff and Núñez propose, mathematics involves, in their words:[91]

1. *The embodiment of mind.* The detailed nature of our bodies, our brains, and our everyday functioning in the world structures human concepts and human reason. This includes mathematical concepts and mathematical reason.
2. *The cognitive unconscious.* Most thought is unconscious – not repressed in the Freudian sense but simply inaccessible to direct conscious introspection. We cannot look directly at our conceptual systems and at our low-level thought processes. This includes most mathematical thought.
3. *Metaphorical thought.* For the most part, human beings conceptualize abstract concepts in concrete terms, using ideas and modes of reasoning grounded in the sensory–motor system. The mechanism by which the abstract is comprehended in terms of the concrete is called *conceptual metaphor.* Mathematical thought also makes use of conceptual metaphor, as when we conceptualize numbers as points on a line.

Ordering objects and keeping track of what comes next is a fundamental cognitive adaptation, perhaps shared with other species, but only fostered into mathematics by humans.[92] A sense for numbers is cephalically mediated and is expanded by memory capacity (often with external props).[93] One region linked to numerosity is the inferior parietal region of the neocortex.[94] This region is tied to a number of behavioral functions, including movement and intention.[95] For example, in one interesting experiment (using fMRI), activation of the caudal region of the parietal cortex suggested that spatial representation and tracking events by eye movements are used for numerosity, or simple mental arithmetic.[96]

Meaning and social practice

Peirce well understood – whether under pragmatism or under pragmaticism – that an understanding of our ideas is part of determining their meaning. Meaning is social: it is outside the confines of the cephalic systems of individuals.

Meaning is part of the social fabric of the community of individuals that participate in the culture of inquiry. The users of tools and their products pervade in the social milieu; both are embedded into the practices in which we are steeped and extended across the cultural milieu in which we participate.[97] The social space is, in turn, embedded in meaningful social practices to which the cephalic capabilities readily and effortlessly assent. Peirce emphasized that the social context is the edifice of social knowledge. Knowledge is embedded in communities and not isolated, anticipating the later Wittgenstein. But, unlike Wittgenstein, Peirce understood the practices of science and not just the logician's formalism that they both understood and appreciated.

The importance of why Dewey held out less for inquiry in the abstract, and more for context and meaning, importance and value, was in part because he held on to a form of critical realism. Peirce labored in the abstract, but he also grounded inquiry into something real and tangible. Central to this is the recognition that metaphor is an important part of our cognitive arsenal.[98] Metaphor is a core way in which we navigate relationships between events. Describing everything as metaphorical, as some might, misses the important point of metaphor in the way we explore events. Metaphors are vehicles for engagement and expansion. They are expansive insofar as they are part of the cultural enterprises of inquiry and discovery, invention and expansion.

Laboratory frame of mind and social theory

Critical realism is a form of acknowledgment of the social context, not naïve realism – that things just exist (just look, there it is). Peirce understood that few objects or events sustain epistemological scrutiny of this sort. It was a bad metaphor in which to ground the acquisition of knowledge, the process of inquiry and the scientific ends to which we strive. The cognitive adaptation is that access abounds across the continually expanding informational sources. The hard work of inquiry is the realization that there is no panacea; no myths that a technique and abstract conception can yield axiomatically an expected truth for discovery and inquiry. The hard work of inquiry is the endless labor, the tenacity, as Peirce liked to call it, the toils and musement of the mind – the work amidst the play.

Social theory embedded in informed history of science was very much what Peirce had in mind with regard to understanding inquiry. The sociology of science fits into the larger context of understanding social meaning, social contact, and forms of inquiry and endeavor, as we move away from the grand edifice of abstract unity.[99] No Platonic fantasies in which structure is divorced from social context, and understanding is understated. Social theory is an achievement brought about by discourse and inquiry, the combat and play with ideas, and the continual testing of ideas and how far they can take us.

An anthropology of inquiry investigates our core orientations to events.[100] We come prepared to interpret events, to anchor our projects to objects. The cognitive predilections lean towards objects amidst the cultural advance, or they do not. No neutrality or Archimedean starting point amidst moderate relativity and critical realism exists. Soft areas are pervasive in the knowing process. Gray areas are cultural pillars, and variation in expression is mediated by cultural interests, orientation, and values. The pragmatist always gravitates towards teleological explanations with regard to human behavior, to modern variants of teleological realism with recognition of the under-determination of theory to evidence in many contexts.[101]

Forms of inquiry

Inquiry is not neutral. Even the most banal of orientations, without mounting interests and concerns, reflects predilections or orientations. Peirce understood that seeing is from a perspective, not in a vacuum tied to an interpretative or hermeneutical framework,[102] but anchored to critical realism. Kuhn's insight, and Hanson's at the end of the 1950s, was to point to the institutional framework of science, to see from within an orientation of meaning.

Perhaps we need to think about C.P. Snow in the context of the social milieu. Snow was concerned about two cultures, one having to do with the humanities, another with science. But he lumped all the sciences together[103] – a common predilection to reduce everything to an encyclopedia of science,[104] or a unity of science of one kind or another, or simply a natural predilection to find unity in explanation. A rationality that builds across the disciplines is a worthy goal, as long it is placed in an historical context.[105]

But explanations vary with subject matter,[106] and the main point of a pragmatist position is to broaden the sense of self-corrective inquiry and not to reduce it to one common denominator. The natural and social sciences vary across and within their subject. A plea for tolerance and understanding is a normative goal of three distinct orientations, one housed in the natural sciences, another in the social sciences, and a third in the humanities.[107]

The normative goal, pragmatic in spirit, is to acknowledge the interpretative orientation that pervades our practices.[108] Anthropology uncovers the many disciplines of human investigations amidst the background of a broad view of nature, understanding, wonder, and the senses of the sublime.[109] The hermeneutical orientation rounds out an important feature of our species: the cephalic structure, the cortical mass interpreter with narrow and broad interpretations, the experience that we have that is rich in semantic networks, scaffolded to the many worlds we inhabit.

Indeed, the methods of the sciences vary, as does the subject. As Aristotle noted and Dewey acknowledged, there are many disciplines and variations within disciplines. Competition is rampant, although cooperation is a regulative goal. The practice of inquiry is a cultural evolutionary process in which there is an invisible hand,[110] with clashes of competition and cooperation, dominance, variation, and limitations. Secrecy can be as dominant as cooperation. Egocentric orientations predominate, as in any expression of human endeavor. The Kuhnian framework is institutionalized in settings that are broad based, amidst narrower communities in which there is little settled opinion, and groups may be in or out (e.g. part of the cognitive sciences).

But the Kuhnian sensibility that came to dominate and demythologize the sciences by setting them in a social/historical context was essential.[111] It was a small prophylactic against scientism and a one-dimensional, ahistorical, rationalistic conception of science. This is a chronic danger for a cephalic bent built rightfully on bedrock security and stability in an idealized universe. Formalism lost in a search for foundations can predominate in an intellectual landscape that lands on a lonely, nearly empty bastion. The turn towards the historical amidst a respect for the formal was a welcome thaw and relief.[112]

The acceptance of the sciences and avenues of inquiry continues to turn the dominance of physics as the paradigm of science and discovery, so dominant in the philosophy of science, which began to turn around some 25 years ago to a much broader view of inquiry.[113] This is something that Peirce adumbrated and Dewey lived. Peirce understood science; Dewey lived the life of inquiry, the self-corrective sensibility in which the many kinds of investigations could be engaged.[114]

Conclusion

Nobody, to my mind, is like Peirce at the heart of American inquiry: broad and bold, creative and rich in contradictions, with a laboratory frame of reference, emboldened with remarkable intellectual powers, albeit with endlessly poor social judgment and capabilities. No community for Peirce, an *N* of one.

Surely, he did understand and respected the historical sense of science, as well as the constraints of hypothesis formation (abduction or retroduction).[115] All inquiry constrained by frameworks utilizing diverse forms of statistical inferences and underlying the logics of discovery are very Peircian indeed,[116] and quite modern. His epistemology was emboldened less by the issue of correspondence than coherence and the fruits of labor in inquiry.

What are the inroads, the possible new discoveries amidst some anchor to a possible state of convergence? Inquiry by holism or orientation, in which a large segment of the knowledge is embedded, but not so embedded as to descend into a form of Cartesian space of sentences or beliefs that belies our embeddedness in worlds of adaptation and coherence, in action and contemplation. Peirce was anchored to kinds of objects constrained by ecology and social history. A sense of progress in science marked by insight and discovery, and the rejection or acceptance of hypotheses, is something Peirce understood.[117]

Peirce understood the modern sense of inquiry, of broadened science, yet he remained tied to a laboratory sensibility, and was endlessly bold and theoretical. A critical realism is historical to the core in which context is tempered by artificial and individual variation,[118] and in which the contour of evidence is marked by the machinations of human cephalic capabilities and limitations.[119]

Peirce understood the practice of science, and the scope of progress amidst the fantasy of settled judgment in some very distant, and (perhaps, if he looked more closely) never-ending sense of what is agreed upon, or degrees of agreement and settled beliefs, tempered by habits and action amidst communities with overlapping interests. The emphasis is on others. The orientation is outward amidst creative hypothesis generation and critical testing by oneself and others. Evolution figured for Peirce in establishing a cephalic capability for hitting on the right hypothesis; the background condition for a hypothesis being constrained by evolution and social history, and of a critical realism tempered by the resistance and rejection of hypotheses.

Peirce unfortunately did not get to enjoy what should have been his: a settled and peaceful role in the community of inquirers. Cantankerous and difficult, he isolated himself from nearly everyone he encountered. His legacy is nevertheless a great one.

3
Evolutionary Origins: Oriented to Kinds

Introduction

What constitutes an object is an idea at the heart of what it is to know anything. We take it for granted, but the old adage "no framework, no knowing" permeates the epistemological landscape of coming to know and what can be known.[1] The ongoing debate in Western philosophy between rationalism and empiricism is often couched as a radical separation between object and subject, but perhaps a better way to frame the dilemma is by degree rather than division.[2] After all, the consideration of what something is, of whether it is a member of a category, requires having an idea and determining its consequences, at least to some degree, just as inducing a category from many instances requires having some idea of what the thing is. This is part of the pragmatist predilection: we are anchored to things but within a form of critical realism.

The term "natural kinds" does not originate with Quine's famous essay,[3] but the term gained a lot of mileage as a function of Quine's philosophy in some circles. It captures an important concept, coupled with the empiricist philosopher's idea of an innate "quality space." In fact, the root of the concept of natural kinds is anchored to our biology, to our feel for natural objects, and to adaptation to our surroundings. It is the surroundings of linguistic terms, words learned by association with stimulus events, and not objects alone, that provide the framework for what I have been advocating as a critical realism.[4]

The human philosophical tradition of naturalism is very strong[5] and, in one of its later manifestations, is a core feature of classical pragmatism. A sense of objects is rooted in our cognitive architecture, tied to our basic social and ecological adaptations. We distinguish natural kinds such as fruits or mammals from artifacts or human-made objects easily and early in ontogeny.[6] The induction mechanisms also operate readily within sets of discernable objects. The foundation for discovery is set within cephalic capabilities and suitable social contexts. It is just easier for small children to draw inferences about categories from a wide variety of objects.[7]

Core and newly resurgent features of classical pragmatism include:[8] preserving a strong sense of a landscape and of pervasive real properties in our interaction with objects and people; to sustain a keen sense of the intersubjectivity of inquiry and social contact; to be reminded of our direct connections to a range of incontrovertible objects, and kinds of objects; and to acknowledge the degree of theoriticity and levels of analysis in the knowing process. Resurrecting "critical realism" and never "naïve realism" is a sane, demythologized conception of kinds. Recurring themes in the ongoing process of inquiry and investigations are puzzlement and discernment, awe and dumbstruck confusion, clarity, and some closure.

This chapter is very much grounded in evolution, and the basic human sense of categories linked to objects and kinds, before considering the philosophical history of the rationalism/empiricism debate from Aristotle through the classical pragmatists. I will discuss some ideas about cognitive coherence amid a core sense of objects and our transactions, embodied and enactive cognition within the social milieu in which we are connected with others with social cooperative orientations, self-corrective inquiry, and our evolutionary capabilities.

Evolution, change, and adaptation

Evolution is often popularly used as a synonym for glorified progression, but its key themes of variation and reproductive success are more like biological variables than straightforward pathways. Evolution via natural selection is really about adaptation to niches, the opportunistic use of specific functions in more varied environments, and the development of novel functions for existing structures and molecules.[9]

Conceptions of geological and climate change already flew through the intellectual air that Darwin breathed. Charles Lyell's book, *Principles of Geology* (1830–33),[10] devotes several chapters to both concepts. As well as geographical and climate variation, species differences and even the idea of species extinction had become core concepts, as had the ideas of competition, survival, resource availability, and allocation.[11] All these themes required a notion of kinds of objects, and a predilection for taxonomy.

Biological investigators of the 20th century[12] have acknowledged the continuing importance of these 19th century foundations in the establishment of the modern evolutionary perspective by which we understand the rest of biology. Two factors especially, natural selection and variation in adaptation, remain key ingredients. Climate change and geographic/geologic shifts still shape our view of evolution as well.

At various points in our ecological history, there appear to have been bouts of rapid climate change that resulted in extinction of species (e.g. 2.5 million years ago). There may have been another moment of rapid change within our own lineage that took place about 100,000 years ago,[13] possibly due to a greater preponderance of unstable environments. So evolution as a whole, and our evolution in particular, is replete with both change and stability. That possible climatic change at the 100,000 year mark may have facilitated dispersal of hominoids across the diverse regions we now inhabit, from the Arctic to the tropics, from deserts to rainforests.[14]

Geographic isolation also remains central to our understanding of variation, extinction, and speciation.[15] As Darwin noticed, and as we can still see clearly today, the same species, separated by space and habitat in the Galapagos Islands, produced two very different adaptations of the iguana: a land-dwelling animal and an ocean-adapted animal. Geography coupled with habitat selected for very different forms of adaptation.

Evolutionary theory also puts into context something about our original condition of knowing. The number of genes in the human genome turns out to be much smaller than was thought, and humans share 98% of our genetic code with the other primates. Although no known precursor exists for syntactic competence in such primates,[16] most primates are social animals and exhibit gestural responding, keep track of the social milieu, and have other communicative elaborations which are significant to evolutionary theory and perhaps to our own linguistic competence.[17] In hominid evolution, evolutionarily abrupt changes seem to have taken place in the rapid expansion of symbolic expression, as depicted in Figure 3.1.[18] Moreover, we have some evidence, including fossils, tools, and skeletons, that more than one modern form of the human species populated the earth at the same time, competing and perhaps cooperating for survival (see Table 3.1).

Homo sapiens came to dominate the landscape as other human-like primates became extinct. In other words, the evolutionary perspective suggests that a wide range of human-like primates were existent at the same time,[19] competed, and perhaps interbred.[20] Some 30,000 years ago

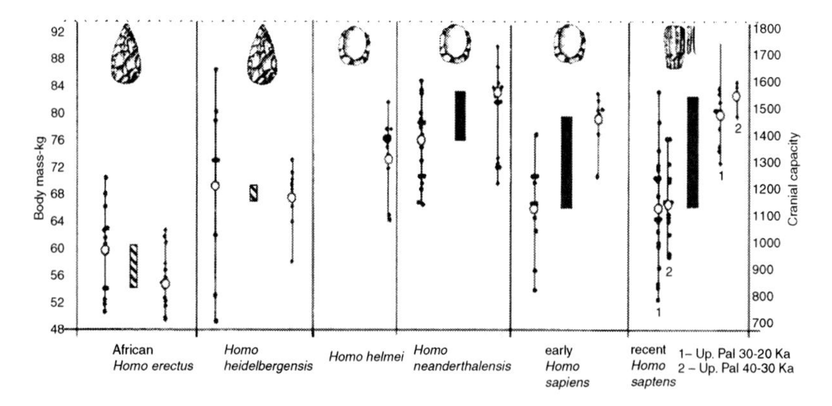

Figure 3.1 A modern depiction of variation in body mass and cephalic expression in hominoid evolution; a number of variants of related hominoids in close proximity[a]

Note: [a]Lahr and Foley (1998).

there is evidence to suggest that Neanderthals and *Homo sapiens* co-inhabited different but overlapping geographical locations in which climate change was a significant factor.[21] From these sorts of interactions, it is clear that evolution has produced a wide range of primate- and human-like kinds, out of which emerged modern humans, as described in Table 3.2.

The development of language – a symbolic representation of the world in which abstract sounds stand for objects, the relations between objects, and the activities in which objects play a part – may have been the beginning of our desire to categorize. From hunter-gatherer to agrarian, to industrial times, humans have tried to lend structure to what they experience and to preserve it in some way. Dwelling together led to the sharing of workloads and efficient task completion, which afforded individuals more discretionary time. With this extra time, at some point, humans began creating visual representations of natural objects in the form of drawings and reliefs, some of the earliest of which have survived in the deep caves of south-western Europe – see, for example, Figure 3.2.

From this simple beginning in these caves, of charcoal on a bare rock face, our ideas about nature have evolved with our culture and our ways of understanding.[22] Early humans originally understood nature through two primary categories: living and non-living, or animate and inanimate.[23] An ancient cosmology based on the four elements (earth, air, fire, and water) eventually gave way to lifeless atomism, ethereal

Table 3.1 Some cognitive and culture features of diverse and related modern species in the context of a common modern ancestor; social features essential for group formation are pervasive[a]

Hypothesized "cultural" properties of hominid taxa and some empirical bases

	Common ancestor and australopithecines	***H. ergaster***	***H. heidelbergensis***	***H. helmei, H. neanderthalensis***	***H. sapiens***
"Cultural" inference		Greater planning depth	Parental care?	Greater planning depth	Strong ethnicity
		Imitation		Ethnic affiliation	Extensive symbolism
		Limited innovation		Symbolic thought	
		Emotional affiliation?		Language	
		Theory of mind?			
Observation	Communities	Significant meat eating	Fire?	Flexible technology	Local networks
	Fission–fusion	Delayed growth	Modern life	Regionalization	Cultural replacements
	Male kin bonding	Complex tool making	History?		Rapid change
	Territoriality	Persistent traditions			Material diffusion
	Political alliances				
	Basic tool making				
	Ephemeral traditions				

Note: [a]See Foley (2006).

Table 3.2 Evolutionary timescale of modern humans

150,000–200,000 BP
Initial emergence of anatomically and genetically modern populations in Africa
110,000–90,000 BP
Temporal dispersal of anatomically modern populations (with Middle Paleolithic technology) from Africa to southwest Asia, associated with clear symbolic expression
80,000–70,0000 BP
Rapid climatic and environmental changes in Africa
80,000–70,000 BP
Major technological, economic and social changes in south and east Africa
70,000–60,000 BP
Major population expansion in Africa from small source area
ca. 60,000 BP
Dispersal of modern populations from Africa to Eurasia

Source: Adapted from Mellars (2006).

forms, and mechanical bodies. Now, lifelike properties infuse both animate and inanimate elements of the world,[24] as we seem to be swinging back to animate nature.[25] One philosopher, who wholeheartedly embraced this view of animate nature, of "nature alive," was Alfred North Whitehead.

Whitehead is sometimes linked to pragmatism and pragmatists.[26] He stated that he understood some of his philosophical "preoccupations" to rescue their type of thought from the charge of anti-intellectualism.[27] But his naturalism – as he long construed nature in terms of "nature alive," something basic to what Scott Pratt has linked to "primitive pragmatism"[28] – teamed with proclivities for which our continuity is made manifest, not by atomism and the mechanical philosophy of inert matter, but by a nature alive and emboldened by our transactions with nature. A form of groundedness to natural objects becomes apparent, as does a predilection to understand them. In the new biology were key features linking him to Dewey, as Whitehead acknowledged.

An "Adventure of Ideas," beaming with abductive moments, heralded the new philosophy of the organism. It was adumbrated somewhat, Whitehead thought, by Locke, the tepid rational doctor.[29] It was the "living person," as Whitehead put it,[30] that we needed to capture. It was in the context of a philosophy of the organism that he thought was the

Figure 3.2 Venus of Willendorf; discovered in Austria, on the banks of the Danube, this figure represents one of the earliest depictions of the body produced by mankind dating back to 24,000–22,000 BCE

essential move. In another text, Whitehead would characterize his own thought as "provisional realism."[31] Whitehead's "organismic philosophy" was very much under the influence of an evolutionary perspective and very close to the organic sensibilities of the pragmatists.[32]

Thus, our sense of nature evolves with our understanding, and has recently taken what is, for many, a gratifying turn towards an appreciation of wild places as paradises discovered.[33] The sensibility of the naturalist pervades the ways in which nature is depicted.[34] It can engender an appreciation for the diversity of species amidst a sense of the unity of nature and the process of evolution itself. At the heart of both naturalism and pragmatism, I would suggest, is this core appreciation.

In the 19th and early 20th centuries, many theorists envisioned evolutionary progress as taking place gradually.[35] They theorized that as animals progressed at tasks, they became smarter and better adapted to their environments. Each stage in evolution was seen as linked to the past, building slowly and steadily from there. In the process, species transformed and new variants emerged. However, we are now fairly sure that evolution is not always slow and gradual, nor is it necessarily progressive. In fact, change might vary radically in rate,[36] indicating a major difference from preceding events, as seems to be the case with linguistic competence, as described above.[37]

Evolutionary theory is filled with gaps and perhaps more speculation than one would find comfortable in other scientific contexts. The gaps, however, fit within the bounds of other evidence within the biological sciences (e.g., genetics) that links humans to earlier species, and evolutionary considerations remain as the background theory to our understanding of the world.[38]

From that first pointing finger or primate hoot, through cave paintings all the way to Darwin, we have come to utilize categories for understanding the changes of living things.[39] Nature's purpose, while not intentional, is expressed in wondrous functional adaptations, which, in turn, shape our perceptions of how the world is and what is in it.

Representational capacities

What emerged in our species was a highly linguistic, tool-using, social animal, with an elaborate diversity of cognitive skills.[40] Early humans developed many forms of instrumental expression and tool use. To go with these, we engendered numerous cognitive adaptations, some broad and some specific. Cognitive expansion and fluidity, in which narrow adaptive abilities expand in use across diverse problematic contexts, have become a signal feature of the human mind.[41]

Language is a unique feature for our species, with its rich syntactical elegance and expansive capacities. Once syntactical language use emerged, our cognitive abilities seem to have increased in great measure,[42] particularly our social discourse.[43] The premium of our evolution is on our social contact and discourse, and our communicative competence and praxis. This social contact and pre-adaptation, in turn, is vital for the formation of basic regulative events that traverse a wide range of reward within the behavioral biology of our central nervous system.[44]

The cognitive capacity for representation probably started in our interest in edible objects: thinking about them, planning for and predicting their use. We do not tend to consider the roots of language and art as being planted in such humble beginnings, but the capacity to represent objects of diverse kinds is central to the evolution of the primate brain. Cognitive/behavioral capacities are, after all, primarily ways to engage the world, to compute probability, and to assess friendly or nonfriendly events.[45]

Our representations are also often social in nature. We are interested in others, and what they do. The representations we make are not simply pictures divorced from life, but active ways in which we engage the world.[46] The outside world, the social world of others,[47] is important to human flexibility, survival, and reproductive success. Representation often has the connotation of something divorced from the object, something that detaches or is detached – though, of course, some forms of representation are indeed that way. To be sure, my use of representation is not that of Fodor, whose nativism is utterly divorced from history, context, body, and evolution.

But representations of others, of those whom we care about, do not divide us from others. Rather, they guide the organization of cognition, in embodied and enactive cognitive and informational systems that are inherent to the organization of action.[48] When we consider others, we want to comprehend their beliefs and desires, the way they are oriented, as well as to what they are oriented, the tools they use, etc.

In terms of the mechanics of this representational urge, the traditional view of the brain sees the cortex as its only cognitive part.[49] However, cognitive systems actually run across the brain – this is what I aim to convey in talking about *cephalic* organization, etc. From cortex to brainstem, the central nervous system is knotted to rich information processing resources.[50] Likewise, our social discourse, transactions, language, eating, and tool use are embedded in the social milieu and not divorced from it. Things like our motor capacity, our visual skills, and our social orientation evolved together.

This dense, interdependent cognitive network we call humanity is really a fairly recent phenomenon. We – modern *Homo sapiens* – only emerged about 100,000 years ago. Agricultural society, the move away from a hunter-gatherer one, developed in the range of 10,000 years ago. The domestication of dogs and the creation of written language emerged something in the order of 5,000 years ago:[51] so amazingly close in time to where we are and yet seemingly so distant from the complexities of contemporary life.[52] The common evolutionary feature through those years was cephalic accessibility.[53]

Expanding cephalic capabilities: tools and cortex

Pregnant within the brain are diverse and rich information processing systems, reflecting expanding cognitive capability. Many regions of the cortex are tied to motor function with diverse forms of enactive cognitive capabilities,[54] and the motor regions are gravid with cortical functions.[55] The wide range of tools, such as those depicted in Figure 3.3, that we have used in adapting to our environment reflects this evolution of cortical and subcortical systems in the brain.[56]

Tool construction and cognitive/motor systems are both essential to the evolution of the brain.[57] Tool use is an expression of an expanding cortical motor system in which cognitive systems are endemic to

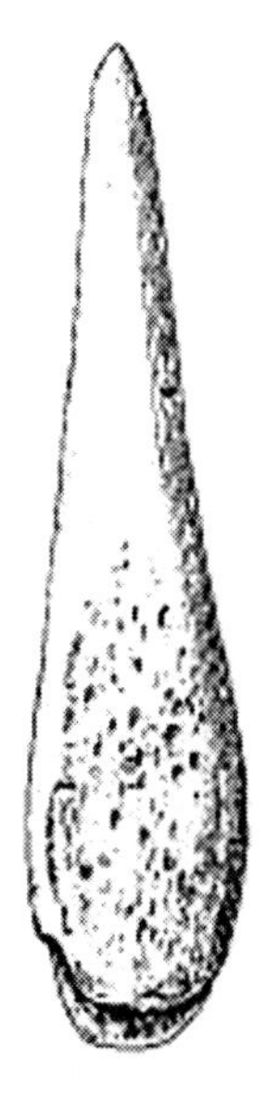

Figure 3.3 Prehistoric tool
Source: Adapted from Mellars (2006).

motor systems.[58] For instance, regions of the brain are prepared to recognize differences between different kinds of objects and their uses, of which mechanical tools are an important subset.[59] Furthermore, frontal motor regions have been linked to the motor features of tool use. Importantly, many species use diverse tools in adapting to their environment and reflect the evolution of cortical and subcortical systems in the brain that participate importantly in tool use, tool making, and tool recognition.[60]

It is the expansion of cephalic function that underlies the tool use that serves physiological and behavioral regulation, as illustrated in Figure 3.4. An expanded motor system with diverse cognitive capacities is no doubt pivotal in our evolutionary ascent. It is not just that the evolution of the cortex or brain is knotted to social function; tool use and other diverse abilities are also knotted to the fact that we are social animals. Thus, innovative tool use has been linked to an expanded brain ratio.[61]

Tool use and its elaboration are reflections of expanding motor capacity, rich with cognitive possibilities. Of course, we are not alone in tool use. Birds use sticks to build nests; chimpanzees and other primates use sticks to catch ants, or even kill prey on occasion. Significantly, the construction of tools by primates such as chimpanzees is often tied to a social context. In their tool-using, we can see the first glimmer of our ability to use objects as tools facilitating the expansion of our social

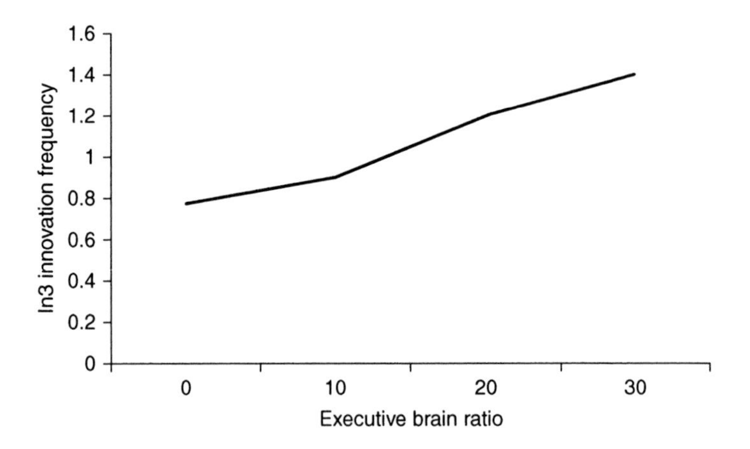

Figure 3.4 A graph illustrating the ratio between brain size and tool use
Source: Reader and Laland (2002).

milieu, for it is the expansion of cephalic function that underlies the tool use that serves physiological and behavioral regulation.

Anchored to kinds of objects

Aristotle wouldn't have put it this way, but he would nevertheless recognize that our social and moral existence is anchored to kinds of objects. Aristotle was a great taxonomist of what is, and his fascination with the body and animal kinds flows in great detail from his pen, limited though it sometimes is in description and understanding. For Aristotle, categories and objects are at the heart of knowing anything. Determining kinds of objects, their life history, and their development is the fundamental way in which we learn and know about the world. Plants, in particular, Aristotle described in much detail. While some of his observations are fanciful or even just plain wrong, they remain valuable for their reflection of what was seen in the classical world's conceptual lens.

Aristotle's taxonomy of the plant world also reflects, in some ways, the kinds of observational categorization used by children. And so we are taxonomic animals from our earliest infancy. We categorize objects[62] and come prepared to discern events from that core orientation and perspective (e.g., the solidity of objects).

Key categories about kinds of objects pervade the epistemological landscape. Cognitive predilection underlies action and inquiry.[63] Categories of understanding converge at every step in our intellectual development.[64] Symbolic and computational systems permeate all levels of human understanding. Generic categories of kinds, both naturally and culturally derived, are operative early on in ontogeny.[65]

An orientation to biological objects seems to be a core disposition in our cephalic organization.[66] Table 3.3, for instance, shows one set of relationships that reflects our orientation to taxonomic categories of living things.

We readily accept this sort of categorization and look for objects in our surroundings that fit into it.[67] Ecological factors are essential to understanding our cognitive capabilities.[68] Our preparedness to note relationships and make categories is reflected in many forms of folk biological discourse (e.g. avoiding foul meat, searching for warmth and security, connecting with others in common bonds, and expressing ourselves symbolically), even though specific categories and ends may differ across cultures.

Table 3.3 Taxonomic categorization[a]

Superordinate		Animals			Foods	
Basic	Cats	Dogs	Horses	Watermelons	Grapes	Peaches
Subordinate	Collies	Irish Setters	Terriers	Red Grapes	Purple Grapes	Green Grapes

Note: [a]Waxman (1999).

Diverse taxonomic characterization is the staple of cognitive coherence. But there is wide variation and diversity in our cognitive machinations, and in what we do and do not emphasize. Some groups may even show a devolution of function about natural objects. For instance, Atran's comparison of three- and four-year-old Mayan children and a group of Northwestern University students revealed that the toddlers could name more types of tree than the students.[69] Indeed, we come prepared with a diverse set of cognitive predilections towards objects and nature that allows us to understand, inquire, and appreciate.

So Aristotle's emphasis on objects and categories is very basic to how human beings understand their world. But Aristotle was also anchored to the concept of essence, which he described as "the cause of existence for all things."[70] For him, knowing an object is discerning the essence of it. But he anchored thoughts about essence in objects: "in general, the intellect in activity is its objects."[71] To put it another way, the purpose or function of an entity is as much of what makes it an object as its material constituents.

Aristotle's logic on the diverse forms of knowing varies with the object under study. The objects are centered by their function. But his discussion is also embedded in the Platonic seduction of abstractness at the expense of objects. What grounds Aristotle's discussion of essences, in a critical sense, is the biological focus of his landscape on objects and their function; function is pervasive. He is not inclined to the reification of objects into needless abstractions. In the end, his thinking is about things, about kinds of objects. Aristotle is really an old-fashioned rationalist: reason is knotted to experience, to findings, and not to thoughts in the mind's eye.

Aristotle's rationalism about the essence of things is fundamentally about substances. Objects have natures. For example, "some trees have gummy substances," as he notes in Book 1 on plants. Investigation discerns what they are. It entails both substance and essence. Contemporary inquiry has shown that those gummy substances are nutritionally rich, providing different sources of nourishment, including calcium, for diverse species.[72]

Objects are paramount. Although rationalism about forms and essences is omnipresent in the Aristotelian landscape, that landscape is anchored to events and to objects, and much less to reification. But in that landscape human reason is so knotted to objects and the discernment of their nature that core features inhere the essence of things, of which the temperament of a person is one.[73] Essences are thus not frozen in the depths of a cave but group together core features of a kind of object, thus allowing one to function in the organization of action and the perception of an object.[74]

However, pragmatists do not reify objects, and essence is the sort of thing that is easily reified. Core features anchored to cephalic capabilities are sufficient,[75] dependably related, and justified epistemically. They put the features of the kinds, the objects, in a much more mobile and action-oriented context.

Thoughts about kinds: from Locke to Kant

Aristotelian thought dominated the Middle Ages, but the old rationalism/empiricism debate rekindled in the early modern period. The philosopher John Locke is often characterized as an empiricist, but he still went to great lengths to argue against innate ideas in his treatise *An Essay Concerning Human Understanding*. He still retains some innate space, something like what Quine called a "quality space."[76] Humanistic medical man that he was, despite his protests against "innate speculative principles," some framework for experience is nevertheless omnipresent in Locke's thought.

From Spinoza to Hume and Berkeley, the issue of categories and kinds was often knotted to debates about what is innate and what is learned, what is a property of the mind's imposition and what is received from the sensory apparatus – an intellectual gambit that clearly impacts the issue of objects and our orientation to them. The focus on the mind in and of itself in the Enlightenment period was at times at a cost: a loss of the world, or a diminution of its importance.[77]

Locke, though, reflects more intellectual ambivalence in this regard. He does retain something of Aristotle's sense of being fixed to kinds of objects. This is perhaps most evident in his treatise "On the Names of Substances." When discussing the properties of gold, he asserts that the real essence is the constitution of insensible parts of the body, on which all the other properties of gold depend.[78]

What Locke, at the end of this treatise, begins to emphasize is the importance of reference, about objects, and about tying the epistemological apparatus to things, and to kinds of objects, in this case, gold. But Locke held the view of investigators discovering the "real essence of things" that things are of a certain kind of stuff (precursor to Kripke and Putnam, reference to "water" things in nature).

To use another example, Locke, commenting upon ideas, suggests that blind persons cannot understand color words (although we know now that even those born blind have some capacity to use color words correctly)[79] because they have no experience of color. That is, there is no reference to color in the experience of blind persons. Locke places the issue of kinds inside the more thorny issues of what is innate, what is certain, what is not disputable, and what passes before the mind's eye.

Gottfried Leibniz, on the other hand, in his famous rejoinder to Locke,[80] argued systematically for innateness of categories, that the idea is the form of the object – a form of Platonism in which knowing is about knowing the form of things, that is, the structure of the idea. For Leibniz and his fellow rationalists, Descartes and Spinoza, it is an idea that takes center stage, where clarity of ideas is the measure of knowledge, of truth, of knowing something about objects. Concepts are ingrained, and permanent. Like their predecessor, Plato, Enlightenment rationalists believed concepts do not seem to be learned (similar to a view that Fodor holds), but accessed, uncovered, and revealed by clear reason. Perception and action are set by a cognitive context.

Leibniz opined that confused ideas lie in perception, and a clear demarcation between perception and conception underlies his reasoning about categories. Mathematics, logic, or what Leibniz calls calculus, which he co-discovered with Newton, as a universal form of reason. Leibniz and other Enlightenment thinkers pondering human understanding, categories and knowledge about kinds of objects, would assert the dictum "nothing happens without a reason."[81] But the larger issue of reasonably tying cognitive systems to objects in making sense of the world would mean combining the classical empiricism and the rationalism of two figures: Immanuel Kant, a well-known thinker, and Thomas Reid, a lesser-known figure.[82]

Thomas Reid was a contemporary of David Hume, and the successor to Adam Smith as Professor of Moral Philosophy at the University of Glasgow. His name is often connected, particularly by Peirce, with "Scottish common sense realism." This was an alternative to the rationalism of Descartes, Leibniz, and Spinoza, and to the British associationism of Locke, Hume, and Mill. This eminently sane view championed a sense of being anchored to objects, a real sense of things that is not just abstract ideas but also concrete sensations. While the Scottish school of common sense was perhaps just too common-sensical to gain much renown as a philosophical school, it was an important prelude towards a sane view about kinds, and a concerted rejoinder to the denial of objects and kinds, in favor of a glorified inner mental space of contiguous associations.[83]

Reid demythologized expectations: "... common principles, which are the foundations of all reasoning... seldom, do not admit direct proof, nor do they need it. Men need not to be taught them; for they are such as all men of common understanding know; or such at least as they give a ready assent to, as soon as they are proposed and understood."[84]

Reid's emphasis was on common sense object orientation, and not the "egoism" he associated with Descartes.[85] This was an egoism that took the self as primary, and not the objects and kinds of things which one was trying to cope with and understand. And, while Reid's name and school of thought are largely absent from modern popular knowledge, his influence is deeply felt by modern philosophers, whose focus is towards good enough reasoning that anchors to objects in problem solving and adaptation.[86]

Getting a sense of the basic kinds of objects (e.g. the basic elements and the periodic table)[87] is, of course, an essential part of inquiry. What are the basic elements is one kind of question, amidst answers at more than one level of analysis. Genes are not more basic than molar behavior; neurons are not more basic than cognitive enactive and embodied people; the brain is not more basic than the heart. The pragmatist emphasis is on adaptation and inquiry. Despite lacking the evolutionary perspective of Darwin, Kant – who helped coin the concept of pragmatism in philosophy – got close.[88]

Immanuel Kant, of course, was the great rationalist. When viewed naturalistically, he elevated core conceptions of time and space to their rightful place as necessary tools for actively engaging the world. He recognized diverse forms of representational capacity in category formation that were linked to experience but not derived from it. As he says at the onset of his great treatise on reason, "But though all our knowledge begins with experience, it does not follow that it all arises out of experience."[89]

Key categories about "substance"[90] pervade the epistemological landscape of Kant's knowing process. In coming to understanding objects, causation figures prominently, particularly with regard to tracking objects over time. He was keen to point out that "the proposition that nothing happens through blind chance is therefore an a priori law of nature."[91]

Of course, we now can and do assert that we are prepared to link events, to track events that matter, and that the radical separation of the analytic and the synthetic is less rigid and demarcated than Kant certainly envisioned. Kantian categories are "regulative," as Kant himself asserted with regard to objects and our knowledge, and tied to their use or "practical employment."[92] Move ahead a few centuries and these same Kantian categories are selected and serve as an orientation towards objects.[93] But Kantian moves always tend to the rationalistic, identifying knowledge with concepts, with the emphasis on pragmatic inferences (not representations), rather than being object centered.[94]

Pragmatism, objects, and cognitive coherence

Charles Sanders Peirce liked to quote two figures: Reid and his Scottish common sense realism, and the medieval philosopher Duns Scotus. While Ockham's razor has long been associated with William of Ockham[95] and the elimination of extraneous variables in explanations, the philosophy of Duns Scotus is centered on realism and his "speculative grammar," which is rooted in semiotics about objects and events, and not simply their reduction to sensations or mere simplicity.

Moreover, Peirce suggested before many (e.g. Wittgenstein, Hanson, Quine) that "we have not pure sensations, but only sensational elements of thought."[96] Elsewhere, Peirce wrote that "[c]ognitive systems are pervasive, there are no pure intuitions. Every cognition is determined logically by previous cognitions."[97]

For Peirce, cognitive predilection underlies action and inquiry. What determines cognitive activity and behavior are expectations about an ongoing activity that is itself created from successful expectations. A belief is only discarded when some breakdown in the cognitive-behavioral architecture occurs due to some disruption in expectation.[98]

Peirce's scientific perspective is rooted in our abductive hunches, themselves planted in the constraints imposed by nature. To the traditional investigative methods of deduction and induction, Peirce introduced a third, namely abduction, the genesis of ideas. As Peirce put it, "Abduction is the process of forming an explanatory hypothesis."[99] In fact, since all perception is grounded in cognitive systems, a natural object permeates new inductive inferences. Peirce sought to ground his sense of pragmatism in the concept of abduction, testing an idea to its fullest extent, in order to find the maximal sense of what it means, expanding a list of categories derived from his reading of Kant into his understanding of Reid's Scottish common sense.

William James, a pioneering American psychologist, was also a pragmatist, but his pragmatism was softer intellectually than that of Peirce. James, as a psychologist, was interested in understanding human meaning, adaptation, and utility. He took the meaning of kind-terms to simply be what they are in use. At least it often seemed that way, and certainly Peirce viewed him in that light, thus changing pragmatism to pragmaticism, to distinguish himself from James.

Peirce had a point in his criticism of James, but, being an endlessly cantankerous soul, could barely get along with anyone in the community of inquirers, glorified as he was by them. James, by contrast, was one of those individuals at home with all; everyone was comfortable in his presence.[100] James exuded an agreeable temperament; Peirce was very much the opposite, and this was reflected in the respective arcs of their careers and intellectual influence.

Peirce always prized in his lexicon of scientific understanding that depicting what a thing is – the classification of it, in addition to its use – is an essential part of explanation. Capturing something about kinds, of what constitutes nature, is what makes objects intelligible, and this can be quite separate from simple use. In other words, knowing is knotted to the natural history of the object.[101] Peirce always retained this intellectual sensibility, and his categories reflected this fact.

Behavioral adaptation figured importantly in John Dewey's writings, just as the categories did for Peirce. Dewey believed stability is sought amidst the endless precarious conditions in which one finds oneself, and categories are oriented towards adaptation, discovery, and engagement, not to isolated internal representational space. "The natural and original bias of man is all toward the objective,"[102] toward the real world of nature, or as Philip Kitcher (holder of the John Dewey Chair in Philosophy at Columbia) says, "Few are born antirealist."[103]

What Dewey understood was that cognitive architecture is inherent in the organization of action, an enactive cognition.[104] Dewey's orientation is about cognitive-behavioral adaptation. There is no reification about concepts of objects. They are assumed to be tied to knowledge as a kind of contact sport. Cognitive systems are action oriented and rooted in relationships to kinds of things, and kinds of actions towards things.

Contemporary empiricism thus broke barriers between empirical and theoretical statements. The "empirical significance of the whole science"[105] is understood as "pragmatic,"[106] grounded in a web of interactions of investigators. Quine's metaphors are about kinds, references, and roots, but the depiction is about language and logic. It was a linguistic behavior akin to the later Wittgenstein but, unlike Wittgenstein, grounded in a thoroughgoing naturalism.

But it is a pragmatism that is knotted to building on what Nelson Goodman[107] emphasized as "projectable predicates." How "entrenched" and reliable a term is reflects its encampment in a conceptual framework. In science, the issue is always empirical and testable, and the reliability of a term, such as temperament, is over time, over use, over predication and explanation.

The issue of induction is embodied in the conceptual frameworks of pragmatic use.[108] Similarity, a notoriously and deceptively simple concept, is a defining property of belonging to a kind, a "family resemblance"[109] of which some "standard of similarity is in some sense innate."[110] "Induction itself," as Quine noted, "is essentially only more of the same: animal expectations or habit formation."[111]

In pragmatism, theoretical sensibility is pervasive and underlies traditional distinctions between sensation and concepts,[112] best understood as "theory laden."[113] Of course, conceptual pragmatism is rooted in determining consequences for a common ground for the knowing subject rooted in a community of inquirers for whom judgments are laden with valuation.[114] The growth of science reflects the concept of expansion of our theoretical apparatus, our capacity to see new things. As Hanson put it, "The layman must learn [to be a] physicist before he can see what the physicist sees."[115]

But perhaps the learning part is overstated since evidence in cognitive science already holds quite a lot of expectations about objects.[116] There is understanding, of course, but not necessarily a theory to put it in perspective. The child might know something about kinds of objects and their solidity, for instance, but what makes it up, explaining it, and so on, is the stuff of science. In fact, while children might embody some notion of prelinguistic causation, it is propositions, the stuff of linguistics, that tie causes to theories. Keeping track of objects is at the heart of our epistemological journey, and part of placating our fears is in our representation of the objects in our world (see Figure 3.5).[117]

As Hanson paraphrased Kant several hundred years later, "Nothing in sense datum space could be labeled cause or effect."[118] Hanson also said, "Interpretation ... is operative in the very making of a system."[119] This is a view close to the later Wittgenstein,[120] and others such as Wilfrid Sellars, who held, like Kant, that "intuitions without concepts are blind."[121] In other words, seeing is always based on perspective, and interpretation underlies all human social behavior and understanding. As a result, and echoing Peirce, abduction or retroduction is essential for hypothesis formation, and hence for category formation.

Saul Kripke, in his book *Naming and Necessity*, and also Hilary Putnam in his philosophy of language, resurrected the Lockean view of anchoring interpretation to objects and kinds, as well as fixing reference by

Figure 3.5 Cave art representing forms of objects familiar to the artist

highlighting the naming process. While Kripke and Putnam look at naming and tagging an object as philosophical issues,[122] from our point of view, naming objects and tagging kinds are primarily cognitive adaptations in tracking things over time.

What Dewey described as "the precarious amidst the stable,"[123] the retention of a sense of an object's ontological permanence despite its movement and change through space and time, is thus both an appetitive and consummatory cognitive/behavioral adaptation. While we search for the essence of an object, we also derive satisfaction from capturing a family-like set of features.[124] Family resemblance is for us a reliable way to predict the behavior of, and regulate our own behavior towards, the objects around us. The essence of an object, on the other hand, becomes "expressed in the grammar."[125] Here, the emphasis, as it was for many investigators,[126] is on our linguistic competence at manipulating things and in our transactions with one another.

We are, after all, preeminently linguistic animals,[127] and inherently social and historical. As Wittgenstein noted when discussing perception of a drawing that could be seen as either a duck or a rabbit,[128] perspective matters. Interpretation underlies perceptual coherence. Categories about kinds of objects must always satisfy the condition of conceptual coherence, usefulness, and prediction.

Getting anchored is a most important cognitive adaptation. We are always getting anchored to objects. What concerns us is utilizing the concept of, say, temperament,[129] in a way that picks out a set of properties, which perhaps cluster together in a systematic way, a semantically meaningful way.

As Boyd notes, "Natural kinds are about how schemas of classification contribute to the formulation and identification of projectable hypotheses."[130] Boyd cites the projectable feature in our epistemological lexicon.[131] Reliable predictors of core features about kinds and objects are the epistemological backbone for cognitive and coherent action. Predictive causal properties co-inhabit the conceptual framework of events clustering together.[132] Boyd links a cluster of properties to a form of homeostatic regulation, where the properties are linked into a causal nexus in a "homeostatic unity."[133]

Of course, perhaps a better term might be "allostatic order." Allostasis is a term about change and coherence.[134] Whatever the regulatory term, the context in good epistemic faith is tied to what might be called "promiscuous realism,"[135] or perhaps "critical realism," a broader view of realism.[136] Our cognitive resources, in part, are trying to gather as much information as is relevant for coherent action.[137] Information such as this is also relevant to an evolving science of kinds of objects, our knowledge of them, and the biological factors we want to understand, as Wilson, a philosopher of biology and the cognitive sciences, has nicely stated.[138]

1. Heterogeneous and Specialized Parts
2. Internal Mechanism
3. Information Molecules
4. Growth, Development, Recuperation
5. Metabolism
6. Environmental Adaptations and Construction of Niches

While John Stuart Mill introduced the term "natural kind" to our lexicon,[139] the basic idea was already there and functioning. Pluralism, amidst realism, is emboldened with an expectation of classificatory systems, particular and diverse forms of kind-terms. Even categories like temperament have features that one can acknowledge and therefore keep anchored to a phenomenon without losing a sense of its category, and the cluster of properties with which it is associated. The move is away from a pernicious, overzealous rationalism, in which expectations about essentialism are discarded or toned down to real and rational size, yet anchored with a realistic view of nature.[140] As Dupre remarked, "The term natural kind can be divorced from traditional connections with essentialism."[141]

Reminiscent of Dewey,[142] John Dupre's discussion of kind-talk can be knotted to functional ends of explanatory discussion and desire, amidst diverse epistemologies and methodological functional tools with well-defined ends. In fact, as Dupre noted,[143] individuals have shied away from essentialist thinking in biological classification, and/or tend to be minimal essentialists.[144] Quine emphasized a science that tied together naturalized epistemology, a subset of psychology, and eventually neuroscience; he did not emphasize his behaviorism and innate quality spaces, but cognitive science and neural science.

The discussion, for us as pragmatists, is not about essences, but cognitive reliability, coherence, predictability, and adaptation. Category-based inductive processes have a firm modern basis in modern psychology.[145] Category-based learning about objects is expressed very early on in ontogeny,[146] and reflects cognitive adaptation (kinds of objects, animacy, numerosity, and facial responses).[147]

But one does not have to be an essentialist about categories and kinds.[148] Cognitive adaptation and coherence are bare level requirements for pragmatists and thus not an essentialism about kinds and categories.[149] What is important is the anchoring to objects, the link to cognitive appraisal systems, and the abductive hunches grounding ideas in context and evolution.[150]

We easily get hold of the meaning of an object. Peirce linked this rapid response to an instinct, a hunch, an abductive moment in which we determine what a new object might mean and how to understand it. No simple set of characteristics when experimentally tested can account for the quick conceptual recognition. No Quinian moment or simple empiricist story of stimulus similarity can account for how we categorize, or the range of our categories.[151] Naturalizing Kant within cognitive science is asking about the cognitive conditions for the possibility of knowledge.[152] What is left out of Kant upon naturalizing his thought is the transcendental and the quest for certainty.

One common experimental method that is subject to more than one explanation is habituation. Take, for example, how long or not long the child looks at an object. Both expectations and violations of expectation underlie this methodology. Occluding experiments and unusual events underlie the experimental framework.[153] Basic cognitive capabilities underlie our sense of objects, numerosity, prediction of events, and teleological predictions. Interestingly, one pragmatic theme is the violation of expectations, a feature that Peirce and Dewey emphasized in their understanding of the organization of action, coherence in understanding, learning, and investigations.

But the issue is the kind of categories: how general, how specific, and how culturally expressed. For pragmatist purposes, a cluster of properties, knotted to biology and adaptation, is enough to constitute a natural kind as long as the science can proceed in a meaningful manner. Once one moves from essentialism to reliability in an inductive system, from philosophical essentialism to pragmatism, properties that cluster together and that can be linked to prediction and mechanisms work as a "natural kind." Thus, psychological essentialism is a way to be rooted towards objects, the seen and unseen features of the kinds of objects in our world (an orientation easily noted in children). One can be a critical realist about kind-terms, as we are, anchoring a conception of temperament to biology and taxonomic cognitive tendencies, without reifying kind-terms into Platonic space, associated with rigid essentialism.

Epistemological awe and humility: the Platypus

Take, for example, the platypus, an egg-laying mammal found in Australia. A wonder of nature, noted since the 18th century, the platypus was a spur to Western ideas of exploring the islands big and small. The grand taxonomic adventure stimulated by the wonder of the platypus is a big part of the biological tradition, ushered in perhaps as early as Aristotle, but encountering an epistemic steam engine during the 17th, 18th, and 19th centuries.[154] An explosive exploration amidst a quest for endless consumption, conquest, and curiosity for adventure linked the penchant for cataloging, tagging, and theorizing to an economic engine to control, exploit, and consume. Then, amidst a theory or theories of evolution that circulated, came one of nature's oddities, an egg-laying mammal.[155] Our taxonomic sensibilities, emboldened by a theory of natural selection, have come a long way from the frozen taxonomies of the 17th and 18th centuries.[156]

But inquiry is anchored, not necessarily in a representation, but in action and objects.[157] A core feature is a deep sense of what Santayana called "animal faith" and a sense that "the animal mind treats its data as facts".[158] These facts undercut a skepticism that easily keeps objects and events detached or divorced.

So the platypus is put in perspective. In the present era, the platypus can be traced back millions of years. It is something of a vestige; surely there must once have been more of this type of animal, yet they are no longer apparent anywhere.[159] Amidst the endless expanse of the exotic landscapes of the South Pacific unearthed by the explorer, the platypus is understood to have become an oddity via geographical separation and forms of speciation.[160] Evolution favors diversification and myriad reproductive capabilities. The rise of animals and their diversification is one of the wondrous events on our planet, even though the animals themselves (mammals at least) are rather new.

Anomalies like the platypus, pictured in Figure 3.6, are a wonderful reminder of humility. Vertebrates have an evolutionary trend that goes from fish to amphibians to reptiles and birds to mammals. Mammals by definition have mammillary glands and almost always internal egg production with the bearing of live young. Fish, reptiles, and birds are associated with egg-laying. The divergence of the platypus breaks a rule, in the same way that orangutans are social isolates even though they are primates, which are, by definition, social.

What the platypus gives us is a sense of our evolutionary past. Its genomic makeup is now understood and unique; it is phylogenetic history pregnant in the present.

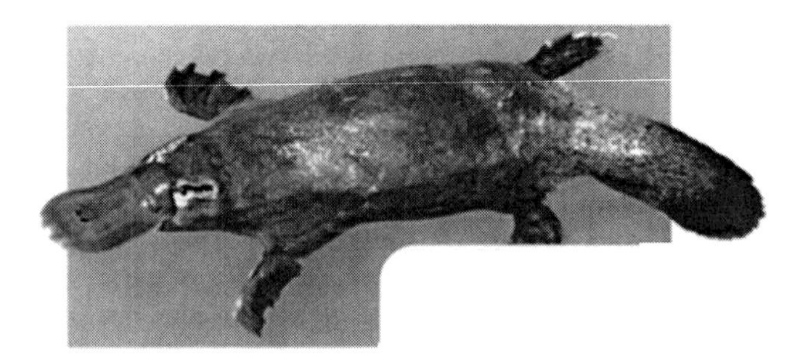

Figure 3.6 A platypus

Thus, the inferences to the best explanations and the more unifying context could be grounded. One feature is that expectations guide action in perception. Unnatural doubt of the kind imposed by Descartes from a pedestal of rationalism, doubt beyond recognition to an interior place of pure subjectivity to a foundation of epistemological security and metaphysical purity, was a vestige useful perhaps as an exercise, but ripe with futility and barren rationalism.

Existential and epistemological exaggerated doubts may play an important role as a methodological exercise under some conditions, but they should never have been rendered the foundation of philosophy. Inferences are tied to action, and foundations need to rest more on activity adapted to circumstance and context. That way, unnatural doubt is displaced with warranted assertions that do not blind and mislead on the epistemological journey.

Conclusion: evolution and cognitive coherence

Pragmatism, grounded within a perspective in which psychobiology is a core theme, takes this orientation to objects as basic to our species' understanding. This chapter is central to this book. Natural kind-terms serve, indeed, as an important cognitive adaptation for grouping features of objects that may be inherently related to parts of biology. The fact that we are born to categorize is not the same as a form of rationalism or essentialism. Realism about categories for kind-terms allows one to acknowledge psychological tendencies about essentialism without an exaggerated expectation of absolute conformity to kind-concepts. This is a pragmatism rooted in the interactions with nature, not something simply for parlor talk.

We just come prepared to group objects, one grouping of which is itself about kinds of objects. And the kind of objects is about a piece of biology and constitutes a kind of feature that pervades an object. Cognitive adaptation lies in the doing of things for coherence of action in complex social environments, diverse ecological conditions, and social communicative functions.[161] Objects for us are always, in some sense, social objects,[162] and our adaptive cephalic systems are expanded in use in social contexts.

Core cognitive architecture is mostly about kinds of objects. While, to be sure, the linguistic competence of our species is itself an innate architecture with rich lexical hooks to semantic representations, language and the evolution of the neocortex is also linked to social behavior, social contact, and social complexity.[163]

Regions of our central nervous system reach out to objects through our peripheral systems, through the construction of objects that allows us to expand our sensory systems.[164] Inventions such as the microscope and telescope are such extensions of our seeing. Pervasive perception/cognitive capacities pervade the central nervous system.[165] We reach out to the objects in space and inhabit them through linking to kinds of objects.

4
Cephalic Capabilities and Medical Decision Making: Endlessly Imperfect

Introduction

When a child comes crying to a carer with a bump or scrape, we kiss it to make it better. The management of discomfort is an ancient adaptation. All human beings seek to ameliorate pain, prevent disease, and treat it once it occurs. Preadaptative cognitive systems underlie this propensity. Shamans, barber-surgeons, and eventually doctors evolved as specialists in medical inquiry.

A Deweyan perspective (viz., cultivation and self-corrective inquiry amidst a genuine community) is endemic to all such medical decision making. The naturalization of human decision making puts medical inquiry in its place, in light of the social progress of our species in the amelioration of discomfort, the promotion of wellness, and the treatment of sickness. On this orientation, medical knowledge is demythologized to human proportions, couched in respect for, rather than the eradication of, uncertainty. It harnesses an experimental spirit to cephalic heuristics, good problem solving, biases, and vulnerability to missed opportunities of discovery – in short, a pragmatic naturalism. Medicine is a good place in which to couch naturalism, critical realism, and social hope, as it is an endlessly imperfect social milieu in which decision making takes place. Furthermore, medicine presents real-world activity and the consequences of that activity in ways that are superior to the many naively simplistic thought experiments of armchair philosophers. Simply put, medicine and pragmatism go hand-in-hand.

American medicine was at first a trade practice divorced from the contours of science. The University of Pennsylvania was America's first medical school (1765), followed by Harvard (1783). At these institutions, medicine evolved as a profession with direct links to science. While physicians and medical researchers are almost necessarily sticklers for authority and perhaps a bit arrogant, self-corrective sensibility has always been an important trait in a democratic society. The culture of medicine was transforming itself and beginning to make itself accountable. This was done in part by markets and capitalism, and the desire to make money, but also because transparency and accountability served as important normative American goals.[1] Legitimation in medical education strongly emerged during 1850–1930, resulting in far greater professional accountability. Two crucial figures in this process were, in fact, brothers: Abraham and Simon Flexner.

While working with my colleague Bruce McEwen at the Rockefeller University in the early 1980s, the laboratory was housed in Flexner Hall, but it never occurred to me to wonder who Flexner was. The hall was named for Simon Flexner, a physician, scientist, administrator, and the first director of the Rockefeller Institute for Medical Research. Simon Flexner believed that science was a universal language that belonged to no one nation. All could participate. His research focused on diseases such as polio.[2] He was interested in virology, and inoculation was a growing focus of his. While he did not find the cure for polio, Simon Flexner was critical in the development of early research and experimentation from 1910 to the early 1950s, when polio ravaged populations, especially children, all over the world. The year I was born, 1952, saw some of the worst outbreaks of polio ever known, but also the emergence of vaccines. A vast public health experiment prevailed and conquered, and modern medicine achieved a major success.[3]

Simon's brother, Abraham Flexner, was less accomplished in the formal sciences, and a high school teacher by trade. Like his brother, Abraham was influenced by the model of science that emerged from Johns Hopkins at the end of the 19th century (he achieved his undergraduate degree from Hopkins). Abraham went on to write the famous "Flexner Report" on medical institutions, which totally transformed medical education in America. He also, like Simon, had an important hand in the establishment of what was then called The Rockefeller Institute, later Rockefeller University, in Manhattan, which was devoted to biomedical research. (Its motto is "Science for the Benefit of Humanity.") Abraham also helped found the Institute for Advanced Studies at Princeton, home to Einstein and other dignitaries.

Besides the Flexners, Dewey and Peirce were also linked to Hopkins; Dewey received his Ph.D. there, and, as discussed earlier, Peirce set up its laboratory in psychophysics where he taught logic and statistics. It was at Hopkins that the notion of medical inquiry informed by science from the laboratory to the bedside, from the clinic to public health statistics, so vital to physician decision making, was developed to orient medicine towards its normative goal.

This chapter is the story of the development of medical inquiry, which goes back much further than the Flexners and the influence of Hopkins-style medical inquiry, of which Peirce and Dewey were very much a part. I begin with a naturalization of epistemological expression rooted in our evolutionary past, discuss the cephalic orientation to objects, and the integration of action and perception. A pragmatist perspective, in which evolution and critical realism link the capability for problem solving and natural objects, healing and medicine, aesthetics and education, is imperfect. Inquiry, as outlined later in the chapter, is demythologized to human proportions without losing sight of the individual experience. The evolution of medical rights, as a consequence of ongoing inquiry, dovetails with social hope, which is a core pragmatist legacy and prophecy.

Evolution, cortex, and problem solving

We have a cognitive predilection, as I have indicated throughout this book, towards kinds of objects and relationships between events through self-corrective inquiry. These predilections include ideas about space, time, and different kinds of objects. Early on, children know something about solidity, depth perception, distance, and different kinds of objects. They know whether something is alive or inanimate. They have notions about agency and about objects related to food ingestion. And children know or can easily learn whether and what objects are dangerous. We come prepared, for instance, to associate food ingestion and gastrointestinal malaise. The learning of this kind of causal relationship is a fundamental, viscerally mediated, specific association. Diverse causal relationships underlie the cognitive arsenal that fuels our reasoning about objects.

Medical problem solving is at once an outgrowth of prepared cephalic expression with an evolutionary background, and an expansion into new domains. Specialized systems are ready to discern objects, to infuse them with meaning, and to look for causal relationships among them and/or events (e.g. relationships, for instance with ingestion of a food resource and illness).[4] Much of our evolutionary past involved finding food, shelter, warmth, and social contact; avoiding predation; and practicing reproductive, social, and deceptive behaviors. All that activity perhaps led to what Peirce called "musement" about what is around us in quieter moments, perhaps when we were safely tucked up in caves or trees between dusk and dawn. There may have been moments to sketch what we were experiencing in our environment. These events were linked to our use of tools and the development of them as we expanded our cephalic capabilities and end-organ systems. These events coevolved with diverse forms of memory capacity that would figure in the development of our technologies.[5]

For example, consider vaccination. How strange it is to inject oneself with the very thing that one is hoping to avoid. Yet vaccination is one of the most successful achievements of modern scientific activity. Not only have numerous problems of disease been solved or evaded through social cooperation, we also have a clear illustration of how the evolution of our cortex in the social context yields initially strange but nevertheless productive musements for activity.

Social cohesion and medicine

Perhaps one of the most prominent features of our cognitive expansion is the neocortical link to social groups. Social cohesion and social formation underlie the bulk of our evolutionary capacity. Many of our cognitive resources are determined from long protracted learning that takes place during ontogeny in a social network, viz., learning from the experiences of others. Neocortical expansion is highly correlated with social complexity and technological innovation.[6] This amelioration emerged with an evolved cortical mantle.

Our desire to provide relief from distress is a major cognitive achievement of our species. Most animals, in comparison to humans, have a limited range of responses to pain. Diverse forms of what would become medical decision making are pervasive across human expression and capability.[7] Folk medical knowledge reflects an evolved predilection towards kinds of objects that we are exposed to and ready to categorize.[8]

Whether acupuncture, herbs, or chanting, the ameliorative factors found in natural objects derive from a basic predilection that is generic to cephalic function, and coupled with both specific and more general problem solving proclivities. These generic practices underlie a basic predilection in our species to avoid and ameliorate discomfort. In other words, a universal inclination is to heal, ameliorate, understand, and discover. Many roads emerged across cultures to reduce discomfort.

Consider acupuncture (see Figure 4.1), for instance. Surely, understanding the mechanisms requires the experimental method, and the development of control groups. But the core idea from many continents is that diverse forms of natural objects are understood to be effective in the amelioration of discomfort.[9] Chinese medicine, for example, rich in historical variety, is an assortment of suggested orientations towards health; some work, some don't. The issue is empirical and experimental.[10] We know now that much of it induces endogenous brain changes that calm diverse end-organ systems.

The predilection to heal is species specific, part of the cephalic adaptative systems. The exploratory discoveries are embedded in the social milieu, and core memory capabilities. Moreover, we have resources devoted to memory (declarative, procedural, and episodic) in which to

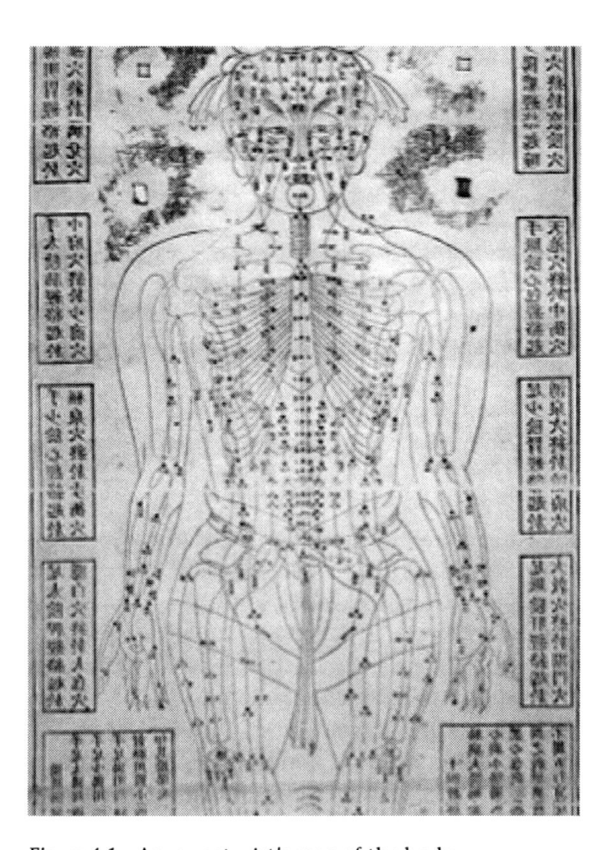

Figure 4.1 Acupuncturist's map of the body

couch the causal relationships we discover. We have predilections about grouping objects into relationships, statistical relationships, and predictive relationships that underlie behavioral adaptation. We also have a cognitive predilection to self-correct, to use the resources we have to determine relationships, and then to shift course towards a new avenue of investigation. This cognitive capacity underlies all forms of human capability.

We are born with an orientation to objects, and a natural predilection to discern and discover.[11] This natural predilection expressed across cultures manifests differently depending upon local context and cultural opportunities. The inclination is to discern ordered relationships in nature. Natural philosophy, then, is not an accidental feature of our species but an essential one.[12]

Children start out as little naturalists, but emerge as more general problem solvers as they explore, wonder, and forge coherent action.[13] The context of medical decision making grows out of this natural predilection. Exploring, testing, and understanding are core practices that take shape early, in the right pedagogical context. With the right temperament and development of habits the development of these core cephalic practices continues over the long trajectory.

The construction of objects, a fundamental adaptation of our species, expanded from our first small tool use – probably barely modified natural objects like sharpened sticks and bashed pebbles – to extensive and deliberate tool design and use. Cephalic expansion set the stage for technological creations, expanding our sensory systems. Seeing by magnifying became an evolving theme as our capacities were extended, and we turned from managing nature towards understanding nature. Tool use and tool making were critical for this development. Figure 4.2 depicts time, the expansion of modern humans across space, and tool evolution.[14]

We went from crude stones and sticks (a feature we share with diverse species like birds, and other primates) to the construction of tools anchored to objects. Tool use is an expansion of cognitive systems as we expanded our social horizon. Galileo could make his discoveries of the starry heavens because of the telescopic lens that expanded sight (see Figure 4.3).[15] Similar discoveries became possible thanks to the microscopic lens or the mechanical devices of Leonardo or Boyle.

Our cognitive systems set the conditions for cognitive coherence in the organization of action that emerged with a propensity to utilize natural objects. A sense of natural history is ancient and is expressed early in diverse intellectual texts.[16] Core orientations towards objects

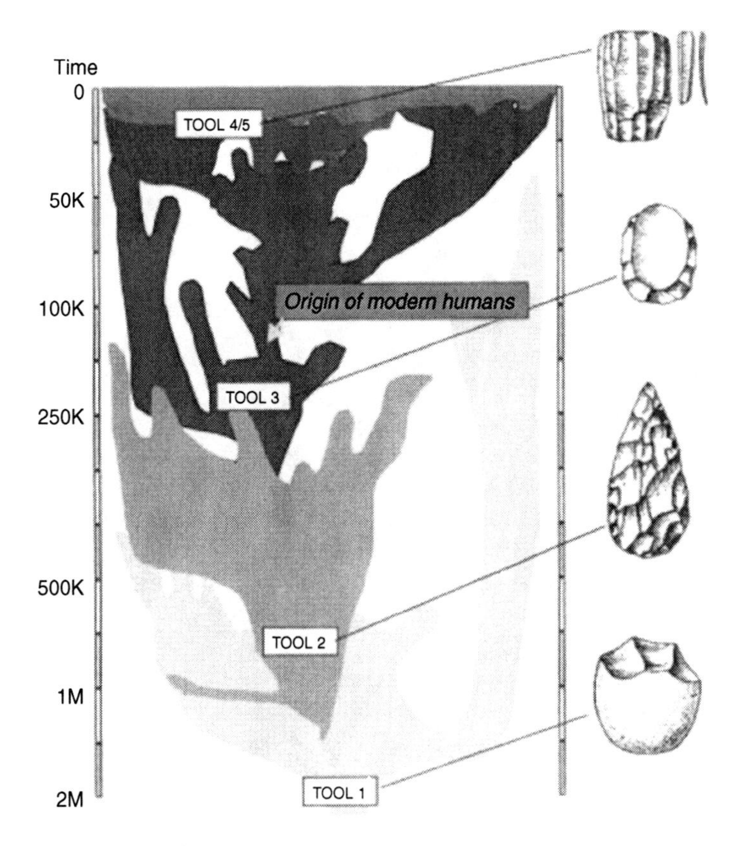

Figure 4.2 Evolution of tool use and manufacture
Source: Foley (2001).

reflect something about their solidity and spatial makeup, causal relationships, and statistical outcomes.[17] This sets the stage for investigations about natural objects, like how to fix, find, and fetch them. In consequence, a set of specialized, adaptive cognitive resources figure early on in neonate consideration of objects. The learning, at first, is innately geared for action; action in which learning can take place – learning essential for coherence in action and perception.[18]

The cephalic capability to determine causal relationships is less abstract and more bodily in function, as are a number of diverse cognitive resources that figure in general decision making.[19] One feature across medical decision making is the search for the various causes of bodily harm and discomfort. Built into tracking causality are probabilistic

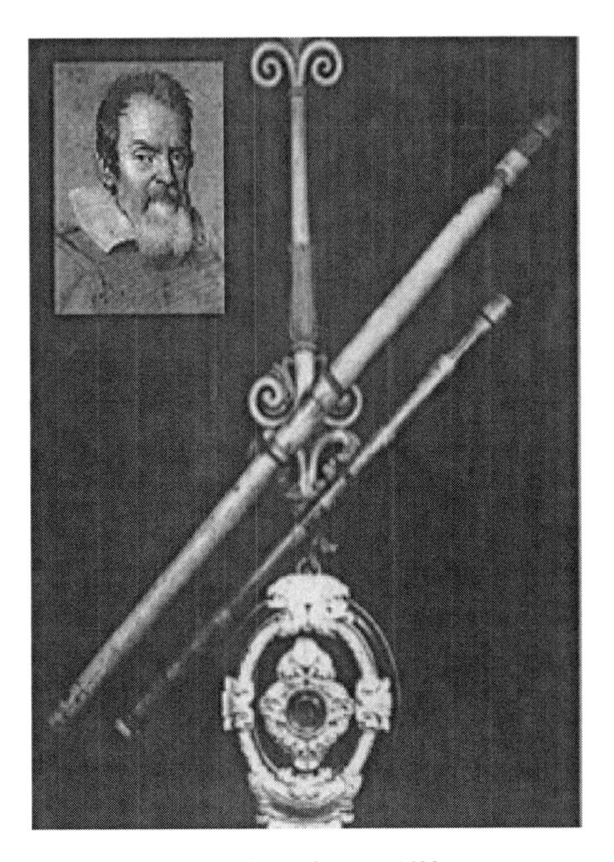

Figure 4.3 Galileo and his telescope, 1609

determinatives of possible harm or safety, rather than perfection in design.[20] Causal relationships are contextualized, grounded in action, and ripe for extension. Preadaptative mechanisms are expanded in use as a predilection for evolutionary competence, expansion, and varied adaptations.

We readily infer mechanical actions, just like intentional actions, early on in ontogeny.[21] The ability to discern, say, a billiard ball's action as caused from outside versus action caused from within, that is self-initiated, is a core cognitive adaptation. We look readily to discern such differences. The roots of this differentiated discernment lie in our observation of natural objects that are foul and dangerous, attractive and approachable, affordable and useful.

Herein lies the context of understanding. The knowing process begins in puzzlement after consternation at something failing to work, after some breakdown.[22] The striving for knowledge is replete with obstacles.[23] Of course, this is almost always the case in medical inquiry. We seek to understand the ailments in the patient. A "laboratory frame of mind" (a Peirceian phrase) grounds problem solving in meaning. Meanings of terms are rooted in situational realism, a critical realism of context, testing, disconfirming, and confirming[24] – a critical realism that logically expands the contours of meaning.

Logic is in the possible connectedness of events or the exclusion of them. Thinking is the "power of following up and linking together the specific suggestions."[25] Dewey notes that, "Thinking is not like a sausage machine,"[26] but a power in "linking together." He later notes that logic is linked to being wide awake,[27] noticing what needs to be noticed, and drawing inferences from the context at hand. In other words, attention systems underlie the logic of inference. Thinking is the bringing together of coherence in action.

Kinds of objects in our lexicon reflect the environment of which we are trying to make sense. Engines of cognitive resources grounded in human action and perception cut across the feel for objects that we discover in making sense of events.[28] There is no sense of where the body begins and the cognitive ends, as Dewey rightly noted in his reflex arc paper. This was a paper that helped usher in a view in which the study of behavior was richly embedded in a sense of the body's own problem solving. In this paper, cognitive systems were inherent in cephalic capability; there was no separation of mind from body.

An inherent propensity to decipher the world into animate and inanimate is part of our cognitive apparatus.[29] Categories, such as alive and well, or decayed and dangerous, underlie this predilection towards objects, which itself holds up the origins of medical decision making. It is not surprising that universal expression of herbal medicine, natural products to heal wounds, to delay devolution, and to promote health is a feature of the human condition. It is as basic to us as language and tool use.[30]

As Dewey noted, cephalic expansion into intelligent problem solving is a basic hominid expression. Ideas and problem solving are tools, not Platonic entities; they serve ends and means in use and adaptation. Evolution, for humans, could indeed be defined as the expansion into more and more novel cognitive performances through increased accessibility to core cognitive functions by our neural architecture. Key features of this cognitive fluidity are the integration of several orientations to coping with the world: social intelligence, technical abilities, diverse expression of natural knowledge, and, of course, language use.[31]

A perspective on medicine

Hippocrates, a Greek physician, who lived during the age of Pericles,[32] is best known for his allegiance to the patient. He was dedicated to the eradication of disease, and to its management and understanding. His treatises on medicine, on managing disease, and what came to be known as the Hippocratic Oath, are testimonies to his influence on medicine, the primary feature of which is the sacred relationship with, and responsibility of the physician towards, the patient.

For Hippocrates, observational science was the primary mode of investigation on the empirical side. But the focus of the inquiry was towards causation. This predilection to determine causal relationships is a feature of our cognitive architecture, and found expression in Hippocrates' assertion that "to know the cause of a disease and to understand the use of the various methods by which disease may be prevented amounts to the same thing."[33]

The great Greek philosophers, and indeed many of the traditions across every continent (e.g. Arabs, Persians, and Indians), laid out one expression of knowledge. The emphasis was on careful observation and clear reason, and in many ways their influence on classical physicians is most profound. Stoicism, for instance, is a natural predilection, a sense of the balance of nature (see also Taoism). The intellectual air of Galen is steeped in recognizing the balance of nature, and the evolution and devolution of living things.[34]

Galen, the paradigm of the classical physician/anatomist, expanded the concept of surgery, and accurately depicted the anatomy of the spinal cord and other parts of the body. Galen arrived in Rome in about 162 AD. Galen was interested in the structure of organs, and the anatomy of the muscles. His understanding of regulatory systems was vitalistic. He was also interested in preventive as well as curative medicine. Like Aristotle, Galen emphasized whole-body physiological regulation. The heart and the liver, because of their size and suspected regulatory functions, played a prominent role in his understanding of physiology under normal conditions, and also when they broke down and needed tending during pathology.[35] Blood was also of interest to Galen, and bleeding (Figure 4.4) was for him a common therapeutic intervention.[36] Bleeding was used to eradicate disease states, a practice found in many cultures.

Galen theorized a climate of the internal milieu, the secretion of humors that pervaded the temperament of the individual (see Figure 4.5), and the vulnerability of end-organ systems to disease.[37] Indeed, the biology of temperament linked to diverse information molecules, such as hormones, is pervasive, and colors the contours of development, and indeed all of our lives.[38] In fact, temperamental features also underlie medical decision making, whether one is bold or timid.[39]

The orientation of both the Greek and Roman classical world and attitudes inherited from the Middle East, India, and China are all still embodied in medical decision making for many of us in the West. Of course, throughout history, medical decision making is anchored to one stark reality: if you make a mistake, someone pays with his or her life.

Figure 4.4 Ancient Greek depiction of blood letting

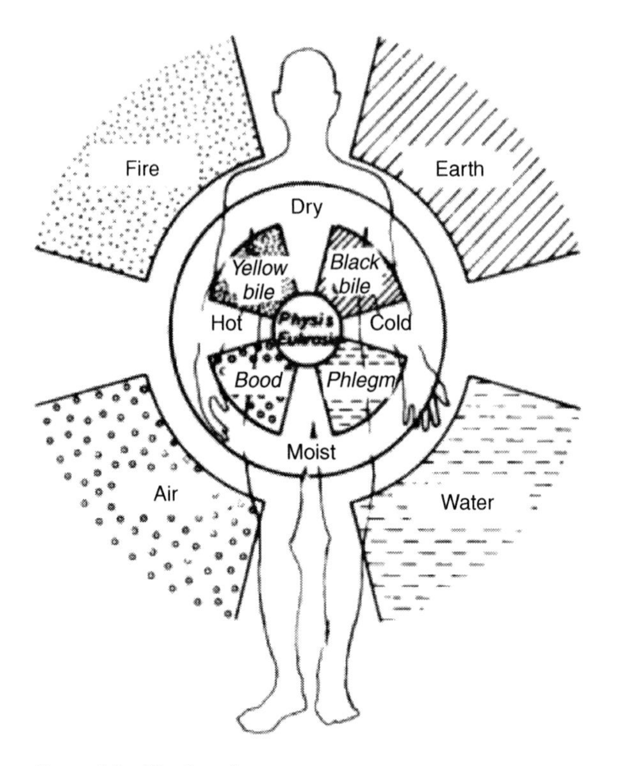

Figure 4.5 The four humors

Cataloging and keeping track

The taxonomic tradition handed down from Aristotle to Galen, through Ray, Linnaeus, and Buffon, along with the experimental approach and the invention of new technologies, were essential for the endeavor of understanding natural diseases.[40]

We are, as I have shown, by predilection, taxonomic animals. The categorizing function of cephalic adaptation is aimed at forging coherence about objects, about kinds of events and entities. Aristotle was a dominant early expositor[41] of this predilection. It is no surprise that early on in our cultural history, cataloging would be manifest in maintaining health and warding off disease.[42] This ability set the conditions for the enterprise of medicine, as nicely depicted in Table 4.1 below. The key features are bedside manner, access to knowledge, hospital care, and social or epidemiological laboratory sensibility.

Table 4.1 Key features of medicine throughout history

	Characteristics			
Kinds of medicine	**Object of inquiry**	**Form and site of education**	**Goal**	**Example**
Bedside	Whole patient	Apprenticeship	Therapy	Hippocrates (c.460–370 BCE)
Library	Text	Scholastic, linguistic, university	Preservation, recoverv, commentary	Constantine the African (d. before 1098)
Hospital	Patient, organ	Hospital	Diagnosis	R.T.H. Laennec (1781–1826)
Social	Population, statistic	Community	Prevent	John Simon (1816–1904)
Laboratory	Animal model	Laboratory	Understand	Claude Bernard (1813–1878)

Source: Adapted from Bynum (2008).

Already present in germ form in the practice of Hippocrates and Galen, these key characteristics continued to inform medical inquiry to the present day.

A rise in experimentalism and in medicine

The breakdown of scholastic architectural argument and observational sciences, inherited and institutionalized from Aristotle and the Arabian/ Persian physicians and philosophers in the medieval period,[43] paradoxically led to new eras of medical progress. In the Renaissance, medicine was regarded as a learned profession, linked to a broader natural philosophy. Its key question was "what was nature signaling?"[44] From representations to coping with disease, the doctor of medicine before the onset of modern science was, as she is now, anchored to this broad array of events. The "learned doctor" was in particular associated with a form of rationalism of thought still abstractly linked to disease,[45] but without the pattern of experimentation, testing correction, hypothesizing, systematization, and reproduction.

Humanism instead focused on translating texts from the classical period, and on an appreciation of the wonder of knowledge. This was a humanism rich in the appreciation of human experience and creativity. Human knowing became a dominant measure; the glory of books ancient and contemporary, a pervading feature; an expression and interest in ancient and contemporary languages as an exploration of the knowing process, linked to a hands-on sensibility of engagement striving for a pervasive excellence, culminating in a "Renaissance person" (e.g. Leonardo). Scholastic frozen thought in architectural structure is replaced with experimentalism,[46] building, constructing, looking, and the valuation of the human knower. The abstract played less of a role, as opposed to the concrete. Tied to the labor of thought and cognitive capability rich in vectors of affective content are states of cephalic function.[47]

An imminent naturalism burgeoned despite the burning ashes of Bruno, and the recantation of Galileo. The heavens opened up through the rise of physics. But an orientation to facts was legitimated, and, in the process, the knowing human mind could be revealed.[48] For poets like Pope, and later Voltaire, Newton's powers were practically magical: nature revealed, all came to light, replete with prospects of rationalism and progress.

In Italy, for instance, this was a flourishing moment of unending discovery. It was an outpouring of human transactions and princely disorder coexisting with human ingenuity and creativity. A penchant for both social cooperation and social deception was manifest on a grand scale, reported in more detail via the new technology of the printing of books.[49] The human condition was made manifest by the cultivation of all things human and on a human scale. Human depravity and bestiality, in the artful hands of Machiavellian social groups, was a fall downward that was not theologically sanctioned but humanely expressed – "human, all too human."[50] In this context, exploration of both human anatomy and the discovery of rich civilizations in China are part of the same humanistic drive.[51]

The Renaissance embodied a revival of languages, the law, the cultivation of universities, and science and philosophy.[52] A revolt against the frozen other, a conceptually rich naturalism, and the taxonomy of Aristotle were present in the sciences, in medicine, in literature, and in discovery.[53] Medicine and mathematics were almost linked through Harvey's burgeoning experimental method, which uncovered the mysteries of the circulation of the blood. Mathematics, biology and physics were linked through the instruments and observations of Galileo and Brahe, and the anatomy of Vesalius and Leonardo.

Experimental spirit infused the Renaissance. Testing and exploring were mainstays of the vocabulary. Regard, for instance, the anatomical figures of Leonardo da Vinci. The important point is that nature was to be uncovered and to be explored. Vesalius, Netter, and other anatomists took anatomical depiction to a new level.[54]

Within a small period in Europe, medical books increased tenfold. The number of medical books produced in Latin from the early 17th century (50 books) is more than twice as many as those produced a generation earlier (20 books).[55]

John Locke is a paradigmatic example of the physician/scientist/philosopher. Locke had the temperament of both a healer and an adjudicator, looking for solutions, wedded to empiricism, yet always acknowledging the innate spaces in which thought and problem solving take place. Locke, to some extent, anchored thought to kinds of objects, and to determining the characteristics of objects.[56] He was by orientation a rationalistic doctor, with a soft heart for the human condition and the plight of others. Unlike some others, however, he was no experimentalist.[57]

A generation earlier, Francis Bacon took up the mantle of the experimental tradition, though experimentation in his time was still largely observational.[58] Indeed, Bacon died a martyr to the experimental method. On a freezing day in London, he stopped his carriage to purchase a chicken from a peasant woman, and spent an inordinate amount of time stuffing it with snow, in order to be able to observe the preservative effects of cold on meat. He arrived home thoroughly chilled and shortly thereafter succumbed to pneumonia.

One of the first great expansions of the experimental method in medicine was the origin of the physiological understanding of the circulation of blood. The measurement of blood had been a core feature of medicine since Galen's time, and with the new mechanical experimentalism that was emerging, there was much focus on this (see Figure 4.6).[59] Concepts of the mechanics or circulation were both metaphoric and real (and as investigators, medical decision makers still straddle from one to the other, often with no clear demarcation). William Harvey ultimately provided a concrete view of the mechanisms and physiological regulation of circulation; but the idea of mechanism was already out there in the intellectual milieu.[60]

Harvey, working on the concept of blood circulation, led medical inquiry towards an idea of mechanism replete with the rise of experimental method.[61] The idea of the *machine* was the empirical feature of his understanding. The experimental philosophy of Robert Boyle and the experimentalists was a precursor to the modern era of science. Instrumentation and measurement were an essential ingredient in his work, along with what Peirce some 200 years later would call the "community of inquirers," and what Boyle himself referred to as the "collective act."[62]

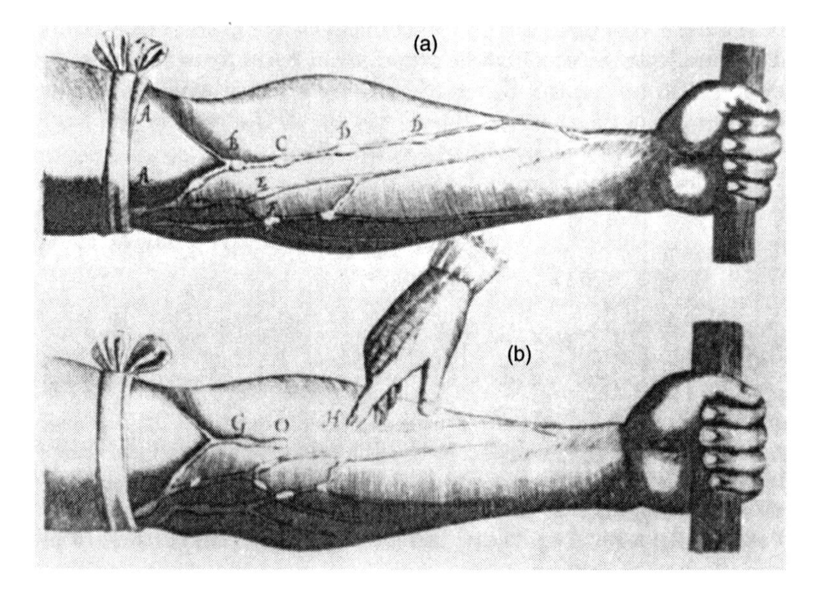

Figure 4.6 Blood circulation in the lower arm

The scientific community of the 17th century was descended in part from the medieval scholastic concept of the *collegium*, a group of like-minded individuals off whom one could bounce ideas. Argument alone, however, was no longer sufficient. Reason had to be coupled with experiment, invention, and instrumentation. Steven Shapin has suggested that truth was dependent upon honorable individuals, but not all were up for the task.[63] As Whitehead noted, social reality suggests that "truth is a faint motive." Individuals had to be perceived to be in good standing and part of a larger culture of a scientific enterprise, emboldened by moral sensibilities and other motivations reflective of the culture and historical context.[64] All this was done with the requisite reputation for truth telling, confirmed and expanded by others – a sense of character for truth, something Dewey saw as essential in education and in medical decision making.[65] Invention is at the heart of the enterprise of science, not simple observations. But invention is a slippery slope requiring truth to prevent avalanches (e.g. Grassi, for detecting motion).

The really innovative conception in the 17th century is its glorified depiction of the machine. Physicians such as Julien Offray De La Mettrie even described the body as a machine, working exactly like clockwork.[66] Intellectual figures as elevated and as disparate as Leibniz, Newton, and Descartes aimed to develop a formal calculus to map out the clockwork embedded in the matrices of geometry and to calculate it in deductive Euclidean precision.[67] This would all change with the idea of diverse forms of geometrical spaces,[68] and eventual divergent senses of time and space, in addition to paradoxes in logic and the recognition of interference effects that compromised certain features of physics.

In the 17th century, however, the clockwork was regarded as absolutely predictive. That was the cardinal feature of what it was to know something, to have legitimate knowledge, in the age in which reason was deductive and glorified. Anything probabilistic was "vulgar."[69] But, importantly, "numerical regularities about disease" were desired.[70] Within a short period, scholars would validate the use of statistical reasoning, for which the search for underlying logic would take place amongst other formal representations.[71]

Physiology, medicine, machines

A new experimentalism specific to the medical field was emerging that would profoundly impact the science of medicine.[72] French physician François Magendie was one of the first figures in this emerging field. One of his students was Claude Bernard, who would consolidate experimentalism within medicine with *An Introduction to the Study of Experimental Medicine.*[73] His emphasis was a rigorous science of medicine grounded in experimental physiology. He sought to understand the normal condition and link it to the pathological, both experimentally and by examining nature as it presented itself.[74] Bernard codified an experimentalism of medicine, forceful and clear. The model, however, was not that of adaptive tissue: there is nothing of nature in his experimental sensibility, the wonders of nature. The focus was on disease, how to fix it, and mechanisms of disease. This is understandably so because he was fighting the demons of vitalism, and other forms of non-necessary events. Either things were necessary, scientific and mechanical, or they were not.

Amongst other things, Bernard made a detailed study of the stomach following hunger. He noted the occurrence of diabetes, and its putative mechanisms. He remained a peripheralist and reflexologist in his perspective. Cephalic capabilities, anticipatory and regulatory, knotted to adaptive behaviors, were not part of his conceptual arsenal when considering adaptive living systems, diseased or otherwise.

Traditionalists like Claude Bernard still envisaged their discipline as a science without probabilities.[75] This would eventually give way to the integration of statistics across the broad array of the sciences, particularly the medical sciences. Bernard, however, though an experimenter in modern medicine and himself a physician, was anchored to experimental medicine with preconceived ideas. In many ways, Bernard still labored under a 17th century notion of the body as mechanism, as clockwork.

Bernard nevertheless understood the diversity of science and method, and that each medical science, and science in general, is contingent upon what is sought after, and what is being understood. In that, he was close to Aristotle: no one method fits all. He shifted his argument from reason to experiment, moving away from the deductive necessity that was required for legitimating arguments. To truly make major advances in medicine, human beings had to be understood not as statistical machines, but as labile, adaptive, and changing over time.

Pavlov and experimental physiology

Ivan Petrovitch Pavlov (1849–1936) was deeply influenced by Bernard, and was firmly in the tradition of experimental physiology and modern medicine. He won a Nobel Prize for his work on the pancreas and the physiology of digestion.[76] Pavlov measured digestive physiology in terms of its reaction to diverse forms of eatable objects[77] and then looked at the role of peripheral control, in particular the role of the vagus, the 10th cranial nerve, a wandering nerve innervating many systems including the stomach, liver, heart, etc.[78]

Pavlov's sense of science was based on broad-based regulation during adaptation and devolution of function during pathological compromise. Pavlov was also a thorough experimentalist, a physician by training and a laboratory manager and leader by experience.[79] One of his best-known experimental protocols is called "sham feeding": one separates what goes into the oral cavity (food) from what enters the stomach and gastrointestinal tract, to determine the efficacy of the oral from the absorption factors on the physiology of food ingestion.[80]

As Pavlov continued evolving as a scientist, he began to consider cephalic systems and thus entered a realm in which the "purposive brain" is paramount.[81] His approach is reminiscent of Dewey's reflex arc conception, where cephalic capability is anticipatory and rich in purposeful expectations. It is not a simple Cartesian reflex at the level of the brainstem (Figure 4.7) that predominates in the organization that underlies behavioral adaptation and flexibility.[82] But, as Dewey and others intimated, cephalic capability is more than reflexive, it is inherently anticipatory. The model is not simply one of reflex to pain for cephalic expansion and evolutionary adaptation.[83]

As Pavlov moved towards understanding cortical control (as opposed to peripheral and brainstem control), he talked about "psychic secretions."[84] The buzzer he used in his famous dog experiment came, by conditioning, to mark a food source for the dog, and a conditional reflex to the buzzer results in the natural reflex to the sight or smell of the food source. Everyone always remembers that the dogs drooled in response to the buzzer, but the crucial "psychic secretion," as Pavlov liked to call it, is not so much saliva but insulin.[85]

Dewey, earlier, criticized the narrow notion of simple reflexes as predominantly cephalic adaptations.[86] Reflexes, he thought, are embedded in a much larger stream of cephalic adaptations and behavioral expression, and thus cephalic organization underlies behavioral adaptation. Cephalic machinations, anticipatory and adaptive in nature, integrate

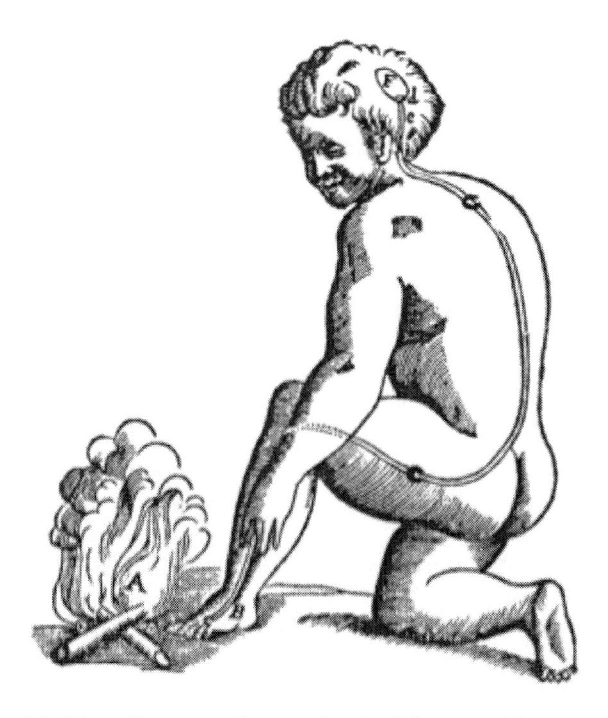

Figure 4.7 The reflex arc as illustrated in René Descartes' *Treatise of Man*

our internal physiology by behavioral regulation.[87] For instance, consider the rich innervation of peripheral sites or the activation of peripheral sites by vagal efferent projections into sites along the digestive tract.[88] Diverse anticipatory systems are expressed as we are about to ingest a food resource, many of which are the rich array of information molecules expressed in the brain and peripheral nervous system in the absorption and utilization of food resources.[89]

Pavlov called it the "cephalic phase." By this he meant the anticipatory systems in the utilization of resources required by the body, such as the secretion of insulin before the utilization of the resource, and the regulation of insulin secretion by time of day in which a food source might be available.[90] Thus, anticipation of food is but one of many information molecules vital for the acquisition and utilization of food resources. The context of food passing through the oral cavity facilitates the anticipation of utilizing and distributing the vital food resources and therefore, in this context, insulin is secreted in advance of absorbing the nutrients.[91]

All our evolutionary adaptive systems conserve and redistribute resources from end-organ bodily organization and cephalic expansion. The seduction of the sensory, laden with cognitive content, pulls us here and there by objects of our invention: processed foods, sweets, and fats to which we are gluttonously vulnerable, resulting in devolution of function.[92] Diverse hormonal systems that regulate energy balance act in the brain to facilitate attention to eatable objects, some good – many not, especially in our culture, with its deluge of consummatory seductions.

Insulin is now a common, and well known, peptide hormone. Realization of its function was a profound medical discovery.[93] Glucose in the urine, with its dark and protruding smell, has long been noted.[94] Its association with some factor in the blood began to emerge at the end of the 19th and early 20th centuries. Scientific understanding of blood glucose and the role of insulin became clear in the 1930s, and their relationship to diabetes became obvious a short time thereafter.

Insulin is, in many respects, an information molecule.[95] Insulin has a primary role in glucose regulation. It is secreted from the pancreas, but insulin-like factors are found in other tissue. The biological origins of insulin are millions of years old and are part of a large family of peptide hormones with common phylogenetic ancestors.[96] Common growth factors similar to insulin can be found in a large array of taxa.

Evolutionary trends show the diverse ways in which we store and restore bodily resources to sustain ourselves for the short- and long-term, within limits. While, for example, the secretion of insulin from the pancreas is vital for the regulation of glucose in the blood, insulin is itself a growth factor and can generate and sustain diverse tissues in the body (e.g. brain).

Normal insulin secretion monitors glucose and stabilizes or renders viable bodily tissue function. Anticipatory insulin secretion, cephalically mediated by sights and smells and tastes and time and context, is vital before the ingestion of a food source.[97] As soon as a food source is detected and is needed by the body, insulin is secreted to utilize the resources and distribute them to the end-organ systems that need it. Our evolutionary history is towards storage and use in a world with restricted resources, however, not towards excess and gluttony.

With the onset of endless supplies of relatively cheap and easily obtainable junk food, the exaggeration of this phenomenon is a devolution of function. Chronic exposure to food resources and exaggerated insulin secretion results in decreased function, and, ultimately, the breakdown of tissue. Type 1 or juvenile diabetes is well known and, while the ultimate cause is still unknown, it may be triggered by a virus or caused by injury to the pancreas, resulting in a loss of insulin production. Insulin injections restore function. With careful control, individuals can live a full and long life. One colleague of mine is a marathon runner, in his 60s, and has been a diabetic nearly all his life.[98]

Type 2 diabetes (adult onset dysregulation of glucose and low levels of insulin) is in this sense a devolution of function. It wreaks havoc in the regulatory systems in a culture in which junk foods are omnipresent, in which people are vulnerable to over-ingestion and chronic over-stimulation of cephalic regulatory systems. Diverse hormonal systems acting in the brain trigger a cephalic drive of exploration, and a readiness and a propensity to respond to diverse eatable objects that can be misperceived for their regulatory richness – a devolution of function.[99]

As a food source enters the mouth, insulin is secreted in preparation to absorb and redistribute the glucose to which all food is reduced by the digestive process. Foraging a food source and its utilization are therefore intimately bound. The appetitive phase, the search, and the consummation are common categorical distinctions used by Dewey and many other investigators[100] in linking these events to many avenues of animal and human adaptation.

The discovery of insulin and its use as an ameliorative factor is a profound example of the growth of knowledge.[101] The social cooperation and competition in the research are themselves features of our primate urge to share and compete, explore and understand, test and retest. The knowledge produced is collective.

Stem cells, progenitor cells for all possible cells, may turn out to be the answer to diabetes. The stem cell may potentially be used to clone insulin, or other growth-like factors, transmitting them via vectors that can then be incorporated into devolved cells damaged by the wear and tear of secretion and use. Use of stem cells in this way may take time. Genomic research to cure diseases has been oversold and has not materialized since it was first envisioned. We know now that Galen was not entirely correct when he stated that knowing the cause of a disease was to know its cure. But we now know most of the structure of the genes, and we have characterized the genetic fault lines for diverse pathologies, or features of devolution (e.g., cystic fibrosis, diabetes, etc.). We have the idea of how to cure disease – push out the defective gene with a viral vector and replace it with a functioning gene – but getting it to work is quite another thing. The modern, Pavlovian experimental approach may eventually give us the answers we seek.

Pain and social knowledge

The amelioration of pain is both a regulatory adaptation and a social phenomenon. It derives from part of a community of educated or educable individuals. The social milieu is essential to the advancement of the cause. Direct observation was the most authoritative means of discovering the secrets of medicine, coupled with mathematical representation.[102] The perspective of nature at the heart of modern science also shifted dramatically during the enlightenment period.[103]

A body in pain was familiar then as it is now, but dying was closer. It occurred at home for the most part, not isolated in a hospital. Life, as Hobbes put it well before Darwin, was "nasty, brutish and short."[104] Darwin emphasized the survival part, but before him the breakdown of the world through bodily disease and pain was a part of the material of everyday life in the world, sometimes exaggerated during famine and war.[105] No one except a detached rationalist would make the body inert, the mind alive and separate them radically. Medicine's job, as Bernard would later say, was to minimize disease.[106]

Pain is a biological adaptation. The urge to remove it is only one response to the problem of disease. The rise of surgery, which causes damage in order to heal, is another (Figure 4.8). The surgeon is one of the pillars of the medical profession in its origins of the discovery and eradication of bodily discomfort.[107]

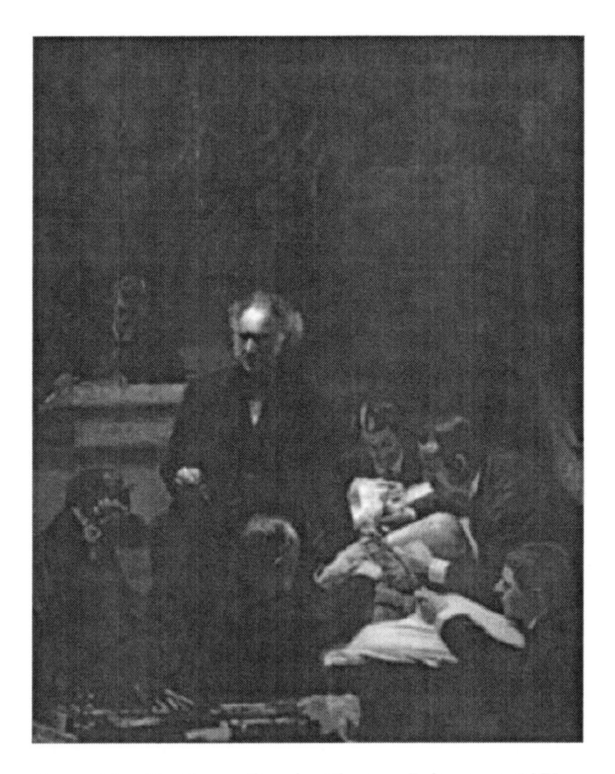

Figure 4.8 *The Gross Clinic* by Thomas Eakins, ca. 1875

Of course, pain transcends social context, just like many other medical events for which the physician plays a fundamental role in management and amelioration. But the social context is as pervasive as is the experimental.[108] Medicine, in Bernard's view, was the science of disease; but medicine is, in fact, more than that. It is part of the larger biological field, in which disease and its understanding is as essential to knowledge as is the normal and well-functioning physiology. Core concepts are development, regulation and devolution of function. These were all concepts familiar to 19th century physician scientists such as Bernard or Hughlings Jackson, the great British neurologist.[109]

Hopkins and medicine

While Pavlov was working in Russia, German science dominated the sensibility of medicine at Johns Hopkins. The first real integrator of modern science, the German approach was a combination of the theoretical, historical, and experimental.[110] German intellectual influence in the sciences is made transparent in Johannes Muller, the great physiologist (1801–1858), whose influence reached to Henle, Schwann, Bois-Reyond, Helholst, Virchow, Remal, and Haeckel, and most fields of the burgeoning biological sciences, ripe with anatomy, physiology, physics, and evolutionary theory. The rigor of German medicine was based on the physiological sciences, a physiology emboldened by biology.[111]

Muller's followers discovered and explicated the loops of Henle in the kidney, Schwann cells, and the study of viruses. They made an attempt to apply the understanding of electricity to physiology, made contributions to the science of perception,[112] and even developed a specific view of evolution. German medicine was grounded in science, the science of the day, and set a high standard of excellence. Thinkers such as William James travelled to Germany motivated by a desire to study with the great scientist physicians of the day. The study and practice of medicine at Hopkins is based on the German approach.

William Osler, pathologist and Hopkins professor, is sometimes called the father of modern medicine.[113] *The Principles and Practice of Medicine* (1892) was his major text, written during the great era at Hopkins when Osler and three others were extraordinarily productive – a time reminiscent of the four horsemen at Notre Dame (although that, of course, was football). A Canadian, who initially trained in Europe, Osler held appointments at McGill University and the University of Pennsylvania School of Medicine before becoming the first chief of staff at Johns Hopkins Hospital, and one of the first professors of medicine at the Johns Hopkins University School of Medicine. Osler ended his career in the Regius Chair of Medicine at Oxford University. Physician, humanist, and historian, Osler was a true Renaissance person, and Anglican to the core in religious sentiments and responsibility. *The Evolution of Modern Medicine* is one of his best-known treatises.[114] Osler's lectures on the "Medical Anatomy of the Brain" were legendary. Pathology, his specialty, focuses on the deterioration of tissue. It was a major criterion for the dissection of disease, and for tracking objective changes in the devolution of function.

Besides Osler, the remaining "horsemen" at Hopkins during this period were Halsted the surgeon, Welsh the educator/administrator, and Kelly the gynecologist. They were immortalized in John Singer Sargent's painting. Sargent, himself the son of a doctor, depicted them as the four immortals, as they were regarded. William Welsh was the teacher of Simon Flexner.[115] At Hopkins, this pathologist was the dean of development of academic medicine. Atwood Kelly, head of gynecology at

Hopkins, also a surgeon, was unlike most such medical specialists in that he was not much interested in obstetrics, but in the conditions of the female anatomy.[116] William Stewart Halstead was an amazing surgeon and a thyroid specialist. Parathyroidectomy was one of his surgical feats, which influenced many, including Curt Richter, a laboratory scientist and artisan craftsman at Johns Hopkins for 60 years in the middle of the 20th century.[117]

Simon Flexner,[118] much influenced by all these individuals, wrote about William Henry Welsh and the Heroic Age of Medicine. It has been some time since the word "heroic" has been uniformly presented in the context of medicine, but it is certainly appropriately applied to these seminal figures. Teaching and research in medicine found welcome valorizers in the Flexner brothers, endowed with the Rockefeller and Carnegie models for transforming American medicine.

Abraham and the Flexner Report

With a Hopkins undergraduate degree and originally teaching high school in Louisville, Kentucky (his hometown), Abraham Flexner, the nonscientist supported by the Carnegie Foundation, was saddled with the task of cataloging the state of medical education in America. It was as much a social critique as a scientific one. It was more critical journalism, perhaps, than anything else. The Flexner Report appeared in 1910,[119] reporting on 131 medical schools in America and 24 in Canada, totaling 155. Many were ill equipped, and all had uneven standards. Few were regarded as truly outstanding. The Flexner Report suggested that most medical schools at that time should not exist, encouraging, as they did, an almost charlatan approach to medicine. The report stated that all were in need of a sense of excellence and accountability. Indeed, in 1910, few medical schools even required college as a prerequisite for entrance.

Interestingly, there are many parallels between Dewey's philosophy and the Flexner Report. Dewey's emphasis on progressive education, the importance of science and the scientific approach in everything, are all elements of Abraham Flexner's approach. His approach went so far as including a sense of the student, taught as an apprentice, and brought into the culture of science, by example, towards a sense of lifelong learning as the normative goal.

Under the influence of John Dewey's larger sense of philosophy about continuous education, pragmatism, self-corrective inquiry, and the progressive stance, patient care and medical orientation finally found a welcome in America.[120] The emphasis went from undergraduate education in medicine to graduate-level medical school, anchored by the rise of biomedical learning and the practice of biological sciences. A continued rise in graduate education amongst medical students fueled the learning expansion in the profession.

John Dewey's approach to the experimental method emphasized its link to Darwin and a philosophy of change,[121] adaptation, and the retreat from certainty.[122] The cultivation of education is one of intelligence and problem solving, broadly understood and endemic across all areas of the human condition. Authority is an achievement through experiment and self-corrective inquiry, but it is never a certainty. Dewey naturalized the philosophical quest to know and made it part of the empirical sciences, while he broadened the empirical sciences to take into account philosophical issues. Reflective inquiry pervades all inquiry as well as our democratic ideals. Our notion of progress, medical or otherwise, is in backing a philosophy of change, ushered in, in part, by the biological sciences. With the decline of necessity came a realistic sense of freedom of expression amidst the constraints of culture and our biology.

Dewey on aesthetics, medical science, and evolution

At the heart of classical pragmatism of the sort emanating from Dewey is the continuity of art and science. Aesthetic sensibility is pervasive across the human knowing experience.[123]

Halstead at Hopkins influenced a wide array of surgeons, including Harvey Cushing.[124] Cushing, one of the leaders in neurosurgery in this country, was a student with Halstead, and from him learned much about the removal of tissue. Cushing syndrome, associated with adrenal pathology, was first described by Harvey Cushing, himself.[125]

Cushing, a major founder of neurology in the U.S., was not always so forward thinking in his personal opinions. He believed in limited space for Jewish students, for instance.[126] Nevertheless, his work on the hypothalamic pituitary adrenal axis helped define a whole field of inquiry. Cushing was also a bit of a Renaissance person. A fourth-generation physician, he would write a renowned biography of the life of Osler, which won a Pulitzer Prize in 1926. The aura of Hopkins would define his orientation as a researcher and a physician.

Besides being a biographer, Cushing was a gifted artist, much like William James. Cushing would later embrace what would become the first art department at a medical school. But a German immigrant at Hopkins, Max Broedel,[127] would make the greatest contributions to artistic medical depictions. His work harks back to Leonardo da Vinci. His depictions are endlessly graceful. In them, there is no separation of mind from body. Through his artistic renditions of medical pathology the body is up close and personal.

Art, as Dewey so well understood,[128] was part of the experience of inquiry. Art as experience was not separate or for the parlor, but endemic to inquiry: "Esthetic cannot be sharply marked off from the intellectual experience."[129] It was essential for the reconstruction of aesthetics to the knowing process and the expansion of human experience and human knowledge.[130] Aesthetic features are at the heart of the contours of basic research experience. The elegance of the design to the clinician's eye, history-laden and dependent but building on the visual sensibility, is what defines our experiences and defines our species.[131]

Art, medical or otherwise, heightens the vital sense of learning, ripe with an organism's intensity. It permeates a wide bodily surface, cephalic in expression, with directed attention. Art like this is a form of experience, permeating the embodiment of knowledge. Art as experience heightens the vital sense of worth in the context of human discovery; humans have understood this for ages. Art may, in fact, be built into our biology, stretching back to cave paintings and our desire to represent our surroundings, to make sense of our world.

Dewey emphasized the instruments that helped the physician explore and that also helped the artist expand our horizons. Problem-based medicine was linked to empiricism, with any tool that helps decision making as the aim and ideal of the experience for the physician. The aesthetic experience focused on the form of the body, elegant, depicted in normal and pathological conditions: the "esthetic forms in terms of completeness of relations."[132]

Technological advance underlies Dewey's conception of the ascent of the epistemological engine, medical or otherwise. Tools are embedded in his theories of seeing, knowing, orienting, and perceiving. The tools facilitate the search for some control in understanding. Knowledge is endemically instrumental in use and can be aesthetic in scope, and practical in use.[133] The point is that there is no separation in these cognitive functions tied to the organization of action and perception.

In other words, epistemological advances rested on technological advances, so that medical or any other advances became a form of decision making. Artisan hands are essential and endemic to a culture of discovery and discernment, to medical prognosis and medical advance, and to patient healing or soothing. The practice of science rests on the tools of discovery; the telescope, and the microscope, seeing through the lens of the technological to advance the sensory systems.[134] Art, in the sense of *techne*, runs through the veins and pulse of science. The reconstruction of others, for instance, represents one small glimpse of the pervasive features of art in science (see Figure 4.9).

The open-ended sense of medical or other technological innovation is something Dewey well appreciated, and underlies technological change, a change that comes fast, punctuated by historical moments,[135] and having far-reaching consequences and applications.[136] In medicine, two key factors are omnipresent: clarity (in some instances endless gray, with some adjudication in many issues), and technological innovation in effecting the resolution of a problem. No predictive or control factors can predominate in the investigative process in terms of the larger range of consequences. Nevertheless, understanding the full meaning, contextualized from its idealized end, requires an expansive notion of inquiry and investigation; the technological embrace is at the heart of the medical advance amidst the aesthetic component inherent in the innovation, with the endless tie to correction.

Art is endemic to human knowing,[137] and that is the point. There should be no mythological sense of art, just the importance of enriching our experience as the human knower. Art, unified in the expression, is endemic to medicine, is part of the valuation of events. The impulse to know is part of capturing the kind of illness, the expression of health, and the development of a function.

The expression of form is a constitutional feature, an architectural expression – from tool making, to responding to objects seen, to objects

Figure 4.9 A paleoartist working to reconstruct models of human ancestors

understood, is vital to art in the medical field – of the body, of the evolution of form as it is the devolution of form in disease and normal aging. These considerations always, to be sure, must be understood in light of history and social context, the normal and pathological themselves both having an historical context.

Historically, medical practice is particularly prone to dominance relationships, since it is so authority driven in the U.S. However, it does not last if it is not good practice. Decision making not grounded in good science – slow as it might be, imperfect as it is – won't last. That is the saving grace of medical science, faint, pallid, and imperfect as it is.

Statistical reasoning

C.S. Peirce, as I indicated in Chapter 2, writing the first treatise about probability in the U.S., while criticizing deductive methods, helped instantiate the science of probability into the logic of decision making. Peirce's contributions to understanding the logic of statistical reasoning are immense.[138] Advances in statistical reasoning in biology, from Pearson[139] to Fisher, changed the landscape and offered a chance to embrace probability as inherent in biology and medicine. A sense of objectivity long held within physics now changed when statistical theories questioned the inherent probability of events, moving away from the Newtonian view.[140] The metaphors in physics changed towards a probabilistic sensibility. In the feeble sciences, such as economics, probability was desperately embraced and understood to provide a logical apparatus for clarifying inferences and reasoning, to compute probabilistic relationships, and to ground statistical inductions. A logical apparatus was needed for computational activities, and the form of statistical relationships took material shape.[141]

Thus, the rise of statistical reasoning took on a pejorative sensibility in the hands of the deduction models that dominated from Euclid and mathematics: statistics evolved from noting state records and cataloging diseases that impacted human and animal populations. Although it was seen as a lesser science, a science of chance, nevertheless "taming chance"[142] became a wellspring of activity at the heart of critical reason and epidemiological thinking.

Human biases and heuristics, and medical decision making

Our cephalic and adaptive responses aim at conserving and storing when we can. We evolved a set of cephalic adaptations. Toolboxes of problem solving capabilities, as envisioned by Peirce and Dewey, realized in the later part of the 20th century and the first part of this in terms of naturalistic predilections, embedded in adaptive systems and diverse forms of reasoning and arguments.[143]

Central to these concerns are the vital sense of meaning in human expressions. Ethics, for example, is an extension of moral sentiments.[144] Ethics is not identical to morality or moral sentimentality. Rather, ethics is the reflective inquiry into moral sentimentality. Valuation is a fundamental part of our experience and cephalic expression. Forged against the enrichment of human social contact and discovery, hard fights in the trenches amidst moments of ease and delight, the meaning of human activity captures the worth in things. This is the value of cephalic expansion.

The heart of the discipline is that decision making is fraught with inherent "framing effects," or biases in reasoning.[145] This is neither good nor bad in itself, although "you are biased" is usually considered to be a pejorative statement, meaning that you see only one side of something. A more appropriate way to consider bias is to see that our perspective provides our orientation, the direction from which we see things.[146]

It is important for today's clinician to be able to weigh evidence, recognize biases, and reach an informed decision. This is a central aspect of clinical reasoning. Cognitive biases can be quite persistent and are therefore unlikely to be totally eliminated. Instead, the goals should be to help physicians become sufficiently aware of the influences of cognitive biases to avoid being fooled by their own perspectives.

The decision sciences literature is rich in detail. It depicts the diverse ways in which we overvalue or undervalue variables (e.g., whether hormone therapy will result in breast cancer), depending upon our orientation to events, the starting points or frames of reference from which we begin. This also holds for our memory of events, by which we associate rewards or hedonic value. For example, we may overvalue (due to a biased frame of reference) whether we would be happier living in the state of Washington versus somewhere in the Southwest, regardless of whether we actually live in either place.

It is necessarily true that we see things in relation to a background framework, and so we always see from a particular orientation. The question is whether the orientation can be detected, is testable, can be criticized, challenged, experimented with, replicated, and expanded. In this regard, scientific inquiry must be seen as an orientation like others. Biases permeate any endeavor and must be recognized, examined, and taken into account in decision making. Moreover, the recognition of biases functions through what social scientists call "sensitizing concepts." That is, the biases themselves provide a context that helps us understand how we decide.

A formal part of the decision sciences is quite rigorous (e.g., decision-analytic trees and the logic of decision making), and these tools have proved useful in some real-world contexts. Decision science is closely tied to statistics and psychology, and is an attempt to make transparent the decisions that we make. But it is misleading to think that the decision tree is anything more than a tool that assists us. The decision sciences nudge us to acknowledge the biases that inhere (and that may or may not be a bad thing), and then to work to prescriptively correct them.

There are diverse forms of biases. One of them is the "status quo" effect: we have a tendency to stay with what we know, particularly when something is ambiguous.[147] Another form of bias is that something is better if it is made by the body than if it is synthesized. Endogenous estrogen by nature, defined by this view, has to be better than anything we could produce in the lab, although in fact right now there is no real evidence to indicate whether this may or may not be true.[148] Then there are wide ranges of biases that reflect the frameworks that we impose: our statistical over- and under-estimations, the limits of what we are able to imagine, and diverse forms of correlation confusions,[149] to name a few kinds of inferential biases in our thinking.

A realistic way in which to understand our many intellectual predilections is to recruit them pragmatically to enhance human decision making, including, but not restricted to, cost–benefit analysis of the decisions that we make. One also needs a framework in which our appraisals are to be understood. Such a framework includes: the values that inhere in our judgments about events;[150] the social cooperative behaviors as an extension of cephalic capabilities; and how we understand our value judgments.

Valuational judgments are at the heart of our appraisal systems, the cognitive capacity that reflects our interests, and the cultural milieu in which we live.[151] As Jonathan Moreno so nicely put it, "the pragmatic naturalist understands that the knower and that which is known are in the same matrix, just as the inquirer is within nature and is one of its entities along with the object of nature, not outside of nature or fundamentally disconnected from the object."[152]

The origins of decision science are bound to statistical analysis, defined as the determination of the utility of a decision in the likelihood of considering the outcomes.[153] Regardless of whether we are physicians, patients, bus drivers, or rocket scientists, we are vulnerable to many kinds of errors in reasoning[154] and need to work to self-correct for them. Heuristics – orientations that seek less than perfection as a starting point – are close to the pragmatist position. Gigerenzer, following Herbert Simon, looks for ways in which we adapt and satisfy ourselves with the best we can do with mechanisms that serve us well. Some of the heuristics are shown in Table 4.2.

One part of the cognitive revolution has been to highlight the diverse forms of vulnerabilities: to over-exaggerate statistical relationships, to miss judgment of our expected pleasures, to risk misdiagnosis because of frequency of experiences, to be particularly vulnerable to mismanaging proportional relationships in time-dependent judgment choices.[155]

This litany of intellectual foibles stands alongside a set of cognitive predilections. Here are several:

Ambiguity Bias	Stay with what you do for lack of certainty
Dominance Bias	Narrow scope, single mindedness, ignores facts and observations
Status Quo Bias	Stay with what is already the case
Recency Bias	Unduly influenced by recent occurrences
Omission Bias	Resisting action
Naturalistic Bias	What is natural is right
Intertemporal Bias	Inability to choose a reward and outcome

This list of categories is finite, but long, not simply modular and isolated but integrated. The organ metaphor, as used by Noam Chomsky, is useful here.[157] Kidney function or liver function are separate and discrete,

Table 4.2 Heuristics that are likely in the adaptive toolbox

Heuristic	**Definition**
Recognition heuristic	If one of two alternatives is recognized, infer that it has the higher value on the criterion.
Fluency heuristic	If one alternative is recognized faster than another, infer that it has the higher value on the criterion.
Take the best	Infer which of two alternatives has the higher value by (a) searching through cues in order of validity, (b) stopping the search as soon as a cue discriminates, (c) choosing the alternative this cue favors.
Tallying (unit-weight linear model)	To estimate a criterion, do not estimate weights but simply count the number of favoring cues.
Satisfying[156]	Search through alternatives, and choose the first one that exceeds your aspiration level.

Source: Adapted from Gigerenzer and Todd (1999).

but also integrated inextricably with other bodily systems; so, too, for cognitive capabilities. Language, the organ of decision making, is tied to memory and other cognitive capabilities, which necessarily alter core features of cognition. The hallmark of cognitive evolution, therefore, is integration and novel expression of diverse forms of capabilities.

The key here is coping with uncertainty, demythologized human reasoning capability, medical and otherwise, without tossing rationality aside. There are better and worse forms of decision making. Medical decision making forces online decisions in real time. As it is now taught in medical schools, bias correction focuses on the cognitive orientation that we discern good reason from bad and recognize the vulnerability of systematic mistakes.[158]

Evidence-based medicine

Evidence-based medicine (EBM) arose from epidemiology, which deals with the logic of research design, the clarification of studies, and the broad assessment of data and the story it tells. In a real sense, the term "evidence based" is a term that signifies scientific legitimating. The idea that experts would grade information, make it available, and then back it up by whether it meets criteria, harkens back to many attempts to legitimate inquiry and place it on a firm footing.

What is different between evidence-based knowledge and medical decision making has been characterized in Table 4.3.[159]

Table 4.3 Differences between evidence-based knowledge and medical decision making

Issue	Evidence-based knowledge	Medical decision making
What are the core problems?	Insufficient knowledge	Combining data and values consistently
	Keeping up with the literature	Judgment limitations
Solution	Critical appraisal	Decision tree
	Answerable clinical question	Sensitivity analysis
Formal models	Unimportant	Essential
Decision processing costs	Low	High
Utility assessment	Largely ignored	Core research issue
Cost-effectiveness analysis	Largely neglected	Major application
Decision psychology	Largely ignored	Core research issue

Source: Adapted from Elstein (2004).

EBM is a more recent attempt to give a scientific foundation to medical decision making. Its legitimacy derives from scientific rigor in research and the use of research findings. EBM is part of the larger culture of evaluating the glut of information that is available and turning that information into informed choice. Information must be parsed and made manageable. Teams of experts assemble to evaluate information, to decide how well studies are designed and how strong the evidence is. EBM seeks to find rigorous scientific grounding for medical decision making by having the decisions strictly follow the science and codifying knowledge.

Evidence-based material provides physicians with the ability to consult evidence-based analyses of findings of randomized control trials. One well-known version of EBM is the Cochrane Library, which evolved to provide health care decisions based on a sound, logical footing and good science.[160] The goal is to systematically review the scientific evidence, codified in terms of levels of evidence. The Cochrane Library relies on principles that include collaboration, reduction of bias, keeping current on the literature, access to information, and the sense of inclusiveness and goodwill of those involved. Within a relatively short period of time, several centers both within Great Britain and outside of Great Britain developed. While valuable for generating hypotheses, observational studies provide less convincing evidence than do randomized clinical trials, but valid approaches in this category are cohort and case-controlled studies.

One primary factor in EBM is the value of the randomized control trial, and the grounding of physician suggestions to the patient with the best kind of evidence. The primary argument for developing the guidelines was to reduce harmful biases, to make the evidence more transparent, and to place decision making in the context of self-corrective processes. Physician practices are evolving towards a sensibility that makes space for EBM. The evaluation of the evidence has shifted from individual and semi-autonomous decision making to a more collective professional body of authority. Physicians tend to rate the evidence-based material as very helpful to their practices.[161]

We nevertheless know and recognize that there is still wide variation and inconsistency in the implementation of practice guidelines, and in their effectiveness in promoting better decision making. And we know, at least in the context of obstetricians and gynecologists in the United States, that the evidence-based practice guidelines are perceived as significant and, in some cases, implemented. Keeping track of actual physician behavior, however, is not easy.[162] In some contexts, we also know that if they read the evidence-based documents, obstetrician-gynecologists tend to perform better on self-reported knowledge and practice questions.

The Cochrane studies are the paradigmatic example of evidence-based analysis. The emphasis is on evidence-based material, and the evaluation of different kinds of evidence in the facilitation of medical decision making. Again, evidence-based medicine is a tool amongst other tools and has to be contextualized to be rendered useful, not mythologized as the only mark of legitimacy. When possible, peer review of the reviewer, self-correction, determination of the questions, awareness of the biases – and even consideration of the rules for grading the evidence, including the logic of considering the randomized controlled trial as the most highly valued form of medical research – should be built into the process and translated into health-related policy decisions. Decision science directs our attention to recognizing and accounting for possible bias in decision making. Evidence-based medicine rests on collective choice (asking physicians to accept their colleagues' evidence-based judgments).

Patient rights, pragmatism, and human inquiry

An evolving sensibility of medical decision making has taken place within the context of human rights, patient rights, and women's rights. Pragmatism meets feminism in the engagement of women's rights and fundamental respect across the wide array of human engagement and human value, for which Dewey is one important figure.[163]

The participatory and democratic sensibility of enlightened citizens so essential for Dewey – although it is no easy task and is importantly linked to an evolving notion of human rights and human participation – underlies a patient's right to participate and take responsibility in medical decision making.[164]

One view of human rights and their expansion is tied to the right to know what dangers we face. Participation and access to information are at the heart of the practice of democracy, of a sense of freedom knotted to the knowledge-acquisition process. Also at the heart of democratic practice is the "enlightenment pathway" that is tied to transparency,[165] participation, and citizen literacy and responsibility, something much in accordance with Dewey and the naturalization of ethics, a subset of which would become bioethics.[166]

Cooperative problem solving is part of our cognitive evolution. Harnessed to conflict resolution, especially with regard to social–institutional conflicts, are the joint efforts toward amelioration. Such efforts, like those of a doctor and patient, are rich in social values and humane expression. These efforts have as a normative goal the protection and inclusion into a broader arena of considerations, particularly when it comes to patient respect and patient participation.

The knowing process is always itself, in part, about self-correction.[167] This involves getting a sense of the evidence, as well as gaining some sense of fallibility,[168] particularly with regard to the biases and orientations that figure importantly in decision making.[169] More generally, and historically, a growing sense of rights knotted to autonomy and choice,[170] in addition to social connectedness and cohesion,[171] is a normative goal.

Championed by Dewey was the continuity in moral inquiry amidst the larger sense of investigation. The modern concept of rights is tied to inquiry[172] and has culturally evolved in a liberal participatory democracy.[173] The evolving sense of human rights as something all too easily trampled has resulted in a heightened vigilance, and even suspicion, of medical authority. This can be seen in the development of codified patient rights. Such principles have evolved, in part, for the protection of human subjects in human experimentation (e.g., The Nuremberg Codes, Declaration of Helsinki, President's Commission on Ethical Issues in Medicine). The goal of these codifications is the protection of the human subject, ensuring that when patients undergo a treatment they are maximally informed of what is involved, of the potential consequences and risks, and that they are competent to understand them. Informed consent is now a fundamental feature of biomedical ethics.

At the same time that the world marvels at the apparently superhuman powers of modern medicine, it is also developing a cultural value of moving away from narrow, authority-driven expression. That is, the culture is rejecting the notion that the physician knows what is right, and is moving towards evidence-based decision making within the context of respect and promotion of patient participation and patient responsibility that is combined with an understanding of the social milieu and the cultural biases, which permeate these events.

The promotion of patients being thoroughly informed before they choose, and respect for this endeavor of making the information as transparent as possible to inform choice, while also benefiting and respecting the patient,[174] is a goal of the culturally reinforced covenant or contract between the physician and the patient.[175] Of course, in the real world, "thoroughly informed," like "perfect rationality," is more myth than reality.

We work with what we have. A progressive sense of bioethics sees continuity and looks for it. It seeks to include as many as we can in the conversation about ethics and rights and the technology that expands the biology.[176] The adventure of medicine, like other areas of human inquiry, suffers no boundaries. A progressive sensibility is bound to the ever-expanding possibilities of our inventions amidst the reality of the diverse forms of devolution of function. The focus of science is on diverse forms of methods within diverse ends that underlie the means and modes of action, such as cognitive systems endemic in action.

Pragmatists, such as Dewey, always keep lively ends in front view in the context of the instrumentality required in action in a functioning, moral, medical world. This is a world of consenting individuals emboldened by their knowledge and able to participate in their medical care if they choose to do so, amidst a sense of human dignity and towards moral transparency – towards good practical sensibility and a resoluteness in the search for common bonds.

Informed consent takes place in a context of disclosure, assessment, and understanding. One feature is the attempt to make transparent not only what is known but also the biases that underlie diverse treatments, the orientation of the physician herself, and possible conflicts of interest.[177] In the end, the physician has a moral responsibility to respect the individual, and her right to self-determination.[178]

In an ideal shared decision making process, the physician ought to: (1) understand the patient's experience and expectations; (2) build trust and participation; (3) provide evidence-based material with a discussion of uncertainties; (4) expose biases in the data and orientation; and (5) continuously check for understanding on both sides.

A Deweyan conception of rational choice is very social in its implications, particularly in the ability to interact and have others at a table where there is disagreement. Part of understanding bioethics is to see it as just another branch of investigation.[179] Albeit this is a problematic one for patients involved in evaluating their own course of treatment, and for physicians grappling with the uncertainty of medical decision making.

Rights are less things, as I have argued elsewhere, than they are part of our evolving sensibility, part of our cultural evolution.[180] In other words, rights are part of the expansion of a cultural revolution in which others are included to participate in the culture – a culture, as Dewey liked to think, at the heart of the democratic spirit. Our evolutionary path has taken us towards a bill of rights in the context of medicine and medical decision making that Hippocrates helped ignite.

Conclusion

Medical decision making emerges from a cephalic predilection to reduce discomfort and anticipate danger: two important adaptations. This, coupled with a rising experimental sensibility and self-corrective states, sets the broader stage for problem solving: something endemic to pragmatic naturalism.

The social context highlights what Dewey referred to as the "lived experience," something dear to the Flexner approach to knowledge acquisition. This was an apprentice model of a vision of a research university, as embodied by Johns Hopkins. Peirce and Dewey were part of the Hopkins experimental philosophy that cultivated a laboratory state of mind. But Peirce, an early proponent of statistical literacy, was also a proponent of semiotics. Walker Percy, who had an M.D. from Columbia University but never practiced medicine, would eventually present Peirce's semiotic points to the world with clarity. An understanding of medical decision making emerged in this larger context.

John Dewey emphasized the link between science and social action. The Flexners were breathing that air. A Deweyan conception of rational choices is very social in its implications, in the ability to interact and have others at a table when there is disagreement.[181] Ideals of health linked to a culture of research in medical schools were the Flexnerian ideal, formulated and zealously begun in the first part of the 20th century.[182] Market-driven forces in a capitalistic society drive availability. Medicine is no different. The great capitalists of the first part of the 20th century had much to do with medical reform, as the instances of the Rockefeller and Carnegie Foundations illustrate.[183]

The Flexner model, Deweyan in orientation, was primarily an apprentice model and allowed for considerable diversity. Not everyone had the same skills at the table of science, but biological and medical orientation was a big tent of investigation. The theme was inquiry and self-corrective investigation. The Flexners, all three, were criticized by some for exaggerating the research at the expense of the clinic.[184] Indeed, the culture of research needs to be balanced with clinical experience. Both are necessary in any age, and in this age, with its explosion of knowledge and subsequent demands on that knowledge, both are equally crucial. The way for the physician to stay anchored is to be tied to data, acquisition, and struggle. The way for the researcher to stay embodied and relevant is to be knotted to the clinical issues, to understand them, and to grapple with them.

Medical decision making is also embedded in the larger cultural milieu, as Dewey very well understood. The social context, the consideration of participation, the inclusionary drive to embolden the democratic sensibilities of educated citizens, is a tall order, frail and fraught with human limitations. Part of a modern Enlightenment perspective is to avoid mythologizing, to accept endless human imperfection, yet to be constantly aware of the glimmers of human possibilities. The tough slog of the pragmatic naturalist, amidst profound awareness of social sensibilities, is an important ideal for an evolving culture of human rights and dignity. Inquiry is part of our cephalic capability and our cultural evolution, one knotted to an aesthetics of natural wonder, self-corrective inquiry and problem solving.

5
Psychobiology: A Jamesian Self-regulatory View

Introduction

William James is at the heart of American philosophy, psychology and, to a lesser extent, medicine. His focus on organs and bodies, particularly the brain, remains a viable approach; his strength is his pen and orientation, as well as his good sense. He captures the pulse of the issues of his age as well as ours. His grounding in human experience, in diverse forms of expression, are captivating and engaging, and a tone emerges from the way he sets out the task of psychology, philosophy, and medicine that valorizes human cephalic capabilities.

For James, pragmatic naturalism is up close and personal, as is the core consideration of the individual. James set out to capture psychobiology and a regulatory view of human self-maintenance linked to behavioral adaptations. My view is that James captured many of the dominant intellectual trends, such as Darwin and Hughlings Jackson, Wundt and Freud, which are still part of our intellectual landscape. He was neither an evolutionary biologist, nor a clinical neurologist, nor an experimentalist, for the most part. Yet the sense of these endeavors flows from his pen at every turn. Like Dewey, he integrated the insights of psychobiology into his philosophical orientation, an orientation in which the consideration of the will or effort is a primary feature of our experiences. Central to his view were both the evolution of effort or the will and its devolution of function in pathology and breakdown through the battles of life.

In this chapter, the narrative is simple. James, the physician/psychologist/philosopher, is immersed in naturalism and evolution, and oriented to individual experience and the will or effort. James's insights are integrated into a more restricted arena, involving anticipatory regulation of the internal milieu, a feature of an evolved cephalic system that is less reflexive and reactive yet anticipatory of environmental change. After all, a core feature of pragmatism within the classical period is its future orientation. Outcomes mattered: future behaviors and experiences that are made richer matter. The adaptation is in anticipation of events.[1]

In this regard, I introduce some discussion of hormonal systems in self-regulation in the context of the social milieu, something that James did not emphasize enough. I also discuss the importance of the individual phenomenology in James's thought,[2] and I consider the ramifications of his work for psychobiology and the study of instinct. James really captured the main themes in psychobiology, of which self-regulation and embodied and enactive cognition are themes.[3]

William James's insights into the nature of the individual

Of the entire classical pragmatist oeuvre, James alone spent his time focused primarily, if not exclusively, on the individual. The collective rarely appears in focus in his work; the community or social sense that dominated Peirce, Dewey, and Mead is not a Jamesian concern. Also in differentiation from his contemporaries in the Metaphysical Club, James was descended not from a Boston family, but hailed from Albany, and then New York City, before he migrated to Boston.

The family wealth came from his Scotch-Irish grandfather. Henry James Sr. had philosophical and literary interests, like his famous children (Henry James, the novelist, and Alice James, the diarist, as well as William; there were also two younger sons). The family valued education highly. The individual was elevated and provided the dominant discussion, but human frailty was a core feature of all their work.

America's great psychologist understood the struggle with depression. He would describe the melancholy that he experienced in his letters to his family and friends. Alice, who they all thought was a genius, and to whom William was especially close, was vulnerable to fatigue and was at times institutionalized in sanitaria. William's personal struggles provided him with an abundance of empathy in his public appearances and in his personal letters; he was tender-minded.[4] This may explain why he was so kind to Peirce, despite the latter's denigration of his theories.

There is a thread in American thought and pragmatism that is tied to the romantic tradition (e.g. Wordsworth, Keats, etc.): nature alive. James, like Emerson, Thoreau, and, to a lesser extent, Dewey, is tied to that romanticism, its sense of self-creation and fluidity, its sense of self-reliance and wondrous nature.[5] Although James's health always seemed frail, he seemed heartier as he got older, in contrast to Oliver Wendell Holmes Jr. He watched one of his brothers never really recover from the Civil War, in which he fought on the side of the north. The Jameses, unlike the Peirces, were decidedly pro-Northern in sentiment.

James received an M.D. from Harvard in 1869 and was appointed a professor, first in physiology, eventually psychology, and then philosophy. Regardless of appointment, James was always philosophical. He was interested in the search for meaning, for what is to be valued in life. His sense of the person was whole, not partial. It was a way to exert a moral sense of energies in the wild instead of war.[6] What he endlessly worried about was the loss of the individual in the great monoliths: evolutionary theory being one amongst others.

James made reference to the findings of Darwin and Jackson, Ferrier and Lamarck. He understood that bodily reactions – or simple reflexes, as Dewey,[7] his fellow pragmatist, also noted – are replete with ideation: ideas embodied in bodies in action. Minds are not spectators on the sidelines. This is what James and others called the "idea-motor" reflex.[8] In other words, cognitive systems pervade human action, and human actions are permeated by varied goals.[9]

A romantic, melancholic spirit would always be a part of James, a temperamental feature that made him tend to withdraw into the contours of the self, the inner world (to use his language, not mine). James in this sense was quite different from Peirce, for whom the social tended to be primary; for James, it was struggling individuals looking to gain access to acquire coherence. This is, perhaps, ironic, considering that James was so much more successful socially than Peirce could ever manage.

James was also influenced by the burgeoning field of experimental psychology emanating from Germany. He became, and is still regarded as, a major figure of that discipline. He was eventually elevated into the National Academy of Sciences for his work in psychology, but he would ask towards the end of his life to be known primarily as a Professor of Philosophy.[10] Despite recognizing the brilliance of his own sister, he held diverse sexist predilections about what one could expect from men and women, and their station in life[11] – very much a product of his time.

He was always literary, gifted with the pen and never particularly experimental, though he was empirical and wrote the great treatise in psychology. There was, perhaps, one small experiment in his corpus of experience, all others came from summaries of others. The portrait of the young William James in Figure 5.1 shows a sensitive man with the gift of visual art. He was literary to the core and tied to a pragmatic naturalism – itself tied to the burgeoning visual arts.[12]

Figure 5.1 Portrait of William James

James, psychology, and pragmatic naturalism

While Peirce might be the first experimental psychologist,[13] James is the first spokesman for a systematic view of the mind, body, health, and disease, in a treatise entitled *The Principles of Psychology*.[14] Without a doubt, the treatise is a great book, indeed his best work: reading it today or in 1890 is equally instructive. Mind is placed in a naturalistic context. James, the physician and physiologist, converge with James, the scholar. Like others of his generation, he traveled to Germany to learn the exciting new sciences and to wrap himself up in the exciting German trends towards experimentalism in psychology.

The neurologists of the day fill the early pages of James's first volume on the principles of psychology: Broca, Wernicke, Gall, and Mynerat, to name a few. Insights from medicine, from brain damage studies, and from human/animal performance link the pages. The evolutionary perspective is pervasive. With reflexive functions more phylogenetically mediated by brainstem functions, the newly evolved cortical regions were required for flexible problem solving, and longer-term, less-reflexive problem solving. The dominant view (which James endorsed) in the 19th century, and to this day, is a view of the brain heavily influenced by Herbert Spencer, and gives full scope to the British neurologist Hughlings Jackson. Corticalization of function became a well worked out comparative view of that century and subsequent centuries.[15]

The level of the neural function of ideas was and remains a dominant conception in understanding cephalic expression and behavioral adaptation. It turns out the hierarchy is less rigid, and cognitive capabilities and adaptation are distributed across many levels and systems that are not strictly hierarchical.[16] James understood that cephalic function is oriented to "intelligent action."[17] Intelligent action is ripe with purpose, something James depicted in some detail.[18] James straddled a permissive line between his allegiance to Darwin's conception of biological evolution and his own humanistic inclinations.[19]

Habits well orchestrated, cognitively enriched, as most pragmatists would argue, are the articulated features of intelligent action. Healthy habits serve psychobiological adaptation. Habits are expressed by automatic cephalic expression, as James would characterize in his work. An arsenal of cognitive capability, highly drenched in motor expression, is readily expressible in suitable environments. The issue was control; mind and brain were joined at the hip. For James, the sense of effort and of free will is embodied in cephalic function. The "feeling of effort," a Jamesian expression, is both a biological feature and a moral imperative for human advancement.[20] The will is something that is essential for human well-being.[21]

Kinetic sensibility underlies bodily intentional behavior,[22] as it is tied to motor regions of the brain (Figure 5.2), which we know are both computational and inherently tied to organization and rich in bodily sensibility. The "ideo-motor" conception was something that James grappled with, and it is also very much tied to modern behavioral neuroscience.[23]

The healthy will is knotted to behavioral inhibition and activation appropriate for context and circumstance and goals. Splintered by the blistering area of demands, an effortful will glides into diverse forms of action, embedded in adaptive systems. Importantly, the will or other mental terms are not things but functions tied to adaptation and action. Moreover, the cephalic capability, the instantiation of willed action, is now somewhat known and includes premotor (i.e., getting ready to move), frontal/motor, and parietal cortex.[24]

Figure 5.2 Primary cephalic elements involved in the production of movement

Thus, the will, as James suggests, permeates motor systems: "The will appears in every muscle as irritability."[25] By irritability, I take it he means activation *in potentia*, which might entail motivation and approach behaviors, behavioral inhibition, restraint, limiting choices, and action, all of which reflect cognitive systems that underlie behavior.[26] James outlined a view in which the sense of the will was knotted to diverse forms of action – cognitive systems no less than overt behaviors.[27]

In one of his early books on psychology, John Dewey,[28] like many other investigators to the present day, characterized the will as "directed towards the attainment of a recognized end which is felt as desirable."[29] His discussion of the will, like many others, emphasized "the higher centers" of the brain and their importance in the realization of ends and the organization of behavior.

James, on the other hand, was right to highlight the kinesthetic features of our actions – specifically, the feeling of effort. Cognitive systems are pervasive in our understanding of the organization of action.[30] Importantly, James and others understood cognitive systems as endemic to motor control.[31] Cognitive systems underlie and can be anticipatory of movement. Re-envisioning the motor systems is to suggest that there is no absolute separation of the motor systems from the cognitive systems in the brain.[32]

Just as William James noted that consciousness is not a thing, memory is not a thing;[33] the will is also not a thing. The will serves many functions,[34] and is essential in diverse adaptive capacities, including basic regulatory motivations that are essential for bodily regulation, the ontogeny of behavioral inhibition, and the adjudication of long- and short-term goals. Thus, effort of the will is a product of many neural systems designed to operate and cooperate in the organization of behavior. There is no one system designed for the will.

Now I would suggest that, in hindsight, that same thing can be very misleading. There are different ways to characterize habits, the stream of consciousness, or the expansion of presentation beyond the contours of the readily available. Presentation is a core feature in which there are a number of levels of analysis in which we go from the molecular to the cognitive systems to a characterization of lively experience. This characterization is not passive, but a radical empiricism bent on action. Mind/brain is looking at relationships across levels of analysis and within levels of analysis. Spinozistic monism of the sort that James and others were fond of is just misleading and not helpful.[35] The concept of mind is not a thing, but an evolving concept in our semantic lure of understanding, and it is easily abused. James himself would appreciate this point.

James understood that traditional empiricism was too passive, and that is why he associated himself with action and test-running through human experience. This was not a spectator sport. I would suggest that the pragmatic part is the radical empiricism idea, for radical empiricism moves beyond simple sensation. As James put it, "no one ever had a simple sensation by itself"[36] – this is what Whitehead, the erudite philosopher/mathematician, called "misplaced concreteness."[37]

As adaptive systems, we function largely in anticipation of events, embodied in habits of expectation in cephalic systems that are mobilized. In fact, as James noted, "we actually ignore most of the things before us."[38] Attention is pivoted on what matters. Competition between needs and resources, information capabilities and environmental context, is a recurring rule in a radical experience which engages the world in action – not as a spectator viewing a sensation and then making a judgment. Attention, like judgment, is part of the doing, part of the engaged experience.

James understood that perceptual systems – specifically, the working of them – are largely unconscious and form the basis of attention.[39] James linked his radical empiricism to a great web of human meaning.[40] That is, meanings are relations that are both real and perceived in human activity. One result in modern terms is a virtual embedded world of which we are part. Thus, the experience of attention was something like a variety of problem solving. Attention is not a thing, but a function, a core concept, a lively experience (or not).

James's focus was on the flux of experience and the continuity of the habitual. After all, as he noted, only in the pathological is one fixed, frozen, and devolved of function: "only in the pathological state will a fixed and ever monotonously recurring idea possess the mind."[41] Today we call it obsessive-compulsive disorder, a condition in which regions of a part of the brain linked to habits and action are overactive. Reverberating neural circuits take hold of the individual. Reducing the active sense of effort by individual control would be a more modern way to describe this phenomenon.

James had a particular interest in epilepsy.[42] He saw epilepsy as a devolution of neural function. Hughlings Jackson and other neurologists who influenced James, also focused on this disorder, an uncontrollable firing of neurons resulting in seizures, which was just beginning to be linked to brain function. Epilepsy can, in some instances, induce the perception of auras and religious experiences. Moreover, the firing of neurons in the brain can also result in other states, excessive fear being one of them.[43]

Hypnotic states were also a point to study, much mentioned by the great Charcot, by whom James was greatly influenced.[44] A hypnotic state is a state of suggestibility in which the individual is vulnerable and passive. This is not a good condition for radical empiricism emboldened by a feeling of effort, an effort essential for individual discovery and self-maintenance. But as Perry, a philosophic biographer of James, reminds us, James started with physiology and ended with mysticism.

The second volume of the *Principles of Psychology* focuses on the perception of things, how objects are experienced, and the perception of space – that is, the ways in which we accommodate to the things around us.[45] The book is thoroughly steeped in the facts of the time, and the theories of consciousness James discusses are "in the cortical cells."[46] Decorticated animals can be quite conscious, just limited in scope and ability. James, in this book, stumbles over the unconscious, not ready, and unable to accept the vast area of what would later be called the "cognitive unconscious."[47]

James was always at his best when locating the feel for objects: the expression of a hallucination, the dysfunctional cortical expression, and the devolution of function that can occur from epilepsy. His M.D. capabilities were widely tuned to human mental function and dysfunction. He would have been a completely modern psychiatrist if he had been born in the 20th century.

Reality, for James, is fluid, a reflection of a radical empiricism, not a frozen misplaced empiricism for which isolated sense data predominated the knowing subject.[48] He thoroughly demythologizes empiricism and rationalism and notes no abstract rationality or passive sensations. Concrete problem solving is, for James, a central and dominant theme. The emphasis is on adjustments to contexts based on some instinctual capability that grounds action reflexively, with cortical insight for adjustments.[49]

James expanded the concept of instincts: "they are not always blind,"[50] as he rightly noted. They organize action. As James also noted that memory matters. It expands cognitive capabilities, making them both instinctive and more reflective: "Every instinctive act, in an animal with memory, must be accompanied with foresight to its end."[51] Instincts, for James, are knotted to problem solving. This is a pragmatism rooted in naturalism and realism. Such instincts can be narrow or broad, but they underlie the organization of action.[52] Or, as Susanne Langer would suggest,[53] all behavior in the animal is rooted in some form of instinct, including diverse forms of learning, "... arising from organic sources as impulse seeking expression in motor action and guided to direct or indirect consummation by acts of perception."[54]

James on the emotions

James links a good many emotions to instinct. They are fast, immediate responses. In this he harks back to Darwin. Fear, he notes, citing

Darwin, can evoke a sense of astonishment, the aroused sense that can erupt into freezing or fleeing.[55] James rightly noted that "disembodied human emotion is a non-entity."[56] Emotions, for James, as they were for Darwin, are not features of the irrational. They are rooted in adaptation.[57]

Embodied sense, a common parlance in the last twenty years,[58] resurrects a Jamesian theme of the body inhabiting space.[59] This is the space of social contact with others in which the emotions, whether directed towards danger or safety, are fundamental features of the cephalic adaptive landscape. The emotions are imbued with action tendencies and are bodily represented. They are, however, more cephalic than James understood. I can see a bear and not run, yet I still note the fear. The fear is represented by cephalic sensibility and capabilities. It is a more central state, not a peripheral activation, though the peripheral responses (e.g., heartbeat, sweating, etc.) contribute to an adaptive response (Figure 5.3).[60]

In fact, James was always interested in the psychobiology of temperament, the differences and limits that engulf us, that set the conditions for how we experience the world, the associations we make, the bonds we forge, the falls that we experience, the worries that predominate us, etc. In *The Varieties of Religious Experience,* James framed the importance of temperament in his distinction between sick and healthy souls. Like Emerson, nature for James had a religious dimension.[61] Indeed, we come

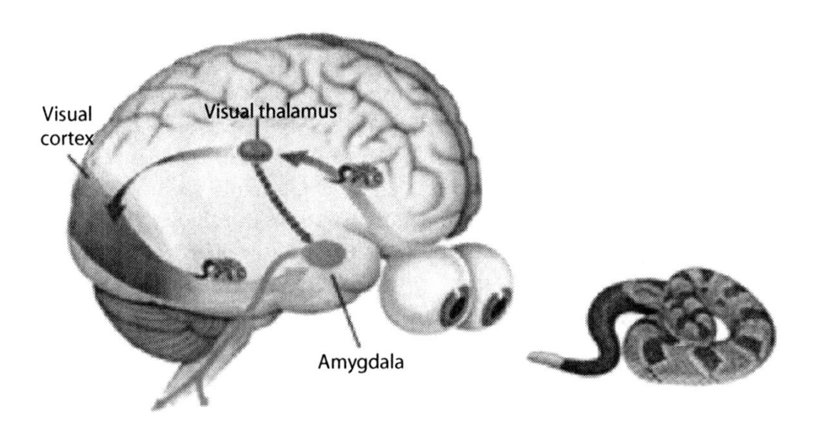

Figure 5.3 Relationship between the visual system and the amygdala
Source: LeDoux (1996).

prepared with a variety of cognitive resources to make religious belief possible, and religious tyranny an all too present reality.

James dedicated his book *The Will to Believe* to his friend Peirce. It is a plea for pluralism and tolerance in expression, a fixture in determining the cognitive basis of action. In it, James presented himself as the softer side of the Metaphysical Club.[62] He softened rationality by embodying it in the sentiments, thereby softening religion by stripping it of outlandish rationalism. He braced for determining the meaning of life through actions that promote the welfare of the individual, demythologizing the argument of determinism and indeterminism, freedom and action, by looking for the consequences of actions, and by limiting abstract arguments far from the experience of a radical empiricist. The will, as I said before, was never a thing for James; just like consciousness and other mental theoretical terms, it is embedded in action, across a landscape of meaning.

Pragmatism is a consequential approach, up front and personal. A decision matters and is judged by consequences. Those consequences can be quite immediate and narrow, while consideration of generalities are perhaps longer term and have more encompassing consequences. The dilemmas of determinism are brought down to earth, to what bears directly on a radical empiricist. James's pragmatism, as he understood it, was broadly humanistic, with a tinge of effort to conserve and embrace a form of naturalism. This was something his student Theodore Roosevelt embraced, along with a strong sense of the importance of biology and Darwinism, couched with a sense of the importance of conservation in our democracy, and the sense of effort: an effort that is at the heart of one's self worth.[63] Roosevelt popularized a very Jamesian sense of the rugged individual, if not his Unitarian soft theology. James had an abiding respect for others. He was tenderhearted, but he could be as limited as any other, as he and his brother eschewed certain eastern European immigrants. Nonetheless, students such as Gertrude Stein and Morris Cohen were becoming more common and more familiar.

Like Peirce in the 1870s at Hopkins, when he was lecturing on how great ideas came from original thinkers, James was trying to understand how Darwin fit into what might be called "Great Men and their Environment." James asserted, acknowledging Darwin's contribution, that much of evolution has to do with "molecular accidents of which we know nothing."[64] The important part of this piece was the worry about "the evolutionary view of history, when it denies the vital importance of individual initiative."[65] The importance of the individual was always sacrosanct for James, and in some ways evolution was a compromise of that fact as it relates to radical empiricism, and an active sense of experience.

A recurrent subject for James is his description of the natural inclination towards religious experience. The power of *The Will to Believe* for me, however, has always been the description of psychological predications, specifically the optimistic and pessimistic outlooks on life. How a pessimistic cognitive predilection to endlessly expect bad outcomes can contrast with a more optimistic view of life with converse expectations is not only a telling fact itself, but also is so well described and understood by James. These are temperamental features that are part of our psychobiological constitutions.

James on lifeworld

In his essay on the present dilemmas in philosophy, when characterizing thinkers, James divided them into two groups: the tender-minded he thought tended to be idealistic and optimistic, while the tough-minded remained materialistic and pessimistic. The tender-minded tended to be monistic, rationalistic, and dogmatic, the tough-minded pluralistic, empirical, and skeptical. James linked our orientation to investigation, and to our psychological predilection, to our temperamental features. Not surprisingly he was instrumental in bringing Freud to Clarke University, but James was never able to accept the vast arsenal of the unconscious. Cognitive systems were also part of consciousness. His radical empiricism, though expanded and extended, was still in the aura of what we could be aware of.

What James understood as the plight of the sick soul, as well as saintliness, were core themes in the *Varieties*, and in his other work. The "feeling of the wider life,"[66] Karl Jaspers would grasp as part of human meaning and health through human connectivity, was the individual with other individuals.[67] James straddled many intellectual worlds as a primary form of engagement.[68]

James understood something about the "healthy mind." This had practical outcomes. He emphasized the functional side, in addition to the phenomenological depiction. James was interested in describing the "sick" or "divided" individual,[69] as well as the whole individual. But for all of James's tirades against Hegelianism, he held steadfastly to a notion of an undivided self – although, indeed, we are many selves, as James sometimes understood. Only Mead, however, really understood the concept of the divided self, as a feature of the extreme in the pathological.

James was part medical person, part neurologist, while mostly an elegant phenomenologist of the sort that many 20th century phenomenologists would acknowledge. Core themes for him centered around the primacy of perception and the body, and the exploration of space, as well as the presentness and completeness of experience.[70]

The focus for James was on what phenomenologists call "the lifeworld" and "meaning context."[71] The events are intersubjective and cephalically shared through endless contact amidst multiple others.[72] As Wittgenstein would later acknowledge, the problem of other minds is no problem; it is a given in our experiences by habit and practice. We have a pervasive sense of others, their shared resources and experiences. It is not so much that it is pre-conceptual, as many phenomenologists would suggest,[73] but simply so much a part of our experience that there is little reflection, no detachment to reflect on. It is simply pervasive to what we are presented, to what we live with. A natural predilection is not only the shared resources that bind us together, that embolden us but also enrich us. In fact, that is how our species evolved, social animals to the core. But, as James understood, the subject is never reduced to the social mass. One of James's strengths was his literary and literal grasp of the living subject, a knowing subject, a surviving subject – an evolving subject.

The strength of pragmatism, in my view, is continuity and pluralism, not discontinuity. The strength of pragmatism is engagement and an enrichment of experience – an acknowledgment of it. Thus, for James, the emphasis is on experience and its description, on the diverse ways in which it can be described, and the heightened respect that it is given.[74] This is a respect that goes to the pre-reflective, to a pervading radical sensibility, a primacy, often perceptual, for which the intentional capabilities are acknowledged. The totalizing myth is replaced with the elegance of the particular individual struggling to understand the perceptual aura around him, focused and engaged. Radical empiricism drew causation close to home through experiences between events.[75] Agency through action was a primary relation to the world. Relations between persons are felt. The sense of becoming felt was a feature for the Jamesian notion of radical empiricism.[76] The world is broad, but intention is pivotal, and a pervading feature for coherent action. An evolutionary perspective on the human condition occurs within a broad evolutionary psychology.[77]

Bergson commented upon James this way:

> People have said that the pragmatism of James was only a form of skepticism, which it lowered truth, that it subordinated truth to

> material utility, that it advised against and discouraged disinterested scientific research. Such an interpretation will never enter the heads of those who read his work attentively. And it will greatly astonish those who have had the pleasure of knowing the man. No one loved truth with a more ardent love. No one man sought it with greater passion. He was stirred by an immense unrest, and went from science to science, from anatomy and physiology to psychology, from psychology to philosophy, tense over great problems, heedless of anything else, forgetful of himself. All his life he observed, experimented, meditated. And as if he had not done enough, he still dreamed, as he fell into his last slumber, of extraordinary experiments and superhuman efforts by which he could continue even beyond death to work with us for the greater good of science, and the greater glory of truth.[78]

James is fundamental to the origins of psychobiology, to the study of self-regulation and adaptation, to understanding the evolution and devolution of function. We will now turn to those fields by looking first at instinct and then self-regulation, and finally addressing anticipatory regulatory cephalic adaptations.

Darwin, James, and the concept of instinct

The concept of instinct figured importantly for Darwin.[79] The issue that permeated biology and psychobiology was the relationship between what is inherited and what is acquired. Darwin was smart when he said, "I will not attempt any definition of instinct."[80] Darwin was always the consummate empiricist; he gave examples of what he thought were instincts (migratory, sexual, and social). He referred to the effects of domestication and how he thought it diminished the effectiveness of instinctive behaviors (e.g., less "broody" fowls, who devote less time to sitting on their eggs).

Darwin noted that species-specific behavioral and physiological adoptions were richly expressed and tied, perhaps, to finding sources of energy, to primary motivational systems, and their release in suitable environments. This was a precursor to later studies of animal behavior and ethology.[81] Mary Midgley, a moral philosopher, has suggested, in citing the importance of the concept of instinct, that "people who dislike the word instinct, can if they prefer often read drive or program."[82]

Darwin, like many others, was less clear about "habits" and their link to instincts, and about new instincts emerging from domestication.

Throughout his writings, Darwin remained a gradualist with regard to evolutionary selection on the emergence of instincts. In the same work, Darwin acknowledged that "instincts are not always perfect."[83] After all, he understood problem solving as not about perfection, but about adaptation. In the *Descent of Man*, he went on to compare our evolution with that of other species and suggested that "the fewness and the comparative simplicity of the instincts in the higher animals are remarkable in contrast with those of the lower animals."[84]

The concept of instinct in Darwin's time, and for a number of years afterwards, held diverse and confusing meanings. Moreover, issues about what was heritable and the battle for Lamarckian transition permeated the debates.[85] There are many related and unrelated ideas surrounding the concept of instinct, but one fundamental feature has always been the dichotomy between the inherited and the learned, or the distinction between innate and learned behaviors. Of course, it need not have been, since learning is part of our innate endowment, as is adaptation to diverse environments. The question is to what degree, and which behavioral systems are being recruited. But that is a modern notion. For when intelligence is part of adaptation, of instinctive behaviors, such distinctions fizzle away. The question is degree, not kind.

James, in an unusually harsh tone, wrote that "the older writing on instinct is an ineffectual waste of words."[86] Some consensus centered on the idea that instincts were reflexive responses to characteristic stimuli.[87] James,[88] though, thought that "instinct is usually defined as the faculty of acting in such a way as to produce certain ends without foresight."[89] An instinct reflected sets of impulses and reflexes. One of James's examples was the reflexes involved in egg laying. Yet, later in his chapter on instinct, when the discussion of memory is linked to instinctive acts, the behaviors "cease to be blind."[90]

Perhaps Susanne Langer is more sanguine in her sanity of expression and reason: "Every instinctive act is motivated by a situation that is deeply prepared in the organism by gene controlled stages to produce massive impulses to muscular action, intrinsically patterned in hereditary forms."[91]

Pragmatism and the origins of psychobiology

Self-regulation is vital to James's notion of health. Adolf Meyer, a Swiss-born neurologist and an admirer of James,[92] is fundamental to the origins of psychobiology.[93] What Meyer emphasized was whole-body activity. Meyer understood psychobiology as "the missing link" between "ordinary physiology and pathology...dealing with functions of the total person and not merely detachable parts."[94] Meyer also understood psychobiology "[to take] life as it is without splitting it into something mental and something physical."[95] In other words, Meyer's psychobiology was interested in "organismal function and behavior."[96] Psychobiology, in Meyers' view, emphasized whole bodily adaptation, and understood it as linking "ordinary physiology and pathology."[97] Curt Richter (1894–1988) would inherit and indeed build onto this total behavioral and physiological orientation.

While he was in Worcester, Massachusetts, working at a psychiatric hospital, Meyer had interacted with James, who, along with C.S. Peirce and John Dewey, had already previously influenced him. He liked what he called their "instrumentalism." Indeed, Meyer was impressed with his new country and its sense of pragmatism. He wrote, "It was the work of American thinkers, especially of Charles S. Peirce, of John Dewey and of William James, which justified in us a basic sense of pluralism, that is to say, a recognition that nature is not just one smooth continuity."[98] It was in American pragmatism and its biological adaptation, problem solving, and whole-body activity,[99] that Meyer located psychobiology.[100]

Meyer's interest in "organismal total function" was passed on to Richter. Meyer no doubt saw in Richter a shared piece of the Old World from which he had emigrated, as well as their shared love for the German language, but also the new-found freedom, the "roll-up-your-sleeves" pluralism, of the country in which he now found himself. Perhaps Meyer looked at the young Richter and saw the outward expression of this point of view. As Meyer put it, "The factors at work in the development of a psychobiological conception had much support in American thought. Pluralism and pragmatism were liberating factors in throwing off dogmatic dualism."[101] A little later in the same paragraph, he described "William James' clear vision of the significance of the pragmatism of Charles S. Peirce and the instrumentalism of John Dewey, and the healthy encouragement given to natural spontaneity of thought and work in the American environment."[102]

Invention and experimentation predominate in a pragmatic orientation. Pragmatism is a philosophy of experimentalism[103] and represents "the open air and possibilities of nature."[104] This open sensibility was an important theme in many of the pragmatists' conceptions[105] and is consistent with what Richter called "free research."[106] Free research is close to open-ended, unencumbered inquiry, something dear to the pragmatists and certainly to Richter.

As I indicated earlier, C.S. Peirce can be credited with creating the first psychology laboratory in the United States, which he did at Johns Hopkins in the 1870s.[107] The laboratory spirit of Peirce and his pragmatism meant that the universe could be opened through experiment.[108] Scientific inquiry was rough and ready, biologically grounded in human and animal problem solving. A persistent feature of this orientation, at least for James, was that "pragmatism is uncomfortable away from facts."[109]

Richter was also uncomfortable away from experiments, and the facts derived from experiments. What hung in his laboratory like a veiled painting of great importance is this quote of Francois Megendie: "Everyone compares himself to something more or less majestic in his own sphere, to Archimedes, Michelangelo, Galileo, Descartes, and so on. Louis XIV compared himself to the sun. I am much more humble. I compare myself to the scavenger; with my hook in my hand and my pack on my back I go about the domain of science picking up what I can find."[110]

Meyer followed the spirit of that quote by orienting his psychobiological perspective to that of American pragmatism. The great *Principles of Psychology* was rich in biological perspective, functional utility, and evolutionary conceptions.[111] Dewey had no less of an influence, and one of his early works, at the turn of the 20th century, was *The Influence of Darwinism on Philosophy*.[112] Knowledge acquisition was now linked to adaptation; the intellectual generation immediately before Richter had struggled to incorporate the new insights of Darwin into current thinking.

These insights swept across Europe and reached the United States. By the end of the 19th and early 20th centuries, the cultural milieu was ripe for the acceptance of the first biological revolution, with a new conception of biological adaptation. Richter would inherit a pragmatism that was tied to the version of 19th century biological thinking that permeated Harvard and later Hopkins through Meyer. These concepts were perhaps already implicitly there for an engineer bent on exploring the biological basis of behavior, and Richter relied on this sense of pragmatism to discern how things worked. Richter never doubted the validity of the scientific method.

Self-regulation: allostasis and homeostasis

A concept of the organism adapting (and coping) with its environment, formulated by diverse thinkers and culminating in Darwin's great works, was part of the experimental framework in the laboratory cultures that were emerging, and was a paramount feature of the classical pragmatism. Importantly so, psychobiology was anchored to self-regulation.

While Claude Bernard was arguing for the fixed nature of biological expression through a science rooted in methodological consideration, others were trying to conceptualize a more broadly organismic sensibility in which to understand physiological adaptation,[113] and, eventually, behavioral adaptation in the regulation of the internal milieu or the social milieu.[114]

The field of psychobiology was relatively new in the late 19th and early 20th centuries. William James,[115] following Darwin,[116] pointed to the role of biology in behavioral adaptation, and Adolf Meyer wrote quite extensively on psychobiology.[117] The relationship between psychology and medicine was already part of the discussion by leading scholars. The concept of psychobiology would eventually take on diverse meanings. Richter would become a major proponent and would come to represent psychobiological research, which emphasized biobehavioral adaptation.

Generically, psychobiology was a way to root the study of behavior within an understanding of diverse forms of biological adaptation. Meyer, who wrote extensively on this topic, emphasized long-term adaptation and organismic responses to ecological perturbations.[118] Behavior, as Richter argued, is about the whole organism adapting to circumstances. "Interest in Human psychology is moving rapidly toward problems of general adaptation involving responses of the whole organism in actual working life situations."[119] Self-regulation of the internal milieu was a fundamental scientific question for Curt Richter. He was not alone in this interest, for certainly the pragmatic naturalism that dominated the Harvard of his undergraduate education, and then later at Hopkins, was pervasive.[120] But he certainly was a major force in providing interest in, and informative contexts in which to consider, the biological adaptation required for bodily health. A key feature in this outlook was the maintenance of balance, generating appetitive and consummatory behavior to restore balance.[121] Walter Cannon (1871–1945), of the department of physiology at Harvard Medical School, was a proponent of this view. His close friend and confidant, as it happens, was Ralph Barton Perry, the biographer of William James, and himself an expositor of a form of naturalism.[122]

These bodily self-regulations include body fluid homeostasis; thirst and hunger; and homeostatic regulation of salt, sugar, protein, fat, calcium, oxygen, blood, temperature, and so forth. Each would figure in

Richter's inquiry. Perhaps the individual to have influenced Richter's research the most was Walter Cannon. In Cannon's words:

> The constant conditions which are manifested in the body might be termed equilibria. That word, however, has come to have a fairly exact meaning applied to relatively simple physio-chemical states, in closed systems, where known forces are balanced. The coordinated physiological processes which maintain most of the steady states in the organism are so complex and so peculiar to living beings – involving as they may, the brain and nerves, the heart, lungs, kidneys and spleen, all working cooperatively – that I have a special designation for these states, homeostasis.[123]

Cannon's critique of James's view of emotions demonstrated that without afferent input and bodily actions, emotions could still be expressed on numerous occasions.[124] Cannon noted that behavior served physiology.[125] For instance, he noted that animals would be likely to ingest calcium during pregnancy when the need for calcium would be great, and that bone and other tissue suffered the consequences of calcium deficiency. Cannon also noted that animals might ingest sodium during periods of sodium deficiency. Cannon laid the seeds for a view of behavioral regulation of the internal milieu. Richter's contribution was to expand this considerably beyond the physiological perspective of Cannon: behavior served physiology in the regulation of the internal milieu. Indeed, the ingestion of calcium is a fundamental drive. Essential during pregnancy and lactation, it is eagerly pursued and ingested, as are, in fact, a variety of minerals.

Richter, in an essay entitled "The Biology of Drive," would assert that "Bernard and Cannon dealt largely with the physiological regulators – responses of individual organs or systems – which serve to maintain a constant internal environment. Several years ago, we found that the organism itself, the total organism, may also play an important part."[126] The emphasis is on the person, whole-body adaptation, and cephalic capabilities. This is what he meant by the behavioral regulation of the internal milieu for which the concept of instinct was the major behavioral category. We now know that neuropeptides in cephalic systems underlie the diverse forms of ingestive needs that are often anticipatory.[127]

What was needed was a vision of cephalic adaptation that described reactive homeostasis to anticipatory responses. This was something that both Dewey was getting at in his critique of the reflex arc article, in which anticipatory cephalic organization is endemic, and that James also was adumbrating in his sense of effort and bodily sensibility and co-inhabitation of anticipatory events – a very modern view about brain structure in which cephalic mechanisms are anticipatory as well as reactive in response to a changing world.[128] The issue is not constancy, just like it is not certainty; it is about change and adaptation. The cephalic resources are embedded in that.

Feedforward systems emphasize anticipatory systems, laden with cognitive/motor systems. James and Dewey diluted the difference between the two, without the mistaken view of identifying cognition with movement. Cephalic systems nevertheless are dynamic. Motion is fundamental, a point emphasized by pragmatists and their sense of experience and adaptation.

The degree of corticalization of function corresponds to the expanded connectivity of cortex and diverse brain regions that underlie behavior.[129] A greater form of regulation is adapting to changing circumstance, one mechanism of which is allostatic regulation. Pragmatism emphasizes the cephalic generators. The brain is in motion, not simply reflexive; it is coherent and purposeful and beyond only reflexive homeostatic responses to deviations of set point emphasized by homeostatic frameworks. Allostasis is a cephalic anticipatory regulatory function that goes beyond homeostasis by anticipating new stabilities through change.

Corticalization of function is perhaps best represented in the context of social intelligence and social regulation of the internal milieu. The utilization of others in the management of our needs, which have to do with things like safety and closeness, as well as the satisfaction of other basic needs such as human contact, is essential for survival and reproductive fitness.[130] Some of the known connectivity implicating neocortical tissue in anticipatory regulatory control is depicted in Figure 5.4.[131]

In other words, our evolutionary ascent is inextricably linked to social complexity, to tagging the beliefs and desires of others as well as forging both cooperative and non-cooperative behaviors that reflect larger forms of regulatory competence. The ability to be patient is therefore an evolutionary trait. The greater the range of cooperative social behavior, the greater the modulation of the internal milieu by social contact, in addition to the greater ability to exhibit patience in order to further future preferences, a feature that James understood. No doubt one feature of this ability to restrain behavioral responses is in corticalization, a core idea since the 19th century.[132] Within this shift to cortical capacity is the restraint of behavioral expression.

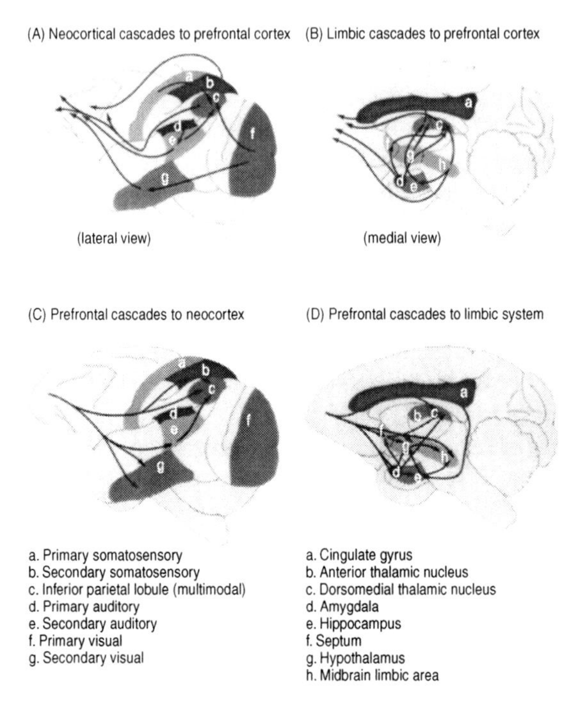

Figure 5.4 Innervation of multiple cephalic systems
Source: Sterling (2004).

The central nervous system aids conservation and maintenance by generating the behaviors that serve physiological viability. Importantly, the central state of the brain orchestrates the behavioral responses in anticipation of and in adaptation to environmental events. One major factor in the evolution of the nervous system is the organization of physiology and behavior to orchestrate anticipatory responses to events that precede a reaction to them.[133] Cephalic innervations of peripheral physiology (e.g., insulin secretion in anticipation of a food source) increase the flexibility and variability of physiological and behavioral responses.

Evolution favored a number of selected behavioral and physiological mechanisms designed to maintain internal viability. It is also apparent that the central nervous system plays an important role by superseding local physiological function in the maintenance of bodily tissue. In other words, the central nervous system is helping orchestrate bodily viability. This adds a new level of flexibility to regulatory events. The central state reflects the balancing of a number of factors essential for maintaining physiological and behavioral viability.

Darwin emphasized social contact as essential in our evolution. Most of the pragmatists recognized the essential feature of the social context for self-regulation. This is particularly true of Dewey and Mead. James, the psychobiologist, would have seen the importance of this towards our survival and instinctive anticipatory regulation.

From the point of view of the brain, two kinds of regulation are taking place: the regulation of the internal milieu, and adaptation and regulation of the social milieu. Both require a brain with diverse physiological and behavioral regulatory systems. In both cases, anticipatory mechanisms underlie adaptation both within the individual and within the social milieu. The greater the degree of social contact and social organization experienced by a human, the greater the trend for cortical expansion.

Fast social judgment underlies human decision making, and is linked to amygdala function.[134] Social attachments are footholds by which we are connected into the larger world and anchored towards others. Regions of the brain long linked to reward and social perception (e.g. the amygdala) are knotted to social judgment more generally, and for getting the "gist of things," or quick heuristic judgments.[135] Such fast judgments underlie many social decisions.[136] Diverse forms of cognitive adaptation emanate from a prepared cephalic system, underlaid with sensory/motor systems pregnant with the ability for getting the gist of things.[137]

Conclusion

William James straddled the dividing lines in the sciences and the humanities. His universe was embedded in psychobiology and naturalized epistemology, while staying outside a narrow scientism, on the one hand, and disembodiment on the other. His discourse and discursive interaction was continuous but not blocked into chunks, something he eschewed. An endless aesthetic and a concern for his intelligent companions populate the pages of his letters. James was a humanist looking to understand self-regulation, psychobiology, and social contact. Regulatory concepts figure more in Dewey[138] than James; but both looked to link self-regulation, instincts, and expanding cephalic capabilities, and both looked beyond narrow and merely reactive homeostatic models towards anticipatory cephalic adaptations.

James was ultimately housed in biology where he embraced the evolution of brain and end-organ systems. His mildness of character allowed him to bridge communities as well. He was not an experimentalist, and there is little data in what he presents. He was the student of the human subject, our adaptations and our devolutions. He never detached the body as a vehicle for exploring terrains.[139]

James was housed squarely within psychobiology. He had an emphasis on human contact and evolutionary considerations, and moved away from narrow notions of the central nervous system function. His view of the brain was quite modern in his emphasis on cephalic adaptation, bodily sensibility, humoral factors in animal contact, and avoidance.

Part of what makes James so compelling is his feel for the human condition; something that Peirce knew about James but thought of him as soft minded. He was just not an experimentalist; he did not have grand social views. He was housed with a focus on the individual.

James understood and suffered. He focused his attention on human frailties, understood something about it himself, never left that behind, and, despite embodying the negative biases of his times, reached out to others outside the confines of the Boston Brahmins. He understood something about human virtue and human strength, and had the glorious pen to capture the human condition.

William James is at the heart of American culture, of which pragmatic naturalism is just one part. The possibilities imagined by James, amidst the recognition of human frailty, human disease, and devolution of function, were quite apparent. Evolved cortical function, expanded regulatory systems are one thing. Their breakdown, the loss of human social contact, and the breakdown of a will in action or in realizing beliefs are omnipresent in our condition.

James does not gloss over the human condition – the possible and the actual. Many of the discoveries I have just mentioned, a small fraction within psychobiology, are very much consistent with his corpus.

6
Social Smarts, Moral Sentiments, Social Constructs

Introduction

More than any other classical pragmatist, George Herbert Mead was fixated on the importance of the social milieu, the social act, and its evolutionary and psychobiological underpinnings. John Dewey, however, was the pragmatist most concerned with the moral context of social actions. In this chapter, we will start with the pragmatist approach to social acts and then move on to its implications for our moral life.

A propensity for prosocial behaviors is competitive with the rest of our predilections. A consideration of the prosocial instincts embedded in cephalic bodily responses towards others, amidst a frail sensibility of human kindness, as well as diverse cognitive predilections embedded in human/animal sentiments, sets the stage for moral progress, social contracts, and diverse cognitive rules. We know from a number of studies that human survival depends on simple bodily contact, a need that underlies our psychobiology. We are compelled to reach out to others. In doing so, we discover what contracts and which rules best serve our interests and hopes.

Human dignity in small drafts of expression is frail and ephemeral, but real. It is something beautiful and awe inspiring to behold. Idiocy often results from the fall downward and the trend away from meaningful social contact, and meaningful relationships. Cephalic expansion in suitable environments underlies these events.

Let's turn to George Herbert Mead and, again, John Dewey. Throughout this chapter, as in the rest of this book, I couch the discussion, where possible, in classical pragmatism. Mead, like Dewey, emphasized an evolutionary perspective in a social context, and broadly understood psychobiology from a pragmatic perspective. I will follow a discussion of

Mead's orientation, followed by a modern sense of what underlies social adaptation and the evolution of the brain. I then examine the link to prosocial expression and finish with a consideration of the social construction rooted in this cephalic capability.

George Herbert Mead and the social act

Mead held a chair of philosophy at the University of Chicago for the first part of the 20th century. He was a leader in the naturalization of philosophy, and a social reformer in theory and somewhat in practice through Hull House, an embracing settlement house for the immigrants and working people of Chicago.[1] The labile expression of the human condition at Hull infused them with social hope through a conception of social adaptation, social connectivity, and social meaning.

Mead is also a core classical pragmatist. His father taught theology at Oberlin College in Ohio (where Mead studied as an undergraduate). His roots, however, were in New England, as he was descended from a string of New England ministers[2] (a history perhaps underlying his recurrent sense of social justice and pragmatic practices). Mead went on to study at Harvard, and from there, to Leipzig, Germany, to study with psychologist Wilhelm Wundt, whose experimental laboratory sensibility was racing ahead of others. In Berlin, Mead worked with Dilthey and Ebbinhaus, amongst others. From the start, he was imbued with the empirical stance.

After a stint at the University of Michigan, Mead migrated to Chicago, where he rose to become the chair of philosophy at the University of Chicago at its origins. The Chicago pragmatists, of which Mead was a prominent figure, came to represent a distinct movement in America: socially minded, naturalistic, open-ended, and beaming with broad Mid-western space and New England ideals.

A primary concern for Mead was what he called the social act, by which he tried to develop a naturalized conception of action. Mead understood action to be an adaptation, which builds on perceptual capabilities that capture coherence and that can be explained in broad physical terms, rooted in Darwin and the physiological sciences. Mead called himself a social behaviorist: intriguingly, John Watson, one of Mead's students, always claimed he did not understand him.[3]

Mead was grounded in Darwin and the advances in experimental psychology. He understood behavior as enactive, and human experience as embodied cognition. Biology and physiology were his first languages, though reading Mead's work never involves looking at any data.

Rather, his perspective comes from the background of naturalism. The core feature of his work is the social gesture. It is the way we mirror each other, by cultivating a collective self in which we find each other, learning about each other in social acts. Meaning arises through communicative action with others, is social in nature, and is tied to our competence. The self, as Mead understood, is a social object with an evolutionary cephalic capability for human experience that is embedded with cognitive embodied and enacted expression.[4] For both Mead and Dewey, the scope of their philosophic understanding is social, with individuals preserved and participating in communities.

From the new field of experimental psychophysics, and more generally experimental psychology, Mead derived a lifelong interest in the experimental approach to space perception, to temporal relationships, and to Darwin. This interest provided resources for how to ground theory in human adaptation and performance. Mead's philosophy, even when somewhat obscure, was always embedded in the natural sciences, with a conception of who we are in the natural world, given our evolution and our capacities. He looks to understand the mechanics emerging at the end of the 19th and start of the 20th centuries, and, importantly, to move beyond the narrow mechanical view of the 17th century. Whitehead serves a purpose for Mead. Whitehead's broad "philosophy of organism,"[5] with nature alive and broad, and human purpose open-ended, inspired Mead to see in human progress a thing of great possibility. Humanistic and evolutionary naturalism was for Mead, like Dewey, a dominant and recurrent theme – something that was very much part of the psychobiological perspective at the University of Chicago.[6]

Since functional discourse predominates in Mead, the references to self and time serve as a collection of perceptual experiences of others with an attunement to our own temporal predictions. In fact, the discussion of the present is couched to capture something of the experience of the present, a sense of coming into being,[7] stretching temporal experience by the attunement to objects of perception that are rooted in bodily experiences with others.[8]

Mead would often say "that the social process relates the responses of one individual to the gesture of another,"[9] and that "mind arises in the social process."[10] With this, we see his affinity to both Aristotle and Darwin, who both naturalized the social process inherent in our species. Semiotic capabilities – interpreting signs – are naturalized in the broad array in which animals, like ourselves, are endlessly interrelating with our surroundings, interpreting events, and relying on cognitive capabilities. Mead always stayed close to this conception.

Animal communicative processes were, of course, of great interest to Darwin and Dewey. These processes were always part of Mead's thought. The difference is that Mead wrote about them so abstractly, in a way too divorced from fact, and highly conceptual. This was even the case with the appetitive and consummatory phases of experience. Nevertheless, the communicative experience and the social milieu – a richly semiotic world embedded in an evolutionary context but fully integrated into human experience – permeates Mead's orientation.

The social act for Mead is the confluence of diverse social factors in an individual's process of making sense of experiences in the present, extended by anticipatory factors into the future, and beholden to the past.[11] His sense of the social act was also directly connected to objects in the world, in a way that advocates the critical realism I advocate throughout this book. Mead writes, "Every act...is moving on from its physical objects to some consummation."[12]

Mead was orientated to the "act," which always occurs in a social milieu. Of course, post-Mead, the social act is quite rich in the modern language of mirror neurons; social imitation in ontogeny and its development through formed connections with others; cognitive capabilities that align us with others from brute imitation to higher levels of intentional acts.[13]

The philosophy of the social act holds that the social act is embodied in a subject making sense of experience, that it is rooted with others, through a broad-minded self, not oneself competing for expression.[14] The self is experienced bodily.[15] The subject is amidst often competing habits that forge into a coherence in a common place. The interpretation in a framework of Darwin , biology and experimental psychology really emerged from Mead's studies in Leipzig,[16] reinforcing pragmatism's ties to psychobiology.

Like Dewey, Mead had a naturalized view of the individual and the development of the state. Like all classical pragmatists, experimentalism and fallibilism were, for Mead, normative ideals that are essential for human inquiry. What interested Mead, like other classical pragmatists, was characterizing human adaptations, cognitive capabilities, invention, and discovery in science. For Mead, the experimental spirit was a force of nature.

The importance of Mead and his "social act" lies in the pervasive intersubjectivity of the integration of biology in the consideration of the social. Like James, the "present," for Mead, is extended into the future as a lure, and into the past. Cephalic evolution allowed for knowledge as "a process in conduct that so organizes the field of action that delayed and inhibited responses may take place."[17] This allowance results from the greater cephalic capacity for both memory and anticipation of the future. With the evolution of greater memory of the past, in both its detail and quantity, our present is more richly informed. Likewise, with the evolution of greater anticipatory function in the cortex, our present is informed in such a way that we are not only pushed by the past, but pulled into the future. Mead's embrace of an evolutionary perspective allowed him to understand a living process within the biological and cultural milieu, while warning against "narrow provincialism or patriotism in science."[18]

Mead was surely thinking about the brain science of his day; however, his conception of the brain, its structure, and its function were limited by the atavistic opposition between an active organism and a passive mechanism. For instance, he writes, "[T]he structure of the central nervous system is too minute to enable us to show the corresponding structural changes in the paths of the brain."[19] A page earlier, he writes, "The cortex is not simply a mechanism. It is an organ that exists in fulfilling its function."[20] What remarks like these illustrate is that, due to the limits of scientific and technical methodology, conceptions about mechanism and teleology are subsequently poorly understood or articulated. Today, we have the benefit of better techniques that have afforded us subtler or more nuanced conceptions of teleology and mechanisms that evade the atavism of yesterday. To these contemporary insights, we now turn.

Social cooperation and cephalic expression

Social cooperative behaviors were an ideal for the early critical realist, many of whom were oriented to pragmatism. They are also a core feature, in addition to deceptive behaviors, in our evolution and the corticalization of function. Mead was also responsive to the empirical contours and was an early proponent, like Dewey, of a cognitive form of behavioral adaptation. Cognitive systems underlie all of human action.[21]

Mead did not write with the eloquence of James; he was more like Dewey in the density of the writing. Plus he was less within the culture of science than Peirce in that he did not readily utilize recent developments in the sciences as Peirce did. Nonetheless, running through James, Dewey, and Mead is the idea of cortical function and evolution.

For over a century we have known that cortical function is knotted to more elaborate cognitive capacities. In fact, the 19th century was dominated by this realization.[22] From a comparative perspective, the frontal cortex, for instance, has expanded quite a bit in the primate brain. The frontal cortex comprises almost a third of the human brain, and it is clearly expanding well beyond closely related species – see Figure 6.1.

One third is quite a large area, and the common comparison of cortex size between primates is always revealing, and instructive.[23] But we must not limit our understanding of the brain to its size alone. Equally important is the organization of the cortices, particularly with regard to specific functions. The degree of corticalization of function is nicely related to the anatomy of the visual systems. What stands out is the visual system in our species.[24] We are visual animals. Thus, the expansion of visual cortex, as depicted in Figure 6.2, is quite large in humans. The visual system, while exceptionally large in humans compared to other primates, is not organized in a significantly different way. The size of it, though, is telling when it comes to the integral role vision plays in our social engagements with one another. We communicate in many ways thanks to vision, from seeing each other's facial expressions to silently signaling with our eyes (the whites of our eyes are noticeably larger than other species, and enable us to direct one another's gaze).

The amygdala is phylogenetically ancient, as discussed in the previous chapter, and expressed in all vertebrates. It is located under the temporal lobe and is almond shaped by tradition, hence its name. The boundaries of the region, however, are highly conceptual and therefore can be subdivided in a number of ways. The amygdala plays a role in diverse adaptive behaviors essential for social function, including

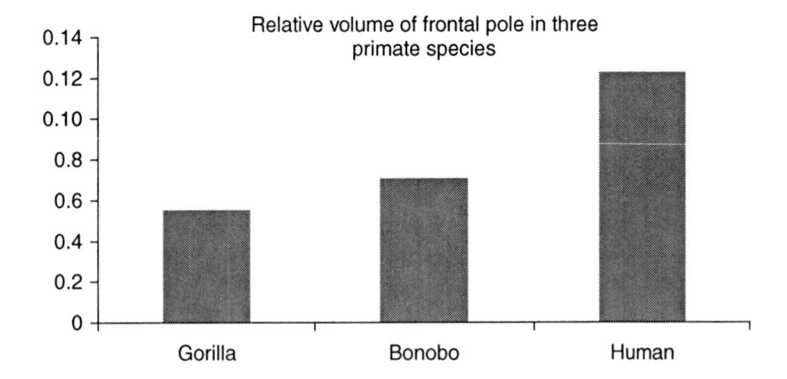

Figure 6.1 Relative brain sizes of primates

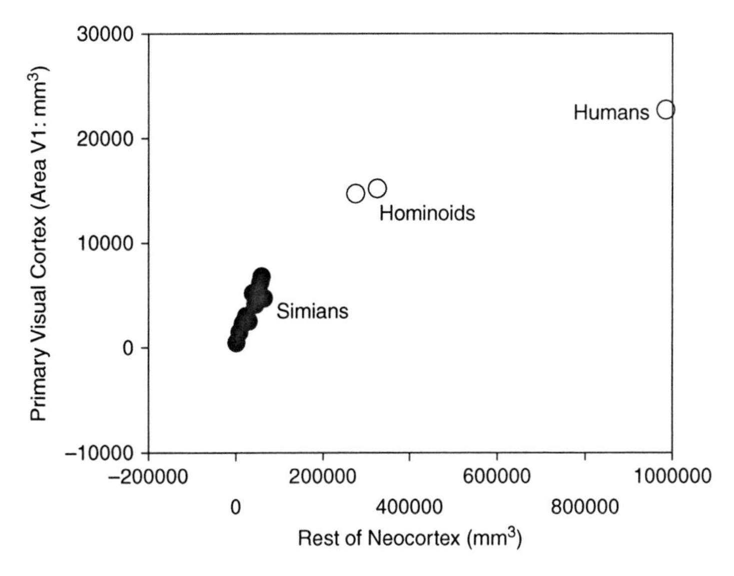

Figure 6.2 Relative sizes of the visual cortex of primates
Source: Barton (2004).

approach and avoidance behaviors, as well as appetitive and exploratory behaviors.[25]

Regions of the amygdala essential for social attachment and avoidance also demonstrate significant changes in humans: for instance, enlargement of the lateral amygdala, which is closely tied to neocortical function.[26] The largest nuclear region is the basal lateral region.[27] In one comparative study of apes and humans (including humans, chimpanzees, bonobos, gorillas, orangutans, and gibbons), investigators found that the size of the lateral division of the amygdala expands quite a bit in *Homo sapiens* compared to the expansion in other primates (see Figure 6.3).[28]

A premium is set on cognitive evolution, which amounts to an expression of diverse cognitive/behavioral adaptations coupled with cephalic expansion. As Mead noted, "The social process relates the responses of one individual to the gestures of another and is thus responsible for the rise and existence of new objects in the social situation, objects dependent upon or constituted by these meanings."[29] With this context of social interaction, survival of each other is at stake: the greater the social cooperative behaviors, the greater the infant survival (see Figure 6.4).

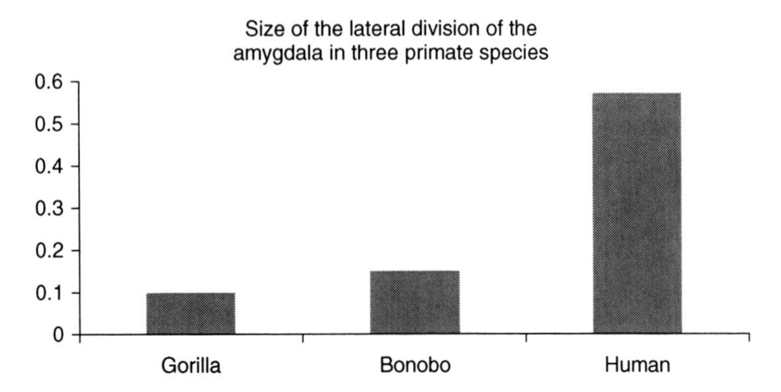

Figure 6.3 Relative sizes of amygdala of primates

Source: Adapted from Barger et al. (2007).

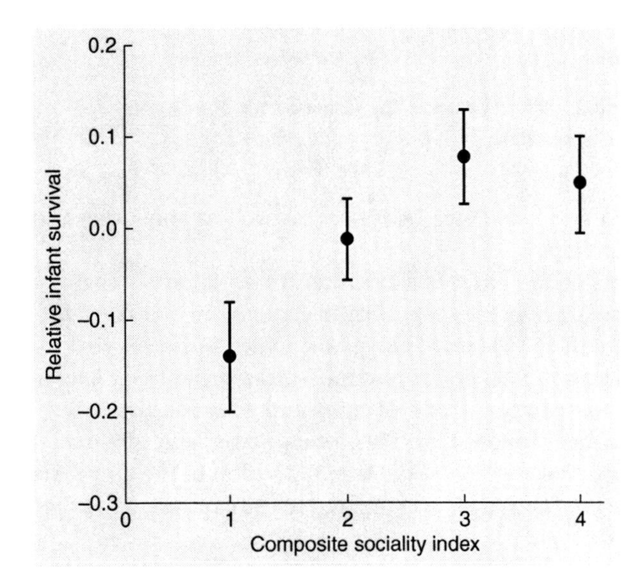

Figure 6.4 Relation between infant mortality and sociality

Source: Schulkin (2011), Adaptation and Wellbeing.

Diverse factors underlie the link between corticalization of function and both social and ecological factors in primate life: life span, group size, terrain adaptation, detection of predation, approach behaviors, foraging behaviors, etc.

The naturalization of our intelligent adaptation manifests itself with the fact that we are pedagogical animals, while core cephalic capabilities and the social context set the conditions for our continued social evolution. As social animals, we are oriented towards diverse expressions of our con-specifics, which root us in the social world: a world of acceptance and rejection, of approach and avoidance towards one another and towards social and ecological objects rich with significance and meaning.

In non-primates, a broad-based set of findings demonstrated the link between social complexity and larger brain size.[30] The metabolic investment of larger-brained animals in expensive neural tissue indicates that this high-energy organ provides compensating social benefits. Interesting correlations have been suggested between neocortical size and social cognitive skills.

Detection and deception amidst cooperation and social prediction are common occurrences that utilize diverse cognitive systems. In some, presumably unusual and unstable, environments, a cognitive expansion, a change in brain development, led to tool use and social cooperative behaviors that proved fateful for our development.

A diverse set of cognitive functions underlies the great range of behavioral capacities that we display. On the biological side, the cortical function underlies the great array of social and communicative cognitive-behavioral systems. In other words, cognitive capacity is knotted in primate evolution to neocortical expansion. The flexibility that marks the behavioral repertoire of our species is a feature of corticalization of function.[31]

Our evolution is tied to social groups working in unison across diverse terrains. Omnivory is a core feature on the ingestive terrain, and social sophistication is another core feature. Key abilities include discerning the wants and the desires of others, along with cognitive adaptations such as recognizing the kinds of objects that are useful or affordable and avoidable,[32] coupled with a wide array of inhibitory capacities. All those abilities contribute to social cooperative behaviors.

This cognitive competence, so social in its roots, also reflects rapid brain growth during critical periods in our evolution, which may have figured in the dramatic expression of our social intelligence.[33] Cultural variation in decision making, for instance, is anchored to core cephalic predilections (see Table 6.1). Perhaps this resulted in a brain oriented to change, rather than stability. The world is social for our species. The self is embedded in the larger linguistic/social community.[34]

The degree of cognitive competence and social gesture, bipedal organization, communicative engagement, diverse tool use, and pedagogy are clearly linked to an expansion of the range of social contact.[35] Contributing to this expansion is the entanglement of all these with

Table 6.1 Components of human social interaction

Component	Description
Attribution of intention ("mind reading")	Crucial precondition, but is by itself nowhere near the abilities needed to generate human interaction.
Mirror worlds	Mental computations that allow us to simulate the other simulating us.
Gricean intentions	Intentions that drive behaviors whose sole function is to have an effect by virtue of having their intentions recognized. Allows for high-level communication.
Cooperative human interactions	There is no point in getting into competitive conflict resolution without presuming cooperation.
Observable practices	Taking turns, sequencing templates and repairing templates. These are ethological proclivities to account for multimodal communication.

Source: Adapted from Levinson (2006).

each other. For instance, bipedality afforded animals like us greater use of our hands for other projects, from communicative gesturing to tool use – both of which are involved in pedagogy. Our social evolution is linked to both the cooperative and competitive. Two variants of chimpanzee reflect this. The bonobo is more peaceful,[36] more inclined to share with one another, and less fierce than its more well-known and larger cousin, the common chimpanzee. Bonobos, unlike the common chimpanzee, are not dominated by male aggression. Some have suggested that a difference in oxytocin levels, facilitating social attachment instead of social aggression, is a contributing factor.[37]

Depicted in Table 6.2 are some common themes in our cognitive development, particularly that of social development (it is adapted from Tomasello and colleagues).[38] What marks us off from other primates, and probably very early on in the history of our species, is our vast social milieu, with our endless social acts. We have evolved a set of specialized social-cognitive skills as a product of an inherent need to form social groups, even more so than other primates (see Figure 6.5).

Our evolution is knotted to language, social competence, standing upright, expandable limbs, and long ontogeny (i.e. a long childhood).[39]

Table 6.2 Human cognitive development

Infancy: understanding others as intentional
1. Following attention and behavior of others
2. Directing attention and behavior of others
3. Symbolic play with objects
Early childhood: language
1. Linguistic symbols and predication
2. Event categories
3. Narratives
Childhood: multiple perspectives and representational re-descriptions
1. Theory of mind
2. Concrete operations
3. Representational re-description

Source: Adapted from Tomasello et al. (1993).

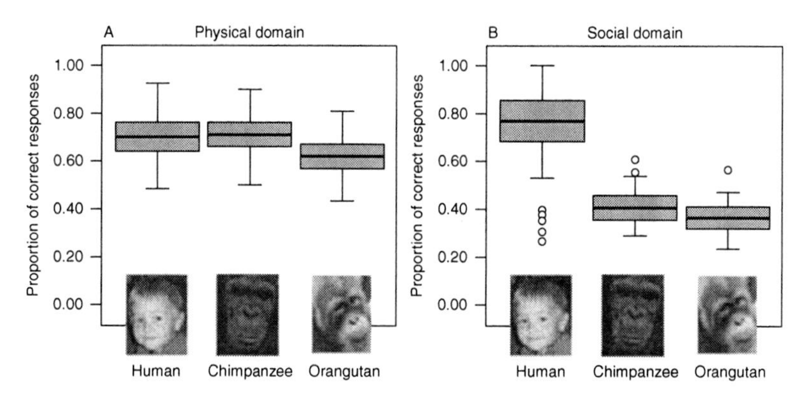

Figure 6.5 Comparison across primate species between physical performance and social performance

Source: Adapted from Hermann et al. (2007).

Notes: Box plot graphs demonstrating that humans perform on a similar level as other primates on the Primate Cognition Test Battery when assessed at the physical interaction level with inanimate objects, but surpass the other primates in the social domain, which involves analyzing other animate beings' intentions, perceptions, and knowledge.

It is also knotted to endless linguistic competence and creativity, which underlies cephalic expansion, along with other cephalic capacities.[40] One cephalic capacity essential for social discourse is a moral sense – the communicative competence essential for pragmatists[41] – for transforming high ideals into practical outcomes, into consequences that matter.[42]

Moral sense

John Rawls, the great moral theorist of the 20th century, linked human happiness to the fulfillment of rational plans, as had others before him. From Aristotle through Thomas Aquinas to Dewey, happiness was linked to reasonability, to the development of habits that promote some notion of the good, a good beyond simple hedonic sensations or calculus. Dewey always labored to link moral decision making within the contours of decision making, and to a moral life. That is, it is not enough to make good decisions; one must also live in such a way that both embraces the consequences of the decisions one makes and also cultivates moral decision making in the first place.

Humanization is of expanding an appreciation of the social bonds amongst us. It is a celebration of what is good about us, however frail that goodness may be. Wisdom lies in keeping track of, without being demoralized by, the good actions as well as the endless transgressions as we trample on one another. As Dewey put it, "judgments of value... have a reconstructive and transforming effect."[43] The link to others is one fundamental adaptation of the larger social selves that merge.

The self, a social collection of cognitive/behavioral expressions,[44] is intimately tied to the moral frameworks that are revealed in diverse cultural contexts. The ethical link to the social is demonstrable. While there may be some core characteristics, there is wide variation, at least on the surface, and perhaps with deep prosocial behaviors. There is something important about the idea of an innate syntax or moral sensibility distributed across the brain.[45] It is, however, also tied to a lot of other cognitive functions that figure in the social milieu.[46]

The metaphor only goes so far.[47] While a theme among human nature is its myriad variations, and discontinuity in our understanding of the origins of language a stable factor, pushing the metaphor of the syntax of morality, analogous to the syntax of language and the language organ, is also misleading. The rules of morality (e.g., the Categorical Imperative) are knotted to emotions of social approach and avoidance, the cognitive systems endemic throughout the brain.[48] And then there are rules that are binding or categorical, that force one towards others; these rules are operative in a prepared social brain. Forming common bonds, for instance, is a normative moral goal that has long been thought to be worthy – it still is.

While social fragmentation is an occurrence of endless regularity, it is in some contexts an aberration and in others an adaptation. Cognitive dissonance is a convenient vehicle to separate behavioral systems; it can serve and it cannot. Moderation and good judgment, as Aristotle and Dewey understood, are often measured in social context rather than as isolated judgments.[49] Amidst the endless ploys of social dominion and temptation, stoic self-control, with a concern for dignity of oneself and of others, is a worthy normative goal.

A dimension emphasized by feminist theorists is the social nature of the human condition. Carol Gilligan's *In a Different Voice* focused on the social part and less the autonomous part of human psychology, something that diverse pragmatists had acknowledged and understood.[50]

Social contact, avoidance, and moral sentiments

Moral sensitivity allows humans to quickly apprehend the moral implications in a social situation depending on context, agency, and the consequences of one's choices, through the active engagement of specific moral sentiments such as those listed in Table 6.3.[51]

Moral sentiments are culturally ubiquitous,[52] despite the wide diversity of how they can be tied to specific cultural and situational contingencies. These sentiments are intrinsically linked to daily social interactions. Anticipated or actual violations of one's own principles and beliefs trigger aversive feelings, such as guilt and shame. Standing up for one's core values, on the other hand, will tend to trigger positive feelings like pride and joy. Moral sentiments are thus strong motivators for human action in a social context.

Moral sentiments depend on the engagement of several cognitive processes, including action and conceptual knowledge, emotion, and motivation. These sentiments require a tight integration among human isocortical and limbic circuits.[53] As such, moral sentiments should

Table 6.3 A partial list of the moral sentiments

Guilt
Shame
Embarrassment
Pride
Indignation/anger (selfish)
Indignation/anger (moral/empathetic)
Contempt/disgust
Pity/compassion
Love/tenderness
Awe/elevation
Gratitude

neither be considered purely "cognitive" or "emotional,"[54] nor a simple sum of "cognition" and "emotion." Moral sentiments can be tentatively clustered on the basis of neurobiological and phenomenological components. The "prosocial" cluster includes guilt, embarrassment, compassion, and gratitude, which promote cooperation, helping, reciprocity, reparative actions, and social conformability. A subclass of those, the so-called empathetic moral sentiments (guilt, gratitude, and compassion), putatively share the attachment component and play a central role in behaviors linked to empathy.

On the other hand, sentiments linked to interpersonal aversion – the other critical sentiments (disgust, contempt, and anger/indignation) – are experienced when others violate norms or one's "rights," and when these others endorse aggression, punishment, group dissolution, and social reorganization. Still, acting in accord with one's own values often triggers self-praising sentiments – among which pride is the prototype – while witnessing praiseworthy actions of others will lead to the experience of other-praising sentiments, like gratitude when one is the recipient of such actions.

Since Darwin, the question of whether our prosocial moral instinct is unique among species has been actively pursued. There is general consensus today that the building blocks of morality have strong evolutionary bases. The origins of morality need thus to be sought among certain primitive motivational-emotional mechanisms that can readily be identified in other social species. These mechanisms can be operationally organized into two broad classes: one linked to approach and affiliation, and the other linked to aversion and rejection. While attachment promotes care, cooperation, and reciprocity toward in-group members, aversion fosters blame, prejudice, and group dissolution. A critical step of human evolution might have involved the functional adaptation of basal forebrain-limbic circuits linked to social attachment and aversion. Intertwining of social attachment/aversion and complex social knowledge through cultural learning thus became a unique feature of human nature.[55]

A *prosocial* cluster including guilt, embarrassment, compassion, and gratitude promotes cooperation, helping, reparative actions, and social conformity.[56] A subclass of prosocial emotions, the *empathic* emotions, does so by invoking a feeling of social attachment (guilt, gratitude, and compassion). On the other hand, *other-critical* emotions (disgust, contempt, indignation) are experienced when others violate norms or one's "rights." Specific brain activation patterns associated with distinct clusters of moral sentiments have recently been described.

Human moral sentiments thus embody intertwined elements of social attachment/aversion and culturally shaped knowledge,[57] for instance moral aversion clustered around visceral disgust.[58] As the philosopher Martha Nussbaum has noted, "primary objects" anchored to an "evolutionary origin"[59] are imbued with moral sensibilities and anchored to moral approach (e.g. magnanimity) or moral aversion (e.g. disgust).

A moral psychology acknowledges the strong and diverse links between social adaptation and a propensity to discover inclinations. Rationality is limited from fast reactions.[60] Well-entrenched moral frameworks Dewey designated to funded wisdom, as they enhance our moral standing. There is little doubt that affect plays a major role in moral judgment.[61] But then it does in everything. Reason is limited and imperfect in just the way that Westerners have conceived affect as being limited and imperfect. This is precisely because reason and emotion are entangled, not ontologically distinct and diametrically opposed. Recognizing this and making peace with it is what it is to be a grown-up by accepting our reality. We need not be offended by this entanglement of reason and emotion. We must accept the demythologizing of our capabilities. We must nevertheless ward against a pernicious devolution of function.

There may indeed be stages in moral development, but they are more labile.[62] The end point is not some Kantian untouchable position untainted by sensory pangs. No blindness by the light, for the light is eminent at all levels in our cephalic systems. A changing historical perspective in moral psychology has landed us close to Darwin and Dewey: Darwin because we are talking about prosocial adaptation, and Dewey because we are talking about good enough moral reasoning. Reasoning of this sort is not perfect; it is certainly compromised, given the competing interests vying for expression. Moreover, this reasoning is social, not autonomous in nature.[63]

A sense of fairness, a concern for others, and an observance of cultural norms permeate human social existence. This social sensibility is the essence of human morality, which emerges from a sophisticated integration of cognitive, emotional, and motivational mechanisms, which are shaped through cultural exposure.[64] Morality is thus a product of our cultural and biological evolutionary history, and represents an important adaptive element for social cohesion and cooperation.[65] By refraining from satisfying immediate self-concerned desires and opting for socially enhancing actions instead, a member of a social group can increase his own reputation and become more likely to be supported by others. Unfortunately, not all members of a social group seek social enhancement or support. Often, if not always, such members suffer from personality disorders.

One set of findings emerges from systematic studies of acquired personality changes due to brain damage, mostly to the frontal lobes. Given the similarities with developmental psychopathy, such impairments in social conduct were dubbed "acquired sociopathy." A review of lesion studies of patients with acquired sociopathy and preserved general cognitive abilities showed, however, that current models of normal social conduct have emphasized the prefrontal cortex (PFC) at the expense of other brain regions.[66] During the past few years, however, a number of functional magnetic resonance imaging (fMRI) studies on normal volunteers have contributed rich material for our understanding of the moral brain.

In one experimental investigation of the neural underpinnings of moral judgment, volunteers were scanned during the auditory presentation of short statements, on each of which they were asked to make silent categorical judgments (e.g., right vs. wrong).[67] Some statements had an explicit moral content (e.g., the judge condemned an innocent man), while others were factual statements without moral content (e.g., telephones never ring). When the moral condition was contrasted to the factual one, the medial frontal gyrus and medial and lateral sectors of the front polar cortex, isocortical regions that are especially well developed in our species, were strongly activated. The right anterior temporal cortex and the left angular gyrus/superior temporal sulcus region were also strongly activated by moral judgments. To be sure, while many may try to construe such results as evidence for a fact/value dichotomy, such a conclusion is unwarranted. Like much of the brain's functions, activity is widely distributed; functions having to do with morality are no exception.

Effects on these cortical regions could not be explained on the basis of overall emotional arousal, as additional analyses revealed. Subsequently, Greene and colleagues probed another important aspect of moral judgment using fMRI.[68] Normal subjects were exposed to moral and nonmoral dilemmas that were structurally more complex than the simple statements described above, imposing a higher load of reasoning and conflict. Moral dilemmas were divided into moral-personal (the agent directly inflicts an injury on another person to avoid a worse disaster) and moral-impersonal (the agent does it in indirect ways, such as by pressing a button).

One instance is known as the trolley car example. An out-of-control trolley car is rushing towards a section of track on which five innocent people are working. In one scenario, the moral actor is required to push another innocent person off a bridge (more personal) to halt the car and save the five. In another, the moral actor must pull a switch (less personal) that changes the track on which the trolley runs – inevitably killing the single person on one track, but saving the five on the other. There is all the difference in the world between these two situations, despite the popular interpretation that these two situations are formally identical (save one, kill five; save five, kill one). The result in the bridge case is greater conflict because it is up close and personal (e.g., should I opt for killing one innocent to save five other lives and suffer the angst of being a murderer forever, or should I abstain from doing so and therefore suffer regret for being responsible for the death of five people because of my own omission?). This is not to say that in the case of the lever puller there is not the possibility of conflict in one's moral experience. Despite the formal identity between these two situations that is only apparent when the meaningful context is stripped away, what is important to recognize here is the lived experience of these situations. Most philosophers and scientists interested in trolley problems neglect this element of our moral lives. Hence their surprise at the results, which basically confirm what pragmatists have long known: the situation matters.

In addition to normal subjects, patients with ventromedial prefrontal cortex damage were also tested. They made more utilitarian choices in trolley-type dilemmas (see Figure 6.6). They also opted more often for costly punishments for non-cooperators in the ultimatum game (see below for further discussion), i.e., they lacked appropriate emotional responses.[69] One possibility is that an intact ventromedial prefrontal cortex is more critical for experience of prosocial sentiments associated with affiliative components (i.e., guilt, compassion, interpersonal attachment), whereas the dorsal lateral prefrontal cortex and lateral orbital prefrontal cortex are more relevant for other critical sentiments (such as indignation and contempt). In other words, reduced prosocial sentiments are compatible with increased utilitarian choices in personal dilemmas, and preserved or increased punishment of others in the ultimatum game. Indeed, the greater the cognitive conflict, the greater the information there is to be processed, and thus the greater the activation of regions of the frontal cortex (dorsal lateral and anterior prefrontal cortex, in addition to regions of the posterior parietal regions).[70]

Other studies have addressed additional key issues in moral arousal, presence of bodily harm, response times, semantic content, cognitive load, conflict, intention, consequences versus means, emotional

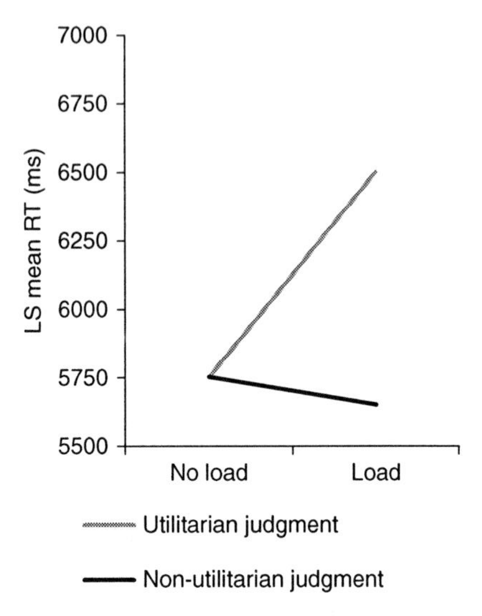

Figure 6.6 Brain function and moral deliberation
Source: Adapted from Greene et al. (2008).

regulating, justice versus care-based judgments, and interacting with theory of mind. These studies have significantly extended our knowledge of the neural substrates of moral judgment and emotions, emphasizing the consistent involvement of lateral and medial regions of the orbitiofrontal cortex, prefrontal cortex, and the amygdala, in addition to other brain regions.[71]

How do moral judgments translate into real-life action? Moral sentiments, which have long been recognized as powerful motivational forces, stand out as strong candidates. Along with sentiments, personal commitment to values also plays a biological role in moral behavior, though their psychological and biological underpinnings are less well understood. Social or moral values can be defined as social concepts which acquire intrinsic motivational salience for a given individual or society (e.g., honesty, courage, social cohesion, autonomy, and benevolence).

One fMRI study, which controlled for the effects of non-moral emotional arousal, provided initial clues on the functional anatomy of moral sentiments. In this study, participants were exposed passively to pictures that varied in their moral content and emotional salience.[72] Activation of the anterior insula, amygdala and subcortical structures were observed for both moral and non-moral but unpleasant stimuli. The prefrontal, medial orbital frontal cortex and posterior superior temporal sulcus (STS) region, however, were selectively activated by moral stimuli. The engagement of the same brain network may underlie a moral sensitivity mechanism, by which certain social situations automatically trigger moral sentiments.

Indeed, damage to the frontal cortex has been linked to a devolution of moral sentiments and moral judgments, sociopathic behavior, and a devolution of empathy towards others. The prefrontal cortex is engaged *whether or not* decisions or behavioral outputs are required in moral scenarios, suggesting that the prefrontal region does not merely manipulate information stored elsewhere in the brain but, in fact, does represent certain aspects of social knowledge or action.

The prefrontal cortex has been implicated in all of these examples of inhibitory function, and there is now strong evidence that it may be exerting its inhibitory capacity on the amygdala in the case of extinction learning. This same prefrontal–amygdala interaction may also be occurring in the attentional domain.[73]

Prefrontal cortex pathology has long been associated with perseverative responding, which may be described as an inability to inhibit behaviors that are no longer appropriate in the current situation. Such responding was typically observed in reversal learning tasks. Behavioral flexibility, or the ability to adjust responses when the reinforcement value of a stimulus changes, is reflected in the reversal learning task. It has frequently been suggested that brain regions such as the prefrontal cortex (PFC) and the anterior cingulated gyrus (ACG) are involved in regulating amygdala-generated emotional responses.[74] The amygdala may be seen as providing rapid threat detection, with the prefrontal cortex providing top-down control when given sufficient processing time.

Brain evolution and behavioral adaptation

Paul MacLean proposed a demarcation among the neo-mammalian, paleo-mammalian, and reptilian levels to account for the evolution of the brain.[75] According to this perspective, the human neural axis is a repository of phylogenetic layers, starting with the reptilian brain ("basic instincts," e.g., aggression and sex), which is enveloped by the paleo-mammalian brain (limbic-mediated emotions), then crowned by the neo-mammalian and the expanded human neocortex. The burden of scientific evidence, however, challenged the validity of such hierarchical, phylogenetically layered models originally proposed by 19th century thinkers and elaborated on by early 20th century researchers,[76] broadening the anatomy of the limbic system and its functional relationships to "higher" cognitive functions related to neocortical systems. The successful development of the broadened view of the limbic brain is partially owing to the discovery of the chemical organization of the central nervous system. Specifically, regarding social behavior, the discovery of molecules such as oxytocin and vasopressin, that exert central roles in prosocial and reward-related motivations, was critical.[77]

Diverse studies have started to demonstrate the role of attachment-related neural mechanisms in humans. Structures of the brain reward system, i.e., the midbrain ventral tegmental area and ventral striatum, along with basal forebrain structures, are engaged when humans look at their own babies or at romantic partners. Furthermore, other studies have provided some evidence for the effects of oxytocin on human social behavior. In a sequential economic game involving trust, Zak et al.[78] found that oxytocin levels were higher in subjects who received a monetary transfer signaling an intention to trust, in comparison to an unintentional monetary transfer of the same amount from another player. Higher oxytocin levels were associated with increased likelihood of reciprocation.

Decreasing social anxiety or fear might also be an important effect of oxytocin, a hypothesis that was strengthened by a recent pharmacological fMRI study. In this study, Kirsch et al.[79] showed that oxytocin attenuated amygdala activation to fearful stimuli. In another study, Kosfeld et al.[80] showed that intranasal administration of oxytocin induced more cooperation in an anonymous economic game by boosting interpersonal trust. In this game (a variety of the ultimatum game, mentioned above), the first player chooses to transfer an amount of money (if any) to another player. The amount is multiplied, and the second player may choose how much he/she will transfer back to the first player (i.e., reciprocation). Exogenous oxytocin administration was associated with increased amounts transferred in the trust game by first movers.

Disgust also plays a central role in social aversion. Proto-forms of disgust, associated with non-social functions, can be found in non-humans. The distaste, nausea, and vomiting that occur following exposure to potentially toxic or contaminated foods and odors have a clear adaptive function.[81] In humans, disgust, and its close relative, contempt, play an obvious role in interpersonal settings.

In contrast to anger, disgust, and contempt are slower to fade out. They tend to "stick" or to become a property of the object, intensely devaluing it.[82] For instance, if you suffer a bout of stomach flu following the eating of canned peaches, you may develop a long-lasting aversion to their sight, smell, or taste, even though you know the peaches are unrelated to the actual illness. Thus, in the same way that neural systems underlying primitive forms of pleasure and social bonding operate in highly complex social situations associated with human cooperation, neural systems underlying aversive responses related to physical properties of odors and foods seem to have been adapted to sustain social disapproval. Indeed, while morality often promotes cooperation and helping, it can also steer hostility towards individuals and social groups.

Diverse regions of the brain tied to social behaviors underlie prosocial sentiments.[83] Prosocial moral sensitivity allows humans to quickly apprehend the moral implications in a social situation depending on context, agency, and consequences of one's choices. These sentiments are intrinsically linked to daily social interactions, and there are several regions in the brain that provide a context for social flexibility.

Recognizing the intentions of others is one critical feature in prosocial behaviors. It underlies the social act in diverse communicative contexts. This cognitive capacity begins early in ontogeny and is tied in to visual sensibility. Eye contact, again recognizing the intentions of others, is compromised in autistic individuals.[84] Recognizing the intentions is knotted to a broad array of cephalic tissue that underlies perspective taking and human moral judgment.[85]

A propensity for prosocial behaviors is small, yet competitive with the rest of our predilections.[86] A consideration of the prosocial instincts embedded in cephalic bodily responses towards others amidst a frail sensibility of human kindness,[87] as well as diverse cognitive predilections amidst human/animal sentiments,[88] sets the stage for moral progress,[89] social contracts,[90] and diverse cognitive rules.[91] As depicted in the Figure 6.7, various regions of the brain tied to social behaviors underlie prosocial sentiments.[92]

The cortex and the adjudication of conflict

All three psychobiological classical pragmatists (James, Dewey, and Mead) emphasize cortical function, inhibition and moral behavior. It is an old theme, still current and valid.

Diverse cortical and neocortical sites participate in the adjudication of competing interests, motivational pulls, and incentive lures, one of

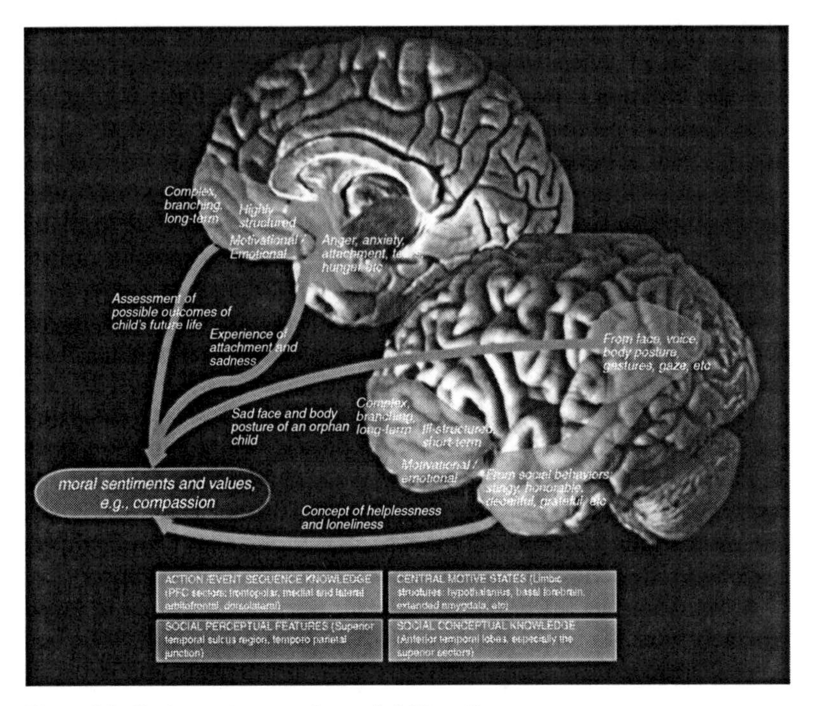

Figure 6.7 Brain anatomy and moral deliberation

Notes: Brain regions implicated in human moral cognition. Cortical regions implicated in moral cognition include the sectors of the prefrontal cortex (frontopolar, medial, and lateral orbitofrontal cortex, anterior dorsolateral prefrontal cortex, and additional ventromedial regions), the anterior temporal lobes, and the superior temporal sulcus/temporoparietal junction. Subcortical structures include the extended amygdala, hypothalamus, basal forebrain (especially the septal region), ventral striatum-pallidum, subgenual region and the rostral brainstem tegmentum.

Source: Moll and Schulkin (2009).

which is moral. Moral sensibility competes with other motives, other interests.

One common view, not necessarily mistaken, but perhaps a bit misleading, is that negative and destructive behaviors come out as the cortex devolves.[93] This view entails that lower cognitive capacity equals less functionality for a civilized individual, and a greater inclination towards base behaviors.[94] Neural systems for control are neocortical; impulses and the less reflective cognitive resources are sub-neocortical.[95]

Control and cortex have become somewhat synonymous, particularly with the frontal cortex.[96] For this, there are many forms of evidence that date back many years.[97] Of course, inhibition and cortical control do provide opportunities, one of which is to restrain behavior. This is the core cortical perspective in light of evolution and the affordance of diverse behavioral options, including essential bodily ways in which to engage and integrate information. Fast-driving visceral representations are fundamental and grounded in cognitive systems and the organization of action.

Dopamine and the organization of inhibition

Our evolutionary legacy is knotted to diverse forms of inhibition and adjudication of competing drives. One way, in fact, to understand something about our conscious sensibility is the broadcast of diverse competing interests perfusing central states, resulting in the inhibition of some interests and the expression of others.

Dopamine expression is essential in the development of inhibition in our species. Aberrations in dopamine expression affect this important cognitive/motor capacity. For instance, children born with Phenylketonuria (PKU) syndrome generate a conversion process that results in disproportionate dopamine expression in the brain. This can be corrected by dietary means (see Table 6.4). If it is not corrected, there are diverse forms of impairment in behavioral inhibition essential for normal development, such as delayed alternation (i.e., there is difficulty with inhibition; delayed inhibitory reactions are comprised), or delay in matching tasks (a task that requires inhibition and is knotted to the development of prefrontal cortex and dopamine expression).[98] This is also a developmental event easily expressed in children by maturation of the brain.[99]

Cognitive control over behavioral inhibition is impaired in this genetic impairment of dopamine.[100] And it is impaired with damage to the prefrontal cortex in macaques. Damage to diverse cortical regions promotes a devolution of diverse functions,[101] including cognitive capacities that encompass many forms of memory and language. Flexibility of cognitive-behavioral functions and opportunity of choice in varying contexts is a key hallmark in our evolutionary cognitive ascent. Damage to diverse regions of the brain, not just the cortex, dilutes these capabilities. Pragmatists who were also psychobiologists, such as James, Dewey, and Mead, understood the relevance of the evolution of cortical control and inhibition.

Social construction, moral sentiments, and biology

We are social animals, and the fluidity of the social and natural is always very permissive. There are several intellectual traditions that attempt

Table 6.4 Outline of biological events resulting in PKU syndrome and its cognitive, and physiological effects

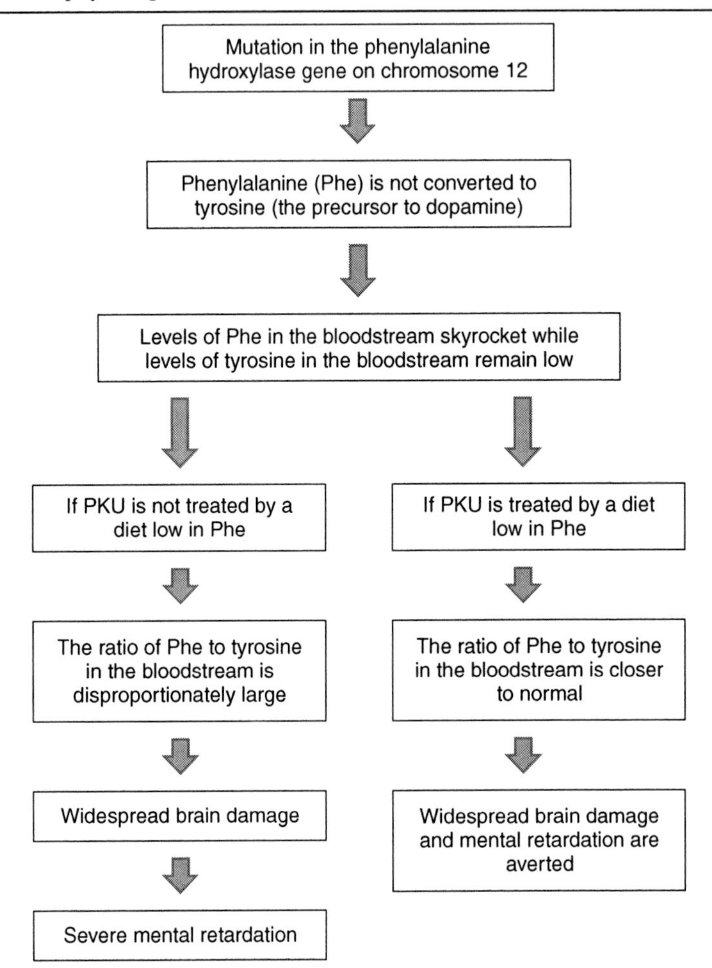

Source: Diamond (2001).

to place the mind in a context broader than individualistic. One is social constructivism. The social constructivist positions – including both those descended from phenomenology, e.g., Schultz, Berger, and Luckmann, and those descended from ordinary language study, e.g., Wittgenstein and Austin – tend to see mental predicates as necessarily involving more than a single individual.[102] Mind, knowledge, emotion are all shared, public, and able to be critiqued – in short, they are sociocultural. They are "forms of life."

The tradition that links Aristotle to Thomas Reid through Peirce and Dewey acknowledges real-world events. As Peirce might say, "The world says no to some of our hypotheses."[103] Aristotle's emphasis on practical reason in negotiating our social world,[104] not just some sensations in mental space, is what we are trying to establish as we tie it to what Dewey meant by "funded reason": those social practices that have warrant and are therefore justified.

Some theorists have given up the faith of Dewey, Peirce, Mead, and James: though there may be no Truths, the world still imposes itself on us and our thinking. It may be naïve to imagine that we have some direct perception of that world. A critical realist, as opposed to a naïve realist, knows that nothing is quite so given as we once thought. But, still, there is room to believe that even social constructions are constructed on and about something, something real. The world may well be lost to Rorty (see the conclusion of this book), but to pragmatic naturalists, with their emphasis on human experience and not simply linguistic competence or propositional knowledge, it is not.

At times, it seems that these theorists believe that if something can't be known with certainty, then belief in it can't be justified at all. They also seem to believe that something that is socially constructed can't also be rooted in nature. It is as if they have failed to see the importance and power of social construction, believing that if something were natural it wouldn't need to be socially constructed. But we humans live in a world of social constructions, even if some of them are also natural. Fortunately, we are not faced with the choice between the world well lost and the world given in a social milieu ripe with meaning, and rich in transactions in which human contact pervades our experiences, especially those of others thanks to cephalic capability ready for contact.

Knowledge grows in a social context. Some knowledge is also about social contexts.[105] Yet sometimes what passes for knowledge, but isn't, passes for political reason. If there really is no straight and narrow, if everything is just conversation, then there is no warping.

Pragmatic naturalists understand that perspectives are public, fluid, and part of the social milieu. Social constructivists and American pragmatists share a strong commitment to the view that knowledge is public. There was nothing anti-biological about Mead or Dewey; somehow, their avoidance of an inherent conflict between constructivism and realism should at least give us hope that a way around it can be found. Pragmatism, with its focus on interactions among individuals, rather than processes within individuals, is comfortable with the social constructivists' ethos. Because both schools see the social world as ever in creation, individuals who come into contact with one another are described as creating – though certainly not *ex nihilo* – a shared reality.

Scientism, after all, is the pursuit of the form of science without substance; it's talking the language of science without rooting it in the real stuff of the world that science is about, in the end. So the biologist knows that in the end all the talk about evolution will have to be cashed in for real bits of DNA, and no amount of discourse about adaptation will substitute. Biology is about real stuff. But in the same way, no amount of deconstruction can do away with the distinction between those who are married and those who aren't. There really is a distinction there, even though marriage is a social construction. The point is to be an even-handed realist here. Social constructions are real; they are real in just the way that marriage is real, in a very important way. But biology is real too. One is not more real than the other. Let us consider, now, a domain where the biological and the social interact, where social constructivists and biologists have taken up very different positions: emotional expression.

The study of emotions provides an interesting place to work out the tensions between the social constructivists and the biological realists or pragmatic naturalists. What do we know? For one thing, there is no denying that emotions and their expression have a universal biological basis. Facial response is a good example. The face evolved to engage others socially, to link approach and avoidance and social communicative functions to their expression. Facial expressions as such are not necessarily indicators of how one feels, contrary to James's view of the emotions (he believed that first I know that I am afraid because of my bodily response to the bear). That does not mean that they are perfect, or that we axiomatically know them all the time, etc. They are merely a subset of emotions that have (roughly) the same expression across cultures (see Figure 6.8).[106] In other words, the James–Lange theory of emotion was too all-encompassing. Peripheral signals, like heart palpitations, are indeed important sources of information. But facial expressions, while emotional systems, communicate emotion towards others and not to oneself, as the James–Lange theory emphasizes. Furthermore, where James's famous story of running from a bear is not – could not be – a story of deception, facial expressions are often used to deceive, either by expressing an emotion that one does not feel or by not expressing one at all despite how one may feel.

Anger is, in part, the perception of an insult which gives rise to an impulse for revenge. Insults are abstract, as is the category revenge. As Martha Nussbaum, in a wise book *Cultivating Humanity*, correctly points out, the cultivation of the emotions is towards wise responses to others,

Figure 6.8 Facial expressions of emotion
Source: Darwin (1872, 1965).

viz., measured responses and fair judgments.[107] Perceiving an insult can also be a matter of apprehending something rather abstract. The faculty for apprehending abstract things is often called cognition. Cognition is also needed, often, to plot revenge. So there seems to be a lot of play for cognition in anger. And, there is also play for social constructions of all sorts. Surely what constitutes an insult (or, for that matter, revenge) is socially constituted.

The social constructivist points out that there are some emotions that are, at least, very uncommon in our culture. The emotion of Amae – a desire to be dependent on a superior and to be taken care of by that person – is very common in Japan, but rather rare in America.[108] Americans, moreover, are likely to characterize it negatively as submissiveness. But the biologist need not claim that *all* emotions are universal. The social constructionist has another kind of fact to offer, and it is that emotional expression is rule governed and normative. One looks sad at funerals no matter how one feels at the moment.

Darwin himself had two views of emotion. One was bound up in the idea of serviceable associated habits, close to that of Peirce.[109] The idea was that with repetition of mental states and actions, primitive forms of the actions began to be emitted as an association with the mental state, even if the full action was inhibited or not expressed. Then the Lamarckian Darwin went on to argue that these products of association were inherited. We, unfortunately, have to reject this Lamarckian account and turn to Darwin's other account, which should make the social constructivist happier.

In Darwin's, Dewey's, or Mead's view, emotional expression evolved not in order to display one's insides to others but as a means of carrying out certain pragmatic, social actions. On that account, emotional expressions are already tied by Darwin to social context rather than to internal states. That culture makes further use of this connection between expression and sociality is not very surprising. One cephalically rich account of the emotions' binding us to social contact is one in which the arousal systems for approach and avoidance systems are nested in broad-based social contact.

The list of human expression is partial. Cultural variation is a key factor in human expression. Some things are basic to any social world and reflect the human condition. Acknowledging social construction is not the same as being reduced to social construction: similarly, the biological turn of anchoring a pragmatic realism to naturalism need not be reduced to a narrow Darwinian view. We are cultural; our cephalic capabilities continue to expand into a virtual landscape in which we are endlessly embedded.[110]

The investigations need constraints. Those constraints can come from two directions, from the biology of the organism and from the social context in which the organism must function. Either way, biology is often needed. On the other side, psychologists have a habit – much criticized – of ignoring the social context and imagining that linguistic phenomena, divorced from action and culture, are really the upshot of internal, cognitive processes. The spirit of American classical pragmatism can be found in both: like the social constructivists, pragmatism focused on social transactions rather than internal machinations, and, like the biologists, pragmatism was rooted in functionalism and the biological.

Our intersubjective sensibility is given some neural meaning through the so called "mirror neurons." Sets of neurons in different brain regions that in diverse ways are common neurons that are recruited whether you do the action, I watch you, or we both do an action together. The learning of moral habits is very much dependent upon the social context and the recruitment of neural systems, like mirror neurons.[111]

Conclusion

Mead, like Dewey and Peirce, was anchored to the social milieu, to forms of social pragmatism in which cognitive capacity is linked to social adaptation.[112] But the pragmatists also had a strong sense of biology and of the dissolution of boundaries between biology and the social sciences. This is particularly pertinent for students of Mead and Dewey. For them, the social was pervasive in our experience, a virtual presence expanding our cephalic capabilities in a connective way.[113]

A moral education for Mead, and certainly for Dewey, is a developmental trajectory, bootstrapping on empirical findings. These findings need to be anchored to moral development. Education nurtures many cephalic capabilities, but one still has to "realize the social nature of action."[114] But a moral education, as Aristotle and the Stoics noted and Dewey reinforced, necessitated the development of character, specifically a moral character in which self-corrective processes are tied to humility, where a nurtured self without a bloated head is a normative goal. Indeed, normative goals flow through the pragmatic processes. Ends are embedded in means, and a "reflective morality" is a normative goal in moral life. "A common faith"[115] is central, in which "intrinsic ties become efficacious in the regulation of a community,"[116] a community of moral agents, in which valuation is emboldened by self-corrective sensibilities. "A theory of valuation"[117] is one in which cephalic capability is pervasive. There is no split of the valuation from the cognitive; there is just a pervasive sense in which valuation is at the heart of our decision making.[118]

There is little doubt about the robust, kindly social attachments diversely expressed in many mammals in nature, something that pragmatic naturalists who were oriented to the all-pervasive social milieu recognized. These prosocial attachments form the basis for moral judgment, from primitive beginnings. Like language, human evolution demonstrates a punctuated abrupt change to what we might call moral syntax. For some time, moral rationalists have suggested something a priori for moral judgment based on the innate strictures that are pregnant in syntax.[119] In many ways, language is the primary metaphor for a "moral organ"[120] – after all, it was Chomsky who originally suggested that language was a mental organ of sorts.[121]

Perhaps Mead and, particularly, Dewey were prescient, following Aristotle and others in linking the development of moral habits with a modern sense of inquiry. Moral inquiry, like other forms of inquiry, needs to be defended and investigated; it is empirical and normative.

Prosocial orientation is an extension of human adaptation and human reason; it is the important continuity that Dewey emphasized.

The utilitarian choice, when and where possible, is the greatest good for the most amidst a consideration of rights and rules and an evolving culture that was respectful of individuals. We must also be mindful of the devolution of individuals in a sense of the collective. Nazi doctors, for instance, were a well-educated class, who nevertheless carried out grotesque experiments and assisted in a genocide of great proportions.[122] The broad-based research arising out of eugenics was of the same educated class.[123] Our generation and nation are not innocent either: we have the Tuskegee experiments,[124] our experiments with hallucinogens and other mind-altering drugs,[125] and our own eugenics history that resulted in the forced sterilization of hundreds of poorly educated, eccentric, and mixed race individuals.[126]

An evolving sense of rights does not mean that they are not historically contingent.[127] They are emboldened in some cultures and exaggerated in others (e.g., a sense of individualism). A moral compass for pragmatic naturalists with social sensibilities rooted in a democratic spirit of informed citizenship is an ideal goal; it is faint but present in ephemeral moments of human history and human cultural evolution.[128]

There is indeed a prosocial sensibility that sets the context for morality, that underlies diverse forms of social, altruistic, and moral senses (repulsion from harm, empathy towards others' pain, etc.),[129] and that is knotted to moral rules of fairness,[130] as well as, within reason, to consequences. It entails diverse forms of cephalic competence. It has consequences even for moral purists.[131] The untenable separation of the sentiments from moral judgment is an intellectual farce, as many have argued.[132] So is the separation of the emotions from cognition.[133]

Dualisms like this were the stuff that Darwin or Dewey never really understood, although they later argued against them.[134] Purism has no place in biology, in the wiggle room that surrounds biological systems. The best way to turn off a biologist is to make rigid pronouncements and rules, and then to provide little content regarding real moral behaviors.

Naturalizing a sense of ethics is grounding us within evolution.[135] There are no simple givens in morality, except perhaps to not cause pain. This is a moral imperative, deeply tied to a long litany of historical cases in which participation is maximized. Valuation relegated to its primary role in the hierarchy of cognitive adaptations is part of a cultural evolution.

7
Evolution, Devolution, and Human Progress

Introduction

Pragmatism has a dark side. In thinkers seduced by a simplistic technological orientation,[1] it can be expressed as a glossy-eyed optimism, a kind of cheap form of "what works" philosophy. Couple this bad reading with a *naïveté* about the human condition, and our short-sighted conceptions, exaggerated self-implosion, delusions, and an overly optimistic conception of human progress are all very much in reach.

Freud, echoing Nietzsche and the classical Greeks, often spoke of the twin human drives toward Eros and Thanatos. Eros represents our tendency to form social bonds, to forge social connections, and to come together. Eros also denotes the urges of Hegel and Marx for constructing secular sacred communities as they brought the religious quest into a more terrestrial perspective.

Thanatos, a death predilection, represents the entropy of things as they break apart, the opposite of coming together. Erich Fromm (a social thinker with whom I studied in Locarno, Switzerland, in the early 1970s) wrote about an "Anatomy of Human Destructiveness" – a darker, self-destructive side of our nature, the downside of repetitive behaviors.[2] Both Freud and Fromm regarded Eros and Thanatos as instincts. While animals persevere above all else, a first premise of most physical systems is that entropy is a fact, systems do break down – all one has to do to prove that ineluctable fact is to look in the mirror. Whether self-inflicted or just the wear and tear of life, the aging process holds no prisoners as the biology withers. Even without a glorified theoretical instinct for life or death in competition, we know that there is both a desire to live, and the undesirable breakdown or devolution of function.

Diverse motivations compete for expression amidst a diverse set of cognitive capabilities, some cohesive, others not. This competition goes on without a super executive that regulates all of the motivations and capabilities. It is less that each system is isolated or modular;[3] it is more that the metaphor of a super-executive conscious or unconscious was, from the first, a mistaken idea. We have some control, and what we mean by *self* competes with the diverse forms of motivational expression, of which social and intellectual progress is one.[4]

For John Dewey, the human condition is best characterized in terms of growth.[5] The naturalism he espoused presumed his corpus of understanding of zeal amidst endless pedagogy, defining the trajectory of growth along with the expansion of participatory democracy and liberal sensibility.[6] His vision was of a glorious community and the cultivation of the individual. This, for him, is the evolutionary trend, the cultural expression of an advance and one in which the technological is clearly part of our cultural evolutionary vector, an educational growth that is lifelong as a normative end.[7]

Technology and humanity are linked to "lived experience," as Dewey liked to say, and are expanded across networks of social meaning.[8] There is no mind/body separation in real-world adaptive events of expanding sensibilities.[9] Mind is knotted to action; ideas are enacted, not simply had.

An idea is one thing, but bringing the idea to life is another. Leonardo da Vinci envisioned a number of flying machines (see Figure 7.1), but their actual realization was not to come for some time. H.G. Wells imagined plenty of bizarre things, and so did Mary Shelley in her *Frankenstein*.[10] Our cultural evolution is beaming forward, putting us at the edge of disaster in both evolution and devolution. In that sense, core human traits have not changed. Our lives are often still brutish; although longer than in Hobbes' time, they remain frail and bright with possibilities and their potential realization. As Mary Shelley understood, our creations may be neutral at the start, innocent; we are not.

An extension of the adaptive use of our cognitive resources is at the heart of Dewey's theory of problem solving and its expansion into every aspect of the human condition. This is nothing but the intelligent use of problem solving, something close to Dewey's pragmatic heart.[11] By our efforts to ameliorate devolved function, using as resources a toolbox of cognitive capabilities, with an expanding sense of memory and culture, we set the conditions and the background stage for human advances.

Indeed, the very concept of human progress itself evolved through the theory of evolution. But, as I said earlier, pragmatism and the

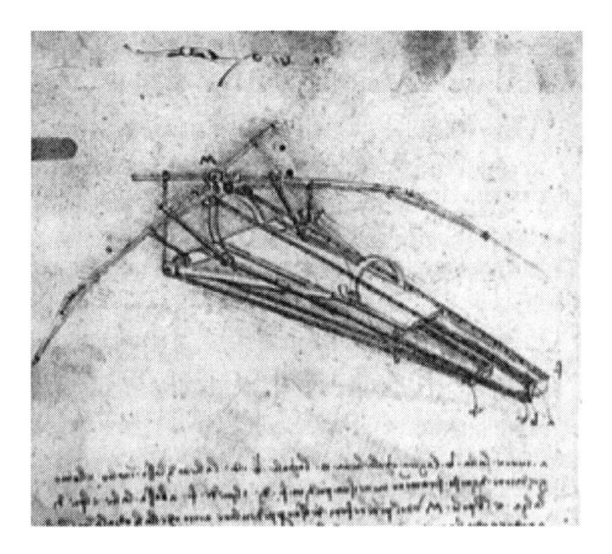

Figure 7.1 A sketch by Leonardo da Vinci of one of his flying machines, ca. 1488

progressive tradition have a dark side. In this chapter, I explore some aspects of that "dark side." I begin with an overview of the concept of eugenics, and follow with a discussion of evolutionary change and the effects of domestication, closing with a consideration of epigenetic change, and the concept of human progress (the latter, something dear to many of the classical pragmatists). This chapter sets the state for the conclusion: social hope or progress within the bounds of reason.

Eugenics, a recurrent theme

Herbert Spencer, a force of major intellectual magnitude in the 19th century, although not much engaged within the 20th or 21st centuries, articulated a conception of evolution sometimes called Social Darwinism. For Spencer, this phrase was a glorious hymn to human evolution in which cultural selection weeds out the weak, the unintelligent, the frail, and the vulnerable, and promotes the strong, the smart, and the savvy. Couple this with a view in which the pinnacle of achievement lies with white Anglo Saxon males, and you have the events of the middle of the 20th century. Indeed, survival of the fittest was a common view held by European and American elites. Spencer's elaboration of Darwinian theory merely gave it perspective and legitimated it into biology. It took very little time for the American Know-Nothings (a nativist political organization formed around keeping immigrants out and supposedly keeping the American "race" pure)[12] to embrace the concept of Social Darwinism.

The concept did more than offer a sop to those who wanted to curb immigration.[13] Social Darwinism would play a pivotal role in thinking about human nature, fuel nativism everywhere, and both provoke and legitimate xenophobic responses to others perceived as different or less worthy.[14] It merged with the dark side of our notion of progress and appealed in many instances to the base side of ourselves in the behavioral sciences and in genetics. Ironically, although it derived from a theory of evolution, the nativism that devolved from Social Darwinism fixed some groups as just that, fixed, with no progress made or possibly made or needed to be made.

My colleague Jonathan Moreno once referred, over coffee, to "the dark side of progressive tradition," a tradition that sparked eugenics, the concept of perfect/perfected people, of weeding out the imperfect. Eugenics and its forms of racism were normative in much of the 19th and 20th centuries. No less a public figure than Justice Oliver Wendell Holmes Jr could talk openly about eugenics without blinking or causing a stir,[15] as did other pragmatists.

Eugenics, the practice of using the theory of Social Darwinism to manipulate human progress, was espoused by American icons such as Alexander Graham Bell, who, in 1923, was an honorary president of the International Congress of Eugenics. Many Americans, as well as Europeans, were swayed by the prospect of eugenics. Although a clear devolution of science and a form of social control, the historical development of nativism gave eugenics a strong normative lure. It resulted in the passage of various eugenically motivated laws in most U.S. states.[16]

Grounding intelligent action and functionalism in the behavioral sciences constitutes the normative goal of inquiry. It is rich in value and linked to diverse forms of realism. Dewey himself always had a broad array of critics (e.g. Lovejoy) for human cultural evolutionary possibilities.[17] Of course, the downward trend, the devolutionary side, is the abuse of the new biology, and just a half a neuron away is eugenics. Indeed, Ferdinand Schiller, a proclaimed humanist and pragmatist, was a founding member of the English Eugenics Society and remained active in it for over 30 years in the first part of the 20th century.[18]

Emboldened by science and supported by organizations like the Carnegie foundation, eugenics gathered full steam under the tutelage of Charles Davenport.[19] This is something they are less proud of, as opposed to the Flexner Report discussed in Chapter 4. Davenport's research, and that of many others at Cold Spring Harbor, a research institute in Long Island, New York, would go on to influence such scientific luminaries as James Watson, co-discoverer of the modern structure of DNA and a major contributor to the field of genetics.[20]

Under the influence of eugenics, depictions of evolutionary racism were commonplace. Georges Cuvier, one of the first great taxonomists, had already described the Europeans as on top of the evolutionary chart, with "Mongolians" and "Ethiopians" (his terms for Asian and African peoples, whom he believed to have smaller skulls and more monkey-like features) on the bottom.[21] This view was quite dominant and pervaded common folklore, eventually evolving into repugnant racist depictions.[22]

Cuvier's theories were legitimated and embedded in romantic and glorified visions of evolutionary progress from the primitive to the civilized, from the subcortical to the cortical.[23] Ernst Haeckel, a 19th century biologist and romantic visionary, depicted core features of animal evolution that captured an evolutionary age with little lasting validity, but much lasting effect.

Haeckel, an artist by temperament rather than an experimentalist, ripped through the late 19th century evolutionary literature with rapacious metaphors of recapitulation using lurid facts (e.g. tails, common forms in embryology, etc.) amidst glorified notions of evolutionary progress and trajectory (see Figure 7.2).

The recapitulatory theme of ontogeny and phylogeny was a common one, and still retains elements of biological validity. But it was overplayed, oversold, and tied into a narrow notion of evolutionary progress, and

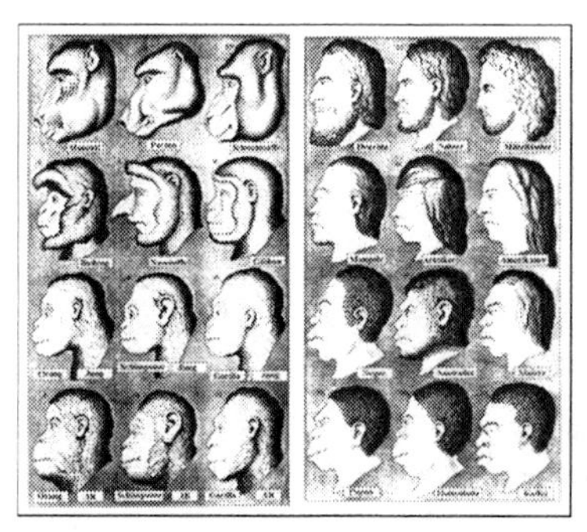

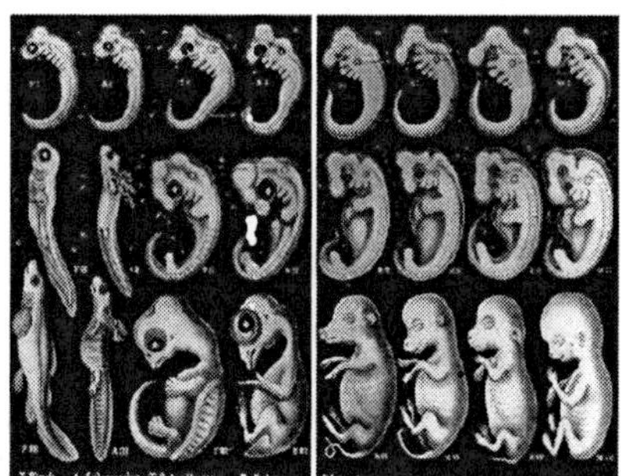

Figure 7.2 Homogeny and recapitulation

Source: Haeckel (1900, 1992).

white male social elitism. Darwin and even Sir Francis Galton, Darwin's first cousin, a polymath who coined the term "eugenics" and popularized the concept,[24] were more careful than Haeckel and made fewer intellectual mistakes. The historian of science Robert Richards has rightly characterized figures like Haeckel and Galton as tragic romantics, who infused a life of controversy into their science.[25] The romantic spirit, not humbled by the endless ways in which we are misguided, mistaken, and embracing wrongheadedness, is no prophylactic against the tendency to run rampant over the contours of the knowable. In the case of eugenics, it ended badly, ultimately in an almost unimaginable extermination effort.

The basic idea of eugenics did not begin with Galton, even though he came up with the term and first applied it extensively to human development. After all, cattle ranchers, for instance, are bred for shape and form and excellence since human beings began to cultivate land, crops and animals. The study of shape and form through cultural practices is a core cephalic adaptation. Eugenics, the idea of maximizing a more perfect state and minimizing a less perfect one, is ancient and an essential part of our cultural history. It even corresponds to a bright spot in the sense of taking hold of our destiny. The humanistic traditions of the Enlightenment were animated by a conception of what we want to be and linked to an experimental spirit by the Renaissance sensibility of experiment, exploration, testing, and comprehension as a worthy goal.

Galton, however – mired in the end in controversy about the validity of his expertise, his experimental findings – turned eugenics towards the worst in us: the Nazi sensibility of the Superman, the perfect being, raw power, unmitigated by softness and social sensibility. This was probably a bastardization of what Galton actually had in mind. What Galton worried about was how to preserve and protect the very best in the British Empire. He did not want to dilute the formation of intelligence. It was Karl Pearson who merged Galton's theories with Mendel's insights about genetic expression to produce a statistical relationship between genes and probability of expression, between innate vectors and actual expression, using regression analysis and other forms of statistical reasoning. Pearson would direct the Galton Institute at the University of London, and write a three-volume biography of Galton.[26]

Before Galton, there was already a cultural or racial implication in Darwin. His original writings on the effects of domestication pointed to biological ways in which to study the cultural selection of some types over others. The problem with this is that the price was not neutral. The level of racial superiority it engendered was off the charts. With the white male of a certain type at the very top of evolution – a common perspective that predates Darwin and Galton (we have already seen it in

Cuvier) – this view was made more pernicious and dominant by Darwin's own theories. For Pearson, the march of civilization put in place a cultivated form of human domestication tied to socialism: the figurement of mental and physical features controlled by genetic/social engineering.[27]

In fact, evolutionary "progress" is far from linear. There is no straight arrow forward, just adaptive radiation, extinction, and diverse forms of stability and breakdown in equilibrium.[28] Both evolution and devolution of function are common features, perhaps even more noticeable in our age, since we live longer and care for those who would not survive in other times in our history as a species.

One disgraceful instance in the U.S. was the exhibition at the Bronx zoo, at the beginning of the 20th century, of an African man in a cage. This was the same zoo that helped save the buffalo from extinction. This shows just how generic racist evolutionary biology had become, its metaphors steeped in lethal injustice.

Domestication and psychobiology

Curt Richter was born in the century of Darwin's discoveries[29] and would be much influenced by them, in particular Darwin's writings about domestication and evolutionary change. During his first trip to the Galapagos Islands, as he recounted in *The Origin of Species*, Darwin was awed by the great variation in species.[30] His observations would eventually lead to his theory of natural selection and the idea that speciation developed through the selective pressures of geographical and ecological constraints. Ideas about adaptation, secondary sexual characteristic expression, and functional fit guided Darwin's thinking.

To understand physiological and behavioral expression, he looked to the ecological niche, the organic conditions in which an animal had to live. Darwin revolutionized the study of behavior by giving it a biological context. This was quickly adopted by those in the emerging fields of psychology and psychobiology, and was soon understood within the evolutionary context of adaptation. But Darwin held a Lamarckian perspective on certain hereditary changes, including use and disuse, and intergenerational effects.[31] Use and disuse, instinct and habits, and the effects of domestication were dominant intellectual considerations for understanding end-organ systems.

Darwin was worried by the price of domestication. He wanted "to estimate the amount of structural differences" that occurred as a result of cultural imposition.[32] The other concern that came to dominate his thinking was the heritable component: instincts were seen as guiding regulatory mechanisms in the organization of behavior.

H.H. Donaldson, a colleague of Dewey and Mead at the University of Chicago (where there were a number of psychobiologists thinking about development and domestication), was a teacher of John Watson, the American Behaviorist and the author of the book *The Rat Data and Reference Tables for the Albino and the Norway Rat.*[33] In this book, Donaldson described the "history of the rat since it arrived in Western Europe,"[34] the animals' life charts, their behaviors, and the significant effects of domestication on their end systems, including the brain, which is smaller in the domesticated rat variant. Interestingly, many decades later, enriched environments were shown to produce improvements in domesticated rats' neuronal structure and learning.

Darwin noted that "in all parts of the world, man has subjected many animals and plants to domestication or culture."[35] He also observed the morphological effects of domestication in a wide variety of species. In Darwin's words, "Our domesticated animals, which have been long habituated to a regular and copious supply of food, without the labor of searching for it, are more fertile than corresponding wild animals. It is notorious how frequently cats and dogs breed, and how many young they produce at birth."[36]

Of human evolution, Richter wrote that "primitive man lived in an environment in which physical strength, endurance and aggressiveness were at a premium."[37] He said about what he assumed was the "growth of community life" that the more aggressive forms of behavior of the "primitive state" became less important as civilization developed.[38] Of course, this explanation of our behavioral biology is now outdated; we know that cooperation, deception, and cunning were just as important in our biological legacy as overt aggressiveness. Richter further asserted that "the wild Norway rat, like primitive man, lives in an environment in which it must constantly be on the alert and often has to fight for its very existence."[39] Of course, modern humans also fight for existence in different but perhaps equally desperate ways. Perhaps the reduction of adrenal capacity (assuming a phenomenon observed in animals whose food options are less predictable and less easily accessible also applies to us) is an important adaptation as we shift our metabolic and cognitive resources elsewhere.

Epigenetics and cephalic growth

Environmental changes on cephalic capability is a core orientation for pragmatists such as Dewey and Mead. After all, if we can control our destiny, why not choose a direction of cephalic capability? Changes in gene expression without altering the underlying DNA is a workable definition of epigenetics, and perhaps lies at the heart of phenotypic alterations in development contingent upon context and circumstance.

Diverse forms of epigenetic theory, in other words, run through the thought of the classical pragmatists. They emphasize the link to Lamarckian theory, and the plasticity required for lifelong learning for self-corrective inquiry vital for Dewey and other pragmatists. We are at a stage in which we can begin to understand some of the cephalic changes necessary for lifelong learning. Behavioral adaptations have long been noted to be fundamental to evolutionary change, and indeed underlie any definition of evolutionary change.[40]

The features of the objects ingrained in the cephalic hardware with epigenetic modification depend upon context and circumstance within the parameters of the systems. Gene expression underlies, for instance, the expression of information molecules like insulin, which is tied to glucose regulation; influencing insulin expression or many other information molecules is one way in which exposure to our ontogenetic and lifelong terrains influences gene expression, which in turn impacts regulatory physiology and cephalic expression.

It is against this backdrop of human adaptation that we need to understand epigenetic explanations.[41] The rich cognitive and social milieu spread the semantic horizons of adaptation and expansion. Jamesian horizons predominate amidst appetitive search engines, consummatory satisfaction, and integrative central states of musement. The cephalic systems expand beyond the central nervous system; meaning, as pragmatists would understand it, is not simply in the head.[42] In fact, the semantic network of science is distributed across the larger culture.[43]

Despite its overall debunking, Lamarckian theory has lingered as a possibility that is less about individual learning and more about populations and their intelligence.[44] It also has some influence on the vectors of evolution. What we have learned is that modification of DNA by epigenetic mechanisms (e.g. demethylation, see below) exists, and genes are silenced or excited by environmental conditions. Both heritable and nonheritable features impact genomic structure via general mechanisms, silencing one set of genes while enhancing others. Epigenetic events impact most forms of development.[45]

Thus, a view for lifelong learning or growth, very dear to the likes of Dewey – viz., the enhancement of experience by suitable ecological/

socio/cultural events – impacts cephalic expression. Animal social contact is one example, amongst others.[46]

We have known for some time, for instance, that enriched environments promote dendritic growth and cephalic expansion, leading to greater problem-solving abilities. The effects are mediated by diverse information molecules. Conversely, impoverished environments have the opposite effect.[47] These events are not confined to a stage in life; but the older the subject, the fewer the neurotrophic factors, and the less the dendritic growth. The epigenetic discussion fits into the changes in neural function from the social-ecological milieu. The experience of social enrichment or impoverishment produces transgenerational effects.[48]

Epigenetics remains controversial, recalling debates between Lamarck and Darwin about acquired traits and evolution, but has grown in acceptance with the onset of molecular integration into broader-based and regulatory biology.[49]

As Evelyn Keller has suggested, our age is the age of the gene.[50] One mechanism that may underlie epigenetic gene regulation is methylation and demethylation, defined as silencing or enhancing the expression of genes.[51] Demethylation prevents transcriptional expression, and one result is the silencing of gene expression.[52]

For example, genes responsible for oxytocin expression, an important peptide hormone that plays diverse regulatory roles from parturition to milk lactation to social attachment, are particularly significant. The same hormone, depending upon where it is expressed in end-organ systems, thus plays a wide variety of roles.[53]

Demethylation of oxytocin expression affected the transmission of these social behaviors to the offspring during development in cross-fostering experiments. Manipulations of demethylation impact the transmission of this social behavior and can even be reversed later in life.[54] The imprinted genes are not permanent; the emphasis on silencing gene expression by imprinting or emphasizing is, of course, an active area of research.

This phenomenon has long-term implications for neural systems. Those rat pups comforted by social contact have greater regulatory capacity as adults in diverse systems in the brain, including neuropeptide and neurotransmitter systems. Importantly, diverse social events can either decrease or enhance the genes and the resultant expression. Diverse forms of social deprivation, poor nutrition, etc., impact the genes that regulate the expression linked to social attachment, thereby promoting devolution of function.

Evolution and devolution

Evolution and devolution are always in close proximity; it is the stuff of adaptive systems. While entropy is a concept from within physics, it applies equally well to biology. The breakdown of biological systems is simply the stuff of aging – the wear and tear – both short and long term. Aging can be facilitated or degraded. The evolutionary part of the equation is related to key growth factors that sustain organs. Thus, the region of the brain tied to memory (hippocampus), while vulnerable to degradation by genetic hardwiring and epigenetic circumstance, also contains growth factors to sustain neurons, and indeed grow neurons. Neurogenesis is one feature of cephalic expression.

Understanding how this region of the brain generates new neurons and what degrades them (for instance, high levels of cortisol) allows us a context in which to study the biological features that underlie neural longevity, an important feature in any species, including us, that is living longer and longer. Perhaps most importantly, adult, not just juvenile, neurogenesis is a phenomenon that underlies diverse forms of learning, including social learning, in a number of species. First demonstrated in song birds, the phenomenon of neurogenesis is now understood to be a widespread occurrence, not all-pervasive in the brain but demonstrable.

Dewey recognized that the brain was "the last physical organ of thought...part of the same practical machinery for bringing about adaptation of the environment to the life requirement of the organism."[55] This evolved cortical mass we call the brain also expresses devolution of civilized function, losing itself in barbaric sensibility. This is the omnipresent vulnerability of our species, which is why conduct and discipline are so important to our continued survival.

The idea of the link between a big brain and intelligence is itself much disputed. Einstein's brain was no different in size relative to his body than are the brains in most of us. Primordial dwarves, with exceptionally small brains, generally have normal levels of intelligence. Bigger brains do link to evolution; the more social the animal, the greater the cerebral cortex is, in most primates.[56]

Nevertheless, most evolutionary conceptions, including Darwin's, were drawn in one direction only. A figure drawn by Darwin shows something of his thinking of the descent and common tree of steady progressive trajectory,[57] though natural selection certainly does not entail this perspective. Like most thinkers of the 19th and early 20th centuries, Darwin understood cultural evolution as a march of progress.[58] Indeed, Darwin emphasized the moral aspect of our evolutionary progress.[59]

Evolutionary change, however, is not always a continuous variable of progressive functions and expression; there are bouts of changes followed by stability.[60] There can, in fact, be features of both punctuated change and stasis in progressive change.

The evolution of our species, for instance, suggests long periods of stability followed by rapid change. However, as Gould notes, "Stasis does not mean rock stability."[61] Abrupt evolutionary change, like gradual change, may or may not be for the better. In this expanded view, great or small changes may occur over a relatively short period, provoking disequilibrium followed by periods of stasis, or a so-called "punctuated equilibrium."[62]

Moreover, a number of related evolutionary and very closely related species (*Homo neanderthalensis*) co-existed and competed with our species at the same time. One such evolutionary description is shown in Figure 7.3.

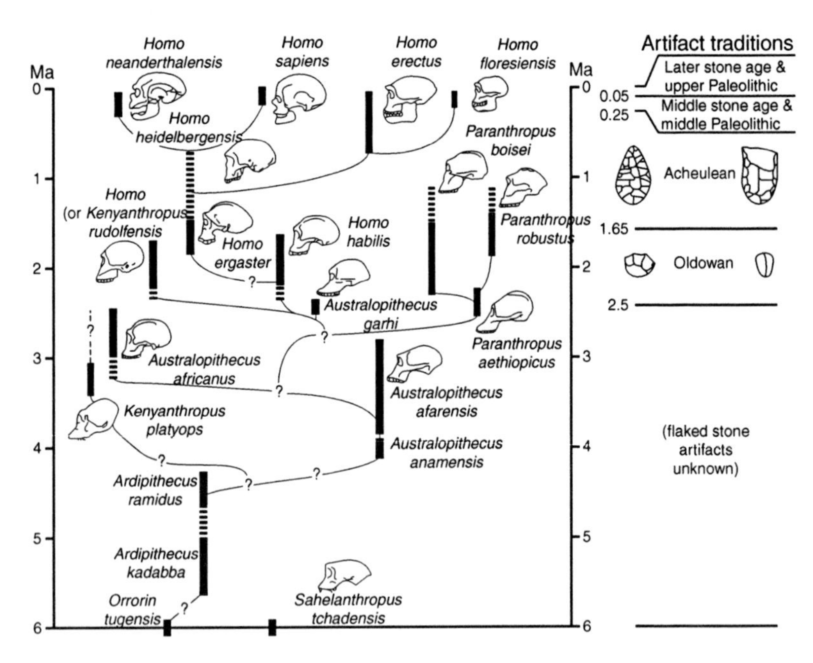

Figure 7.3 The evolution of hominids and their artifacts
Source: McHenry (2009).

Thus, modern evolutionary theory provides a basic heuristic framework within which biological activity is instantiated and viewed from a coherent perspective.[63] That there are discontinuities within this framework may be due to our limited knowledge, or it may be a reflection of the fact that the process of evolution is typified by breaks followed by periods of stasis.[64]

Moreover, not every feature of a biological entity is an adaptation.[65] Adaptive arguments must be viewed with caution, as they often become circular. Evolutionary theorizing requires careful observation amid a core set of concepts (speciation, natural selection, etc.) that have determined the meaning of evolution. It is rich in unresolved disputes, such as whether language evolved from gestures or from vocalization.[66] Our conception of biological evolution is continuous with our conception of cultural evolution, a view held by a number of the classical pragmatists (e.g. Dewey, Mead).

Variants of progress, pragmatism, and speculative thought

A common conception about progress is that it is a feature of evolution, social and otherwise.[67] Progress historically was always a feature by which we could pursue our wants.[68] What underlies the modern sense of progress is the inevitable march of science and the perfectibility of the human condition by the insights and findings of science.[69] The conflict is between two competing conceptions of progress. The Enlightenment conception holds that science necessarily brings progress to humanity. The most extreme version of this view is scientism. The other view is the pragmatist conception that progress results from ongoing and self-corrective human inquiry, of which science plays a major but not exclusive role. The Enlightenment view is absolutistic and determinative (there is an objective standard independent of humans by which to measure progress). The pragmatist view is melioristic (the only absolute is the growth process itself, viz., we measure our progress in terms of whether we are doing better than before).

What the pragmatists demanded, and perhaps too naïvely, was endless flexibility; social context matters. This was very different from positivism, and perhaps closer to a philosophy that emphasized a unity of creative expression across diverse realms of human action, reflective of cognitive pregnancy embodied in human action. Human progress is fundamentally linked to science, but not reduced to it. The larger sense of human meaning is the diverse ways in which we behave and reflect moral sensibility; a moral imperative that necessitates, that one reflects on others, that they be considered in one's actions.

Susanne Langer remarked that "the brain is an organ and like all organs it is built into a greater living whole, an organism that helps to sustain by its special functions."[70] The "living whole" speaks to the pragmatic naturalism that underlies much of James and Dewey.[71]

Cassirer built a deep philosophy of culture on an Enlightenment trajectory (amidst the omnipresent devolution of a culture around him),[72] where a core feature is apparent. This was something he called "Symbolic Pregnance."[73] This is an important component in any form of naturalism, something Langer understood.

Langer, dedicating her book *Philosophy in a New Key* to one of her teachers,[74] Whitehead, was systematic in her sense of capturing nature alive within our diverse forms of adaptation.[75] This is in line with her running theme of keeping her philosophical sensibilities quite empirical.[76]

The march of science, and what later became known narrowly as scientism or positivism, was a march away from religion and metaphysics towards science, preferably physics and chemistry, both grounded in mathematics. Biology, without laws, was suspect and would be for a long time. Psychology was a mere infant in terms of science and legitimacy. Of course, this takes science to be one thing, something the classical pragmatists like Peirce did not accept. But non-Darwin-like statements emerged from the pen of Herbert Spencer, catapulted by Wallace's co-discovery of what Wallace thought were the special laws of human progress, human evolution: no frozen vast "Chain of Being,"[77] just a frail lure forward amidst many forms of devolution.[78]

Evolutionary perspectives dominated the post-Darwinian landscape, whether the topic was population conflicts,[79] competition and evolution, survival and extension of adaptations, or one-end reproductive success. But Darwin was himself a Lamarckian (heritable habits) in some senses, and thus is not always consistent. We know that Darwin had with him on his trip on the Beagle a book by Humboldt and John Hershel, the son of the great British astronomer and noted expository of the emerging natural philosophy. This was seen as something distinct from Aristotle, Bacon, and Descartes. A thoroughgoing naturalistic philosophy was embedded in the new sciences, not just in physics but also linked to biology, a science of natural sensibility and responsibility.[80]

We know from the tragic fate of positivism that science as sense data validity is absurd, a "misplaced concreteness" in Whitehead's terms.[81] Positivism was opposed to a much more social acknowledgment of the various levels of discourse grounded in the sense of which something is useful, to us, as grounded in kinds of objects for us, having a validity of test and expansion.

Categorical features such as capturing the quality, resistance, and relationships of objects and meaning, continuity noted by Peirce amidst generic features of human experience such as the precarious and the stable noted by Dewey, are engaged. Peirce appreciated that "conceptions of philosophy of a primal matter out of which the world is made."[82] No mythical beginnings that were foundational, there is just a long evolutionary trend towards clarity and understanding that is certainly not a continuous function or linear trend.

Contrary to Peirce's conception, there are no guaranteed settled beliefs that merit a truth for all time. There are not Hegelian progressions for those of us in the trenches with laboratory frames of mind, who appreciate speculative thought as a grounding factor in human understanding and capability – although this is something easily abused. We have a predilection towards problem solving built into our cognitive architecture.

In other words, there is very much a tradition within pragmatism that is respectful towards a sense for the speculation that underlies cognitive coherence, towards the generic features that underlie human experience.[83] This is a pragmatism respectful of speculative thought,[84] and responsive to the endless search for meaning, with valuation appraisals.

Since Dewey, the psychobiologists commonly referred to the precarious and the stable as recurrent core features in human experience. The metaphor harkens back to evolution, to biology. Dewey, after he left his early Hegelianism, embraced an open-ended universe of self-reflection peppered with self-reliance and communitarian sensibility. Nature and culture merged into a moral psychology and pedagogy with endless possibilities; or so it seems to some, and utterly naïve and misguided to others. Dewey was also a psychologist, who understood something about the biology of temperament and innate capability – though perhaps he did not emphasize enough our endless stupidity. But then he did stand for social hope and aesthetics of inclusion and self-correction.

Indeed, nature as becoming and responsive to our changes, and nature as eternal and resistant to our changes,[85] are factors in grounding a pragmatism tied to action, to process, and to becoming. This is a nature beaming with new beginnings and fallen degradation. The eternal recurrence of the possible amidst the decline and the degradation that is apparent and all too recurrent. Core abstract features are to be described; nature cut at the joints of conceptual space with a critical realism.[86]

Core themes are a realism tied to experience, with what Whitehead and certainly James understood as experience rich in the feeling of causation. This is manifestly not the passive experience of classical empiricism or the positivism that misplaced sensations for the grounding of knowledge. The dynamic experience of James and Dewey stands opposed to classical empiricism. Instead, the pragmatic conception grounds knowledge in action, in perception, and in an evolutionary context that makes hypothesis formation plausible and understandable, as Peirce noted and many of us have followed.

A sense of the organic, of nature, should not be mythologized as a panacea. That is one of the anchoring points of pragmatism. Good sense, sound consequences, instrumental capabilities are among our hopes. It is simply important to understand and appreciate the organic features of nature.[87]

Evolutionary themes ground perspectives like the one that underlies this pragmatist writing these words. Dewey never lost sight of an integrated evolutionary perspective.[88] That the social is continuous with but not reduced to the biological is a normative goal for our continuing to realize our sense of progress.

Within this perspective is truth making: a context of a critical and well-earned realism that is indebted to the endless appraisal systems that underlie human action and inquiry. The endless sense for the value is inherent in experience and in its link to truth telling and searching.

Whitehead looks to civilize the advances and the dangers, with abject humility and endless social smarts. He has an eye towards protecting us from the advances in technology that exist with the base stable features of our motivations, our search for foundations to hide in, and theological and secular fascisms that are omnipresent expressions of us – in short, what leads to the devolution of cultural expression.

Our notion of progress needs to be anchored to core background sensibilities about nature. It needs to be anchored in a philosophy of nature that captures who we are within it. For we are caught between lofty ideals that transcend with its primordial sense of us within it, and yet there is the constructing of social spaces, in which we take part, that have transformed our surroundings so dramatically these last 10,000 years – not a very long time, but surely a punctuated change in the landscape.

Modern science is to be understood within a romantic conception of our drive to understand our surroundings. This drive is an ancient predilection built into cephalic systems that are linked to beliefs and action. This romantic conception also envisions responsible agents in a democracy as a normative goal, and as a regulative and constitutive factor in human experience, in which essential and conditional factors are integral for understanding human action, human experience, and human creative choice.[89]

Progress is measured in opportunity of choice, rational decision making, moral sensitivity and communitarian bonds. An individual is thus not mythologized in isolation rich with sensations and cognitive splintering, but embedded with others, with a freedom constrained by the social, the biological, etc.[90] Freedom is not a radical choice outside the factors of history and circumstance; it does not rely upon any epistemological conditions divorced from historical context and set into a private space of certainty, like subjectivity or its aftermath into postmodernity. Classical pragmatism offered a way around some of the dilemmas.[91]

Modern science, no panacea, is nevertheless built on the notion of progress, a progress that traces back from natural cephalic predilections towards systematic uncovering of mechanisms, and producing theories and instruments of discovery,[92] to raising an experimental methodology.[93]

But progress is also about the well-being of the human condition,[94] a directive that, like biological evolution, has not one fixed vector or one direction. It goes both up and down. The danger of devolution grows as the destructive force of our technological innovations develop. Evolution and devolution of expression are too close and fraught with endless danger, and dire consequences.

Historical sensibility is not often enough emphasized by classical pragmatism in its forward-looking outlook. Doing so is one small protective factor from absolutism, and dogmatic romanticism. Furthermore, political correctness of diverse persuasions is always an omnipresent virus to cephalic free inquiry.[95] Pragmatists, such as Dewey, always sought to capture the human sciences. James, so responsive to the human condition, studied in Germany with Wilhelm Dilthey, who emphasized the human sciences, and history. For James, an autobiographical perspective was emphasized.

For most pragmatists, myself included, the human sciences are not bifurcated from other sciences; history, autobiography, etc. are part of the human knowing process. Within it are appraisal systems, embodied cephalic systems. The abstract and the less abstract are continuous as we pursue inquiry. Moral judgment and aesthetics are at the very fabric of many judgments in human inquiry. Some forms of self-correction are more readily available than others, some harder than others.

Critical but not revealed or naïve realism is a running theme amidst social complexity, as is a hard-earned "common faith" amongst what is possible – ephemeral though it be.

Utopias are a mythology fraught with possibilities for our cultural evolution as well as our degradation and devolution. Dewey emphasized method, not ideology, with the exception of the cultivation of social involvement. There is no hope for utopia, but there is perhaps an exaggeration of and not enough appreciation of our limits, and our lust for power. Self-initiated group formation and self-reliance are at the heart of human freedom and expression that does not fall into the trap of utopianism.

Again, the key to progress, for pragmatists, and certainly for my view of the integration of science into the heart of human life, is a cultivation of inquiry that does not reduce knowledge to sensation and the simply practical, as James was tempted to do and was scolded about by Peirce. Peirce thought James's pragmatism was indeed a put-down of real pragmatism (what Peirce renamed as pragmaticism), for it was anti-intellectual and narrow to the core, just a cut above positivists.

A pragmatism, however, based on the culture of inquiry, paints a broad brushstroke of what is to be appreciated and mused about. Indeed, musement is a cardinal feature of the life of the mind, not simply outcome orientated, but quite respectful to this accomplishment of imagination. Pragmatism should be a set of beliefs, social in nature and knowledge, that nevertheless defends the expression of knowledge. Pragmatism is not the same as mythologizing knowledge and grounding it in a narrow room of scientism; rather it is rooted in multiple ways of capturing our cephalic capabilities as nurtured by a culture of inquiry.

Some notions of progress do hold on to a concept essential to the human condition,[96] albeit some notions are often materialized into a narrow notion of science.[97] Yet other notions may now be expanded into the larger human community as a normative goal, something sympathetic with Dewey's conception of life. That is, Dewey conceived of a lifeworld brewing with social and moral respect, with individual self-creation and a continued acknowledgement of the dark side, of our devolution.

Adventures and expectations: Whitehead and Dewey

Maybe a return to the historical/poetic vision of Alfred North Whitehead is an imperative, prophetic, and mitigating factor against grandiose expectations. Boldness and adventurousness push us forward, while anchoring to a sense of humility and gratitude amidst the omnipresent development of self-enclosure and disappointment. Whitehead, in *The Function of Reason*,[98] describes the eternal tension between decay, entropy, and the breakdown of events amidst the endless lure of new possibilities, as he describes a love of science tempered with a broader view of the human condition.

Whitehead states in this book the normative goal of reason: "the function of reason is to promote the art of life."[99] He states the obvious difference between survival and "explication of the Art of Life";[100] the goal: "to live well." What might he mean by this? Part of it is an appreciation of the landscape, a natural piety infused with religious ideals, humbly stated. Never by force, only by lure and persuasion does the adventure of ideas go forth, normatively emboldened by cultivated reason.[101]

Whitehead anchors his philosophical meditation with regard to reason in two dominant domains: in reason in the search for understanding, and in immediate adaptation. A lure for reason is the search for novelty, some form of synthesis, a new form of interest amidst the habits of sanity and funded reason rooted in the merely familiar. The competing interests of staying the same and in place, and of searching for the new, gave Whitehead a sense of speculative reason, something he suggested was found in fundamental Greek thought about principles and inquiry into natural knowledge, knowledge of nature. One emphasis that Whitehead suggests is "significance is the relatedness of things,"[102] and, as he put it, "natural knowledge is exclusively concerned with relatedness."[103] This comes remarkably close to Peirce's notion on the evolution of cognitive systems (see Chapter 2).

Indeed, Whitehead's philosophy of organism is rich with the integration of science with speculative reason, the attempt to systematically look for relatedness and to discern the axiology or value in things.[104] Importantly, Whitehead anchored his speculative thought, and his idea of the fundamental desire for relatedness, in contexts in which he is also criticizing the idea of "simple location."[105] He worried about how the abstraction of simple location came to dominate the intellectual landscape; a landscape emboldened by the rise of modern science.

His view is a philosophy of organism, of nature alive, and a precursor of the embodied view of cognitive resources.[106] Whitehead would rightfully acknowledge that the "body is the organism whose states regulate our cognizance of the world."[107] Modern, yet Victorian in manner, a romantic and a mathematician at heart, Whitehead labored in his philosophical writings to provide a civilizing cultivation within the notion of human progress. Ideals lure as fatigue settles amidst the endless disappointments and sense of entropy or decay; evolution and devolution breathe out from his philosophical prose. The normative goal is to acknowledge "the intrinsic worth of the environment," echoing a sentiment that would become common within the ecological movement and with eco-philosophers.

Whitehead understood the culture of ideas, lived it, and understood something about the worth of nature as part of the appreciation and promotion of civilization. This is the second Copernican revolution, just like Darwin's moving beyond the narrow view, towards the social view of others, and the larger view of the ecological context. Civilization evolving amidst the entropy of habits that blind, narrow views that degrade, is the ever present conflict. Devolution of function is the dominance of force over persuasion, certainly as Whitehead understood the engagement of ideas. Of course, the endless vulnerabilities in our ability to reason – viz., to reason well towards living well – is a palpable fact about us.[108]

For Whitehead, and certainly as a normative goal, five characteristics are important for a civilized sensibility: "truth, beauty, adventure, art and peace."[109] I would assume the vulnerability to determine truth, amidst an ethics of nature's majestic beauty, awestruck in an adventure that requires art as an investigatory and peaceful coexistence, are five civilized goals. Whitehead, like Dewey, broadens art and the aesthetic to the everyday, to the practical, in addition to fine art.[110] And Whitehead acknowledged that "we are living in the midst of the period subject to Dewey's influence."[111]

Naïve with regard to the perfectibility of the human condition, Dewey in particular held on to a remnant of both his roots in his Hegelianism, his liberation from it, and his embrace of critical realism and naturalism – holding out for human perfection towards the glorious community. Liberal thought requires both positive action from the government and the negative right of freedom of expression (with minimal restraint or interference from government). The cultural perspective for our condition was both our sense of belonging to the larger culture, and participating in the demonstrative spirit amidst a sense of the individual.

A remnant of the Hegelian spirit in Dewey, naturalized by Darwin, pervades his corpus. Dewey is weaving a view about cultural progress with a demonstrative spirit.[112] This spirit is in the entrepreneur. It is international, moving across porous borders, toward broader social identity and solidarity with the larger human condition. This spirit is less naïve than perhaps his critics thought,[113] and yet naïve enough to understand something of the human condition and its limits and possibilities as a ray of social hope. He dedicated his book on *Liberalism and Social Action* to Jane Addams, the co-founder of Hull House, as a better symbol of reaching out to the less fortunate.[114]

After all, *A Common Faith* is something social in nature (see the conclusion of book). A common faith is connected, rich in meaning, frail to the core but expressed, and to be cultivated by the habits of life; this is what Dewey called "the basal ganglia of society" which can surpass "the cortex of the social body."[115] The liberalism of Dewey "has to assume the responsibility for making it clear that intelligence is a social asset and is clothed with a function as public as its origin, in the concrete in social cooperation."[116] Dewey was the embodiment of a naturalistic tradition respectful of diverse expression in the sciences and of human expression.[117]

John Dewey's corpus naturalizes ethics along with intelligence, human social problem solving, and human growth. Ethical issues, for Dewey, are optimal ways in which to get along, to distribute and to facilitate human excellence and individual expression and individualism.[118] New and celebrated but embedded in communities of social respect, solipsistic individualism is replaced with social contact, atomistic individuals replaced with others who are connected or entangled throughout a cultural network of meaning.[119] It is a culture in which process, not static events, are prized and valued, where solidarity is cultivated but not blind, where psychological dispositions are blended towards human worth, and where worth, as well as valuation, is highly cognitive and meaningful.

There are no pernicious barriers between value and cognition. In fact, all dualisms that separate human activity from cognitive adaptive events are forbidden as relics from an earlier vision of a passive sense of the knowing process, for pragmatism is steeped in action. A true "reconstruction in philosophy," as Dewey noted, becomes less about correspondence and more about adaptation and integration, less about essentialism and more about the features of nature, more about means, ends, and fluidity, and less about stagnant sign points.[120] This philosophy dissolved such dualism into functional ends and means. Fundamental to Dewey is the dissolution of the separation of theory and practice, articulated in his book *Reconstruction in Philosophy*.

The put-down is that there is no rest for a pragmatist like Dewey. Dewey may have come close to this, or not sufficiently emphasized experience without action. Musement is one state noted and celebrated by Peirce; but Dewey's emphasis is on social action. The sense of musement slips in aesthetics for Dewey, since aesthetics, like valuation, pervades human action. The reconstruction about which Dewey proselytized was radical; it pervaded all of human action, an action emboldened by science but also its handmaiden, as the positivists envisioned themselves. Dewey's was a moral philosophy, empirical to the core, aimed at discerning the human predilection and how best to harness it, without a eugenics that looked away from those in need. It fostered a strength bristling with compassion amidst the harness of the human condition towards fragile ends – ends acknowledged to be bound up in means in which tidy divisions are endlessly and effortlessly transgressed.

The cultivation of civilization for Dewey is not so dissimilar to the thought of Whitehead; it is just less formally religious. Truth for Dewey is less a product of justification from the past than it is a prediction and predilection towards the future; his instrumentalism emboldened by a pervasive teleology through enactive and embodied action. In Dewey, there is no mind/body separation. We are a continuous function in which there is no diminution of specious bifurcation that degrades human experience and the knowing process.

Classical pragmatists, perhaps, were often less erudite and self-reflective than they ought to have been, though Peirce was steeped in historical reflection. Historical understanding is essential for us, as it is tied to an anthropology of human expression and variation, and ripe with the possibilities of human construction and devolution of function. Any form of critical realism needs to be embedded in an historical perspective. We are in an age that allows no Platonic transgressions. Core concepts that reach into understanding adaptation prevail. Dewey understood that. While Enlightenment sensibility is a faint glimmer amid endless human limitations and transgressions and base expressions, and often a real limitation on pragmatic thought, with frequently historical sensibilities, it is still a defining lure for social hope.

Conclusion: lifeworld and warranted reason

A consideration of the lifeworld is a core predilection. "The phenomenon of life"[121] is built on a propensity to detect animate from inanimate objects early on in our cognitive development. Within the first few years, children are oriented to distinguish living things from nonliving. In our lifeworld, cognitive systems are knotted to social practices. Social participation is a fundamental and formative part of us. Acknowledging these features is a precondition for the diverse experiences that we have. Scaffolding outside ourselves to access, share, and connect with others[122] is a cognitive adaptation rooted in our getting a foothold into our social world.

Indeed, we look early to determine internal causal forms, the feeling of efficacy as James and Whitehead described, and as Bergson and others understood. What they did not understand was a broader view of mechanism. The choice was not between 17th-century clockwork mechanism and vitalism. Rather mechanism is now better conceived as the means that underlie adaptive activities. The internal generators and effort formed in adaptation and perseverance are fundamental features of the lifeworld of shared and individual experiences. The predilection for a shared lifeworld, social in nature, is a fundamental cephalic adaptation that sets the conditions of a social phenomenology rich in expression.

Nature conceived with vast continuity in form amidst discontinuity forces the embrace of the biological stance, a stance rich in phenomenology, in which pragmatism is understood as part of the problem-solving predilection. The continuity of problem solving moves from local narrow bonds to broader horizons, in which intelligence is linked to cephalic expression, and capability is a welcomed acknowledgment.

Social contact and human well-being are intimately bound together. Pragmatic naturalism is knotted to human social solidarity and human contact. Diverse social cooperative behaviors are a cephalic adaptation. Social cohesiveness, see Figure 7.4 for example, competes with other diverse motivational systems that are all too readily expressed, and are themselves an expression of our endless vulnerability to Eros and Thanatos, to evolution and devolution.[123]

Figure 7.4 Makonde *tree of life*, ca. 1970
Source: Humphrey (2007).

8
Conclusion: Adaptation, Well-being, and Social Hope

The orientation towards problem solving, as classical pragmatists such as Peirce and Dewey understood, is replete with struggle that marks inquiry and its diversity, and what my colleague Gerd Gigerenzer, a director of the Max Planck Institute in Berlin, at a recent conference in Germany, calls the "Enlightenment Pathway."[1] The enlightenment is paved with disappointment and tragedy, with core limitations omnipresent about the human condition as we witness and participate in our capacity for devolution of function in addition to evolution of function.

As human beings, we have not been here very long. It is only over the last 10,000 years that we have evolved agriculture, writing, and reading, expansion of food resources, and then civil government. This is such a short period when measured against the background of evolution, of the planet, of other species, and even of our own species. Of course, grim reminders of species decline are omnipresent,[2] as are the dangers of planetary warming. We seem at times to be in endless peril and devolution, and at other times to face endless possibilities and advancement. Transparency and participation are key features of our experience of democracy, something that Dewey struggled to make clear.[3] A participatory democracy and an expanding sense of rights co-evolved only in the last 100 years with our deep-rooted sense of being knotted to nature. The democratic sensibility helps to constitute a pragmatic form of naturalism.[4]

The sense of nature and the pragmatic sensibility of preservation of nature merge nature and culture. Pragmatism needs to function technologically without being seduced by a technocratic spirit that alienates us from fundamental human social bounds, human social cooperative behaviors, and social competence. Our species' evolution is tied to inventiveness and social expansion, contingent on context, temperamental, and cultural opportunities; social cooperativeness is a core feature amongst other motivations. Broad extension in an age of endless and mind-boggling semantic networks spread across our many terrains is a pragmatic naturalist dream, when harnessed to a democratic sensibility and a bridge towards inclusion.

Our sense of being part of nature does not denigrate the human spirit. It does not eradicate the sense of the sublime,[5] nor the cultivation of humility and awe. Mystery remains; our limitations are a reminder of the hubris, arrogance, and wanton destructiveness and devolution of function to which we are vulnerable. Nevertheless, a sense of value and worth is anchored to an evolving species and to human dignity.

We combine staggering progress with frail humanity and omnipresent bestiality.[6] The "possible lives to come"[7] are rich in epigenetic possibilities, with a social world that fosters human social contact and public legitimacy with an evolving sense of inclusion and refined "habits of heart"[8] – all broadly in extension to a public in contexts of democratic expression and self-correction.[9] In his recent book arguing that violence has historically been decreasing, Steven Pinker, I want to suggest, is in the tradition of Dewey. This is the tradition in which the expansion of rights (e.g. Women's, African American, Gay) and participation in the polity and public can result in decreases in violence. The civilizing process of democracy that Dewey understood may sound naïve, but Pinker argues that it is real – "the better angels" that Lincoln adumbrated.[10]

There is a recurrent theme in inquiry and self-reflection. Our intellectual journey is rooted in core questions that emerge in the recorded and evolving literature: themes about social cooperative behaviors as an ideal, about rootedness in objects and adaptation, inspired by human possibilities and social hope, albeit diminished by devolution and wanton destruction. The answers reverberate as the human species explores its possibilities and limits; there is always the reminder of our finitude.

In this concluding section, I return to core psychobiology, adaptation, and well-being anchored to the process of pragmatism within the narrative theme of critical realism and social cooperative inquiry and social sensibility. I reconstruct the importance of Richard Rorty, a modern neopragmatist and an important intellectual oriented to the conversation about human spirit and its dignity, and place him in an interesting burgeoning literature of social hope and human social civility. He lost contact, in my view, however, within a pragmatic naturalistic perspective, a core feature of what I have been emphasizing throughout, with the fact that the classical pragmatists got something basically right: that hypothesis formation, understood naturalistically, is at the heart of human inquiry broadly understood, that error and change are endemic to the sense of inquiry, that a sense of action and the will are at the heart of investigation in which valuation or appraisal are inherent in the process, and that social cooperative behaviors are at the core of our evolutionary success.

Well-being and adaptation

We have come to realize, as a normative goal, that social hope is our common bond; it is the dilution of differences that divide and harm us. The path of human progress is frail. With glimmers of hope and eternal seduction, the stoic nobility exists amidst the diffidence and difficulties in preserving a broad social compass in which many reach forms of meaningful human happiness rich in existential sensibility.

But there is no panacea. The idea of progress that infused Darwin's conception of evolution, as well as Jackson's and Spencer's conception of the nervous system, has been modified. Corticalization of function does not necessarily mean social advance. Devolution of function is as paramount at times of war and crisis.

Prosocial sensibilities figure importantly and are a constant across cultures, but they compete with diverse motivations.[11] Variation in expression is a constant, but social contact is also a factor across all cultures. The formation of habits sets the conditions for meaningful, lasting social contact.[12]

At one fundamental level, little has changed: "We search for the stable amidst the precarious."[13] The search requires diverse cephalic and cultural resources, and it results in punctuated and gradual cultural epicenters. The human condition remains more precarious, our weapons that much more dangerous, the level of potential destruction that much greater – the precarious shifts towards the more stable by cephalic adaptation. Core needs are always a common function satisfied by food, water, sensual contact, sport, explorations, etc. The diverse motivations underlying these needs are quite broad – as we are broad in potential for expression.[14]

What evolved in our species are long-term social bonds, plasticity of expression, and corticalization of function. As our cortical visual functions increased dramatically, standing up, looking, and forming eye contact began an evolutionary expansion in many primates. Human social contact, representation of objects, and use of objects are core cognitive capacities. Technology is an extension of ourselves, expanding what we explore – a cultural continuation of biological adaptation.[15]

We come prepared with an arsenal of cognitive adaptations rooted in social discourse and commerce with one another and the construction of objects that we use, that is, our tools. Our evolution is knotted to social groups working in unison across diverse terrains. Key abilities include discerning the wants and the desires of others (a core feature of our adaptations), along with cognitive adaptations such as recognizing the kinds of objects that are useful or affordable and avoidable, coupled with a wide array of inhibitory capacities that contribute to social and cooperative behaviors.[16]

Cephalic expansion set the stage for technological creations, expanding our sensory systems. A conception of the experiences of others is one frail but important psychobiological predilection to reduce morally offensive behaviors. Experiments in the laboratory and real life point this out. In experiments where physical contact is required, it is much more difficult to sacrifice one for the many.[17] The more personal the social contact, on the other hand, the more conflict that can emerge. Cognitive dissonance is one mechanism to reduce conflict by segmentation and rationale, justified or not.

Conformity and compliance, lack of judgment, and loss of individual perspective for the inside of another underlie obedience to authority under these conditions. Despite these caveats, empathy is still a vital form of moral sensibility, and a broad array of moral sentiments (e.g. sympathy/empathy, fairness, duty, self-regulation) have been uncovered that underlie the basis for approach/avoidance social behaviors.[18] Sentiments are richly embedded in cephalic information processing and appraisal systems.

Social attachment is primarily a behavioral adaptation. Evolved sets of neural systems are designed to facilitate social contact. Distinct sets of neurons in diverse regions of the cortex are active when one performs an action and when one watches others do so; this is pristinely shown in studies in macaques.[19] That does not mean that there is not overlap in neurons that fire when mirroring others and in performing the action; the emphasis for my purposes is that we come prepared to respond to others. The prosocial systems are tightly linked to moral systems.[20]

While biological determinism is a thing of the past, biological vulnerability and comorbidity is not. There are genes that create a vulnerability to decreases in social attachment and social isolation interacting with other genes for other traits (like aggression), which can do harm to the long-term viability and well-being of an individual and thereby decrease individual happiness. Prosocial moral sensitivity allows humans to quickly apprehend the moral implications in a social situation depending on context, agency, and the consequences of one's choices.

These sentiments are intrinsically linked to daily social interactions, and there are several regions in the brain that provide a context for social flexibility. One virtue, moral and otherwise, is in "deciding together."[21] Recognizing the intentions of others is one critical feature in prosocial behaviors. This cognitive capacity begins early in ontogeny and is tied into visual sensibility. Eye contact, again, recognizing the intentions of others, is compromised in autistic individuals.[22] Recognizing intentions is knotted to a broad array of cephalic tissue that underlies perspective taking and human moral judgment.

A propensity for prosocial behaviors, however, is small and competitive with the rest or our predilections. A consideration of the prosocial instincts embedded in cephalic bodily responses towards others amidst a frail sensibility of human kindness, as well as diverse cognitive predilections amidst human/animal sentiments, sets the stage for moral progress and social contracts and diverse cognitive rules. Diverse regions of the brain tied to social behaviors underlie prosocial sentiments.[23]

Reaching out to others is but one prosocial response that we all share, even though it varies quite a bit across cultures in its expression. The life blood of humanity, after all, is our social bonds and the way we enjoy and manage our solitude.

One ameliorative factor against devolution of function is social meaning – viz., human empathetic social contact, which broadens our altruistic sensibilities towards a wider arena of human beings.[24] Our brains are designed for social cooperative behaviors and social deception, amongst other forms of contact. Social contact, meaningful close relationships, is an important factor in well-being. Sleep, for instance, is linked to meaningful social contact, to social hope.[25]

In fact, one recurrent feature is that lending a hand to others, having social responsivity to others, engaging in acts to placate the response to threat – in short, meaningful social contact – is an ameliorative biological adaptation that promotes trust within the bonds of reason.[26] Less trust amongst us, less exposure to strangers, and fewer family ties result in greater unease. Of course, not all of us experience the world this way. The social milieu and the larger ecological context have always been littered with unease, uncertainty, and endless danger.

Let us not forget this other fact: our cognitive capabilities are rarely consistent and often shortsighted.[27] But amidst these events are meaningful social bonds, and thus well-being.[28] Evolution put a premium on social attachment, gaining a foothold into the world by being guided and connected to others. Cooperative tendencies coexist amidst the competitive predilections, and the desire to live.[29]

Social contact is essential for ontogenetic development, a long-noted piece of epistemological history expressed differently across diverse cultures.[30] Family and group structures forged through meaningful contact are essential for our mental health.[31] Supportive social contact is not an absolute prophylactic, but it is a helpful ameliorative in combating disease and breakdown. Loneliness is a devolution of the ameliorative social factor. Meaningful labor is at the heart of the human condition and underlies recovery and maintenance of well-being. In other conditions that might be favorable to neuroendocrine systems, social isolation exacerbates vulnerabilities across diverse species.[32]

Of course, we know that a sense of well-being can be found in generosity, in reaching out to others; and that it is linked to moral ethos.[33] Aristotle distinguishes the shortsighted well-being of the senses from the well-being tied to a meaningful life. In modern form, the difference between the hedonic treadmill, the computational mistakes and expectations of the decision sciences of the senses, and the longer term of what constitutes a meaningful, eudemonic life is a distinction with which we are all familiar. Diverse cephalic systems are implicated in this process. In the end, trust and social solidarity are essential for human well-being.

Compassion, Rorty, and American pragmatism

Richard Rorty, erudite and deeply educated in Western philosophy and history, has emphasized well-being in his philosophy through human solidarity, and has come to be identified in the last part of the 20th century with a variant of contemporary pragmatism. For Rorty, the conversation of philosophy was reduced to no givens, and certainly not the narrow pragmatism and scientism associated with Quine.

What Rorty had in mind was something quite different. It entailed the end of philosophy as we thought of it: no ideal language, no illusion, and banal scientism, no special philosophical insights into wisdom, no grand metaphysical synthesis. Rorty had three heroes: Dewey, Wittgenstein, and Heidegger. He loved Dewey because he dissolved the distinction between the knower and the known. He valued Wittgenstein for his glorification of ordinary language to eradicate philosophical perplexity. And he treasured Heidegger for the end of metaphysics. All three became part of Rorty's landscape, as indeed they became part of the philosophical landscape. After all, they are three great philosophers.

When I was a graduate student in philosophy at the University of Pennsylvania in the late 1970s, one of the most exciting talks I ever attended was a lecture given by Rorty just as his book *Philosophy and the Mirror of Nature* came out. This brilliant tome concerned the history of epistemology and representational views of nature, something that Rorty utterly rejected. In fact he thought discussions of the mind were a bad invention, a pernicious predilection to be eradicated by modern philosophical therapy. This is not the part that excited me, and indeed I mostly disagreed with him: no biology, no links to kinds of objects – too much mere conversation, too much rejection of core issues in epistemology. There simply was not enough integration of what mattered to Dewey: social hope amid a knowledge and appreciation of nature, our psychobiology and core behavioral adaptations.

It was his knowledge of the tradition that came out in his work that I admired. For his was a serious mind, even though he doubted the validity of the term that is the construal of the mind as representative in which correspondence to truth is a measure of epistemological validity. His version of pragmatism is at odds with my own, in no small part due to his neglect and my embrace of experimentalism.[34] But he sure could generate interesting and important conversation.

Rorty reset the conception of pragmatism as it was known and knotted to the larger, mostly European, world of postmodern philosophy, a world without foundations. Of course, Dewey, who, as Rorty acknowledged, eschewed the search for foundations, but kept something akin to fixing reference to objects, would appreciate the anchoring metaphor and remained steadfast within the biological and psychological sciences. For Dewey, philosophy was just another discipline broadly understood within the sciences, and science was less about representation and more about adaptation, coherence, and advancement in knowledge and learning.

Rorty did, however, keep his hand on an important truth, something the theologian Paul Tillich would have considered of "ultimate concern": the relationship of hominid existence to "social hope." While with one hand Rorty uttered phrases such as the "world well lost," he held out for social hope amidst the endless disappointments of the glaring left that he grew up with.[35] After all, disappointment indicates two important facts about us: the search for security, and our capacity for sympathy. Any conception of hope entails these, dare I say, truths. Rorty became synonymous with pragmatism for many people, especially since he could use rhetorical phrasing and wrote so elegantly of "us pragmatists." Indeed, pragmatism is a broad tent, covering a wide array of individuals. Rorty's is a pragmatism. An important contribution of Rorty's pragmatism is highlighting the transitions between events – transactions with others, in which human growth – is part of the stable of important factors.[36]

We seek the "stable amidst the precarious," as Dewey noted.[37] But Dewey understood the importance of biology and anchored what he said to a broad view of inquiry. This was not in Rorty's arsenal, and without some notion of being anchored to kinds of objects he gave up too much.

Dewey's social hope, natural piety, and humility resonate with a sense of inquiry and turn into the contours of the human condition,[38] replete with a certain *naïveté* (as critics have often pointed out),[39] but less naïve than often understood. Dewey knew the limits of the human condition. For him, the fall into devolution was as plain as the evolution of the human expression. An existential pragmatism – the link to the important insights of Heidegger and Sartre, Kierkegaard and Nietzsche – manifested in the high anxiety (to use a Mel Brooks phrase).

No world well lost, but engaged activities with the use of diverse capabilities cultivated in a cultural milieu that appreciates inquiry and its engagement. Naturalism without scientism or narrow reductionism is broadly understood as responding to the biological sciences, to kinds of objects without foundationalist sensibility.[40] Erudite table talk predominates: it is interesting, well informed, and ripe with empathic vectors rich in human meaning. Its achievement is no small feat but a worthy part of the Rorty ambience.

Social hope, a phrase used by Dewey and resurrected respectfully by Rorty, is what we have to live by.[41] Wise, sane, rich practices undertaken amidst the heartbeat of a participatory democracy of educated participants is a normative goal, if distant, but possible in a landscape of porous borders, calling for nationalism without chauvinism.

No easy chore, social hope is an embodiment and appreciation of the particular amidst the many, an understanding of the old problem of the one and the many without the abstract metaphysical trappings. Dewey sets the stage for sanity, Peirce for inquiry, James for lived experience; these are some core features of these classical pragmatists that Rorty inherited to varying degrees. Unfortunately, the extent of what

Rorty inherited from Peirce was significantly narrowed as Rorty took the linguistic turn.

Rorty himself grew up in a context in which social hope was a calling for political action. The time was ripe for social hope amidst crash, doom, and disaster. Messianic social engineering was a running thought across ideological camps, including fascism and communism. Many of the secular humanists who stood at the steps of City College and then at Columbia with Dewey in NYC turned away from these extreme ideological camps.[42]

Pragmatism and naturalism, instead, found comfort in an intelligence guided by action and by consequences. This was a pious naturalism, embedded in an acknowledgement of the tragic sense of life,[43] with glimmers of social hope and human well-being. There was no grand illusion, no mythological distortion of the human predicament, just the lure of human hope: lose that and the fall downward is endlessly destructive. Intelligence, for pragmatists, was a feature continuous with our evolutionary capabilities that was to be used towards that end.

Indeed, social hope, social liberation, and the breakdown of barriers held prominence in the philosophy of Dewey. His was a social hope for inclusion in an evolving culture, emboldened by individual aspirations and social possibilities, amidst an appreciation that emerges from diverse identities and backgrounds to solidarity towards one another.[44] His was a cosmopolitan social sensibility by solidarity, shared beliefs, and accommodated differences. Social hope requires transactions that embrace the differences amongst us, a conception of the family of humanity – a humanism emboldened by social hope, natural piety, and a broad common faith, emboldened by democratic participatory sensibilities.[45]

This was no easy task to be sure, with narrow group formation a powerful trajectory in our evolutionary success as a species. The solidarity is diminished by other motives, namely the motives to isolate, to stigmatize, to dominate, to depreciate – to hate. Dewey himself investigated the 1930s' Moscow show trials, which obliterated any illusion of a grand ideological erudition of social hope coming from an exclusive, state-controlled organization. There was no block universe as James understood it, as he rebelled against monolithic orders. But social hope is the stuff that ties us together. Our evolution is bound by our social contact, and our sense and concern for each other.

Dewey always seemed to be underestimated, in logic, in policy, and social sense; soft and muddled, not the most aesthetic of pragmatist writers, but perhaps the most sanguine and with the deepest hope about the prospect of the human condition. Moreover, he understood the value of the empirical stance, the continuous synergy of science and philosophy, and the larger historical predicament and precedent. He spoke in one voice as a thinker – an achievement. Dewey had the talent to demythologize while preserving what is worthy and valuable, albeit with diverse limitations.

He radiated a warm yet detached sensibility, anti-authoritarian to the core. What emerged from his writings is a view of demythologized reason, not reason as the glorification of one state, one ruler. That is the lesson of our species' history: frail and endlessly labile, promiscuous and beaming with possibilities, a glorious sense of being connected to others, in participatory labors of self-initiative, self-excellence and self-preservation.

The political side of pragmatism, linked to an expanding cephalic capability and a rich sense of the "interpretative turn,"[46] has been a major part of the pragmatism traced to Dewey. Social hope breathes not through dominance, but peaceful resolution,[47] through understanding, through historical connection, through humility, through semblances of strength. It is a virtue of intelligent action, action historically emboldened but not mythologized, and a conception of reason demythologized and connected to the contours of others. The social hope is naïve perhaps, but it is emboldened with possibilities and ripe with contradictions.

A so-called progressive like Woodrow Wilson, as president of the United States, would initiate and enforce segregation in Washington during his administration, and applaud a racist film about the South;[48] others would celebrate the poetics of democracy.[49]

"Hallowed ground," words of a president (Lincoln) at a place of battle (Gettysburg), looking to heal,[50] and who himself evolved to embrace a broader range of the American experience. Any experience in which inclusion was a primary motivation is testimony to the frail sense of what is possible. But it is frail, and all the talk about the experience of democracy competes in us with many other motivations, many of which are far from democratic, some demonic, effecting the lure downward – a common entropic trend: devolution of function, devolution of possibilities.

Pragmatism, understood as a put-down, can be read as "anything goes," "what works sounds true." This possibility was understood by Peirce, who railed against his friend and benefactor, the softer but wiser

William James, ever loyal to the more lofty but less wise Peirce. But a sense of naturalism and realism coexists amidst a connection to kinds of objects, real history, real consequences, abstract objects. A pragmatism is acceptable for those oriented to the existential fate, and mediated by demythologized problem-solving capabilities and expectations about knowledge, while retaining forms of being anchored to a classical pragmatism in which intellectual honesty is still the normative grounding, although without any guarantee of getting things right. We pragmatists of the classical sort want a form of anchoring without rigidity, a way of understanding social change and action that enhances human experience and conditions and in which scientism does not dominate.[51]

Rorty's "we pragmatists" phrase nevertheless did a disservice to the rest of us. Rorty was deep, compassionate, erudite, and to be respected, but he lumped us together. He gave up the language of nature, an important part of the pragmatist tradition. He eschewed the language of naturalism because of the link to the "glassy essence" that was repugnant to him for the representationalism that he thought it stood for. With the loss of the language of nature came the loss of the very tradition that Rorty was trying to strengthen: grounding action, emboldened by kinds of objects, kinds, or practices tied to the decision making that emerges from empiricism.

Pragmatism, like naturalism, as I have indicated throughout, has always meant more than one thing,[52] not unlike many terms (e.g. paradigm). What is important is to retain a conception of social hope and an embrace of the critical natural philosophical stance, rooted in inquiry.[53] Critical realism has been modernized by a century of discovery, fallacies, and critical common sense, and an embrace of the subject with the endless sense of tacit knowledge,[54] not articulated, but embodied across the practices in which we participate,[55] emboldened by diverse forms of representational systems in which the reactive, the enactive, and the symbolic represent the evolution from reflexes to mimetic capability to a social world rich in the consideration of beliefs and desires.[56]

Social hope requires the intelligence to be anchored in, but not frozen to, things that matter. The variants of pragmatism in the modern era are confusing for many, as it often has an anti-intellectual property, something of course that would have been inimical to Rorty, but which was a badge of merit for some, a put-down for others. There is no denial of truth, it is just more complicated than "one size fits all." By thinking that pragmatism is all the same as a reflex, it has been ripe fruit for cultural and philosophical critics.[57]

Rights and wrongs: expanding participation in an evolving social hope

Pragmatism is profoundly linked to participation in democracy and the cultivation of intelligent problem solving. An expansion of participation through embodying this process to varying degrees underlies the conception of a participatory democracy, as does participation in social groups towards rational ends, where rationality is tied to the clarification of issues, and to the adjudication and the ability to compromise in context without devolution of principle and purpose.

A profound philosopher/sociologist, and a student of William James, W.E.B. Du Bois revolutionized with rational acumen a sense of place and race dignity,[58] amidst an onslaught of noise and danger that permeated and continued to permeate the 19th and 20th centuries. Indeed, the problem of race, as Du Bois noted at the turn of the 20th century, was a recurrent theme, an endless theme springing from historical viruses and myopic tendencies, tendencies built into a psychobiology that can both cooperate and exclude, where groupthink is an endless vulnerability, where dividing and conquering is a predilection, to be curtailed only by social practices.[59]

Du Bois elevated a dignified conception of being amidst racial onslaught. As a student in Harvard's philosophy department, he noted in his autobiography[60] that he was "repeatedly a guest in the home of William James."[61] He interacted in the Harvard philosophical milieu with Josiah Royce, an individual who idealized pragmatism into a glorious community,[62] an "absolute pragmatism,"[63] and was every bit as much a first-rate thinker as them at an extraordinarily difficult time for an African American. Dewey, along with Jane Addams, was at the inauguration of the NAACP in 1909, along with Du Bois, and remained respectful though not fully engaged with issues with regard to race and social change.[64] Dewey emphasized the social dimension and the aim of forging social links,[65] amidst a "productive pragmatism" embodied in technologies with an eye towards enhancing human experience and participatory democracy.[66] An Enlightenment pathway, in which transparency and participation of citizens is a normative goal, aims further towards inclusion, where possible, and the richness of shared human experience.[67]

In the main, race, however, remained a key factor in the 20th century American scene. And in the struggle in the U.S. for human and civil rights, African Americans, women, and Native Americans were pitted against each other, in contradiction and conflict, with both advances and setbacks. Voting rights, Jim Crow laws, and vigilantism fought with a worthy goal of civil rights, the right to participate in the larger culture.

Social hope survived in the narratives, self-exposition, and exploratory texts of Sojourner Truth (depicted in Figure 8.1), sending us in the right direction for broad-based human participation in the culture of democracy and human freedom and expression. Harriet Beecher Stowe wrote a popular novel (*Uncle Tom's Cabin*), of which Abraham Lincoln,

Figure 8.1 Sojourner Truth

a political pragmatist with evolving ideals and an evolving sense of rights,[68] is said to have uttered when he met Stowe, that it helped set the Civil War in motion.

Amidst a sense of rights and social bonding, of "a right of association" and "acting in common,"[69] was a Frenchman, Alexis de Tocqueville, who in the middle of the 19th century documented the Old Regime in France as it finally fell from a monarchy to a Republic, in which common bonds of labor and meaning were at the heart of freedom of expression and social bonds of value. The celebration was about democracy within self-initiatives coexisting with social hope; a social hope of choice, not imposition. Social hope, always frail and yet omnipresent in an evolving cultural milieu, must also be linked to maximizing freedom of expression. The fall into slavery and devolution of human expression through social control is always a half neuron away.[70] Amidst diverse moral sentiments,[71] and a propensity for moral rules, a culture that supports social equity and labor can undergo a cultural evolution ripe with social hope.

Of course, multiple motivational lures compete for expression. One of which is the ability to participate, to draw out rational conclusions in which the many are enhanced but not at the expense of the vulnerable or the minority. This is no easy task when faced with the value of obedience to authority,[72] along with the seduction of authoritarian personality,[73] as it leads to a devolution of responsible action. Yet part of the human trajectory has been the expansion of rights and responsibilities from a single authority across the community.

A common origin of our species emerging out of Africa is the current theory in evolutionary biology, giving us a common past that dates some 200,000 years for our species.[74] Common origins do not speak of superior or inferior, but mark our shared diversification and adaptation to different climates, terrains, food resources, etc.

Despite real genetic differences, we have far more in common than we do traits that set us apart. That, of course, is what made the Gaia perspective possible.[75] Surely that is something Peirce should have appreciated, with his speculations about "evolutionary love,"[76] or Whitehead and his sense of "Nature Alive": evolution, not devolution, of species and living kinds.[77] Another intriguing and suggestive hypothesis speaks to the common origins beyond our species, to the larger ecological context in which we find ourselves, a concept which underlies a number of feminist texts with regard to pragmatism and other philosophical schools stressing commonality and unity.[78] A cosmopolitan sensibility amidst the sense of otherness and difference is an acknowledged sensibility in which others and difference are a real, if frail, social virtue and social connection.[79]

We have elected an African American to the presidency of the United States. That he has received support at home and abroad is telling. This too is a sign of social hope. Social hope comes to mind as something that fires amidst an economic downturn of grave proportions; and yet this historical event is also prophetic in its possibilities for the participatory sense of democracy.[80]

A keen sense of pluralism is an important element in the pragmatic tradition, amidst a cultivated sense of participatory pragmatism, almost a celebratory moment of differences amidst common purposes, a rootedness in nature linked to one another, the one and the many grounded in pragmatic purpose, demythologized amidst the recognition of the tragic elements. Omnipresent is a mature sense emerging from the new pragmatism, a pragmatism without frills, rich in a naturalistic predilection and with a sense of our continuity with nature and other animals.[81]

Perhaps a shared naturalism, or natural piety, was kernelled within the classical pragmatist, and pragmatism much more generally. This naturalism evokes a sense of nature and a progressive responsiveness to others.[82] This is also a responsiveness to tradition, in a shared pluralistic world to rejoice in. This provides a sense of strength in which sacredness of place is turned into the sacredness of others and home – a home big enough for all of us,[83] a democracy that reaches out and a pragmatism that reaches beyond a narrow purview of the included.[84]

The transgressions we have committed upon others are a humbling homily and a reminder of our past, a past rich in mistakes and possibilities. We see this with the progress of a Franklin or a Lincoln, as they are products of their age, whose conceptions of Native Americans or immigrant Americans or African Americans gradually moved forward and became more progressive. The enlightenment pathway is one in which participation in democracy is grounded towards obligations towards others.[85]

The strengths of others were an important element as they grappled with what is right and hard, and the limitations of the times; indeed, as Hegel says, we are products of our ages. An expansion of rights at the heart of the social hope envisioned by Dewey involves an inclusive attitude towards others, a recognition of difference and commonality, amidst an omnipresent bastion of ill behavior and devolution.

Social hope is the ideal of balancing in context both individual freedom and social bonds and responsibility,[86] with an eye for inclusiveness as our cephalic capabilities are expanded. There is no panacea here, there are no miracles, just hard slogging amidst a lot of intelligence, and social practices. We need to acknowledge diverse forms of Thanatos or devolution of function as well as Fortuna amidst re-envisioning the body with little fanfare of Cartesian separation from action that emboldens, while being mindful of our natural continuity with others.[87] The naturalistic urge to forge consequences lies in the participation in group formation fraught with conflict amidst a democratic sensibility, as well as the participation of the least well off being part of the process.[88]

It is a core cephalic adaptation to forge links through problem solving. Intelligence raises us above problem solving toward a rationality not based on deduction from absolute premises but through engagement with others, finding ways to forge ahead. Rationality towards the higher ideals,[89] a "common faith" tied to a "common good" amidst an evolving sense of rights becomes a common expression,[90] an expansion of human dignity, frail and fraught with endless disappointments – these are the alluring ideals that constantly beckon to us.

Our transgressions match and complement our progress. Human beings are demythologized on this journey, not towards perfection but towards peaceful and respectful coexistence. An existence endlessly marked by real danger, real wars, and real Holocausts. Worry remains our common currency, an ontic fact (see Figure 8.2), and a reason to further forge collaborative bonds, linking social hope to memory and forward-looking sensibilities.[91]

The pragmatic sense of progress is particularly the democratic sense of coexistence amidst the use of intelligence for wider concerns, for a pluralism of others, albeit amidst the tragic facticity of our narrow-mindedness and myopic predilections.

Social hope is like the prophetic jubilee that brings many of us together for an ephemeral moment in which some good is achieved.[92] From place to home, with pluralism as its cornerstone, social hope is a natural outgrowth of a continent's evolution, though imperfectly gathered from all corners amidst endogenous people with diverse practices and natural piety.[93] It is a frail homily of possibility, ephemeral except for moments punctuated by greatness. Like the King march on Washington, or the progress of women's rights and equity, social hope sustains us in our pragmatic march along the path of evolution or devolution.

At a time with open-ended possibilities fraught with endless danger and devolution of function, and with possibilities tempered with realism for the diverse motivations that compete and are not simply eradicated with education, what is needed is a sense of an individual

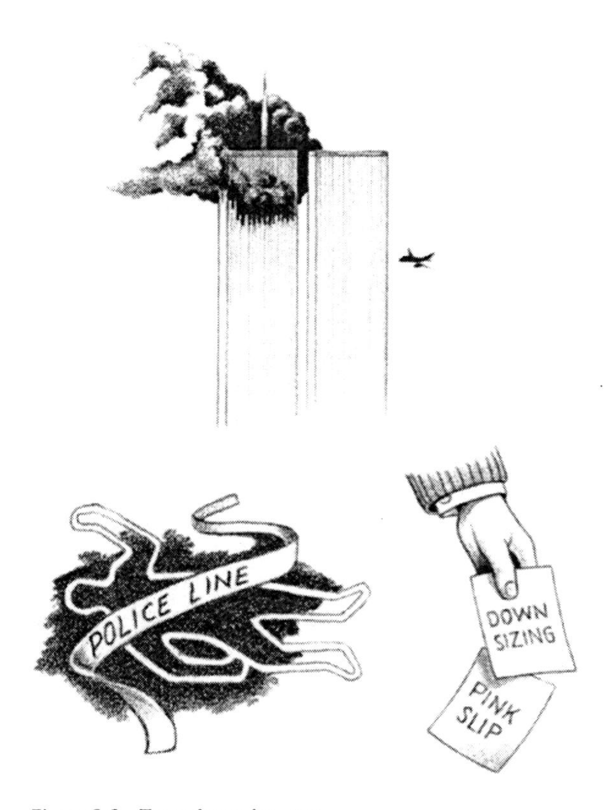

Figure 8.2 Tragedy and worry

tempered by participation in a democracy emboldened by individual responsibility tied to a pragmatic problem solving amidst high ideals for the human condition. That is a pragmatism worth believing in. This is a pragmatism that is historical, evolving, inclusive, and ripe with the struggle inherent in the pursuit of what is worthy of our species – an evolutionary trend with rich causal experiences: no easy task, and no necessary historical vector. That is what makes for exciting and open-ended human expression: it is also what makes life humbling and ripe with terror.

Notes

1 Pragmatism and Naturalism amidst Social Sensibilities

1. Schneider, H.W. (1946, 1963). *A History of American Philosophy*. NY: Columbia University Press; Eames, S.M. (1977). *Pragmatic Naturalism*. Carbondale, IL: Southern Ill. University Press.
2. Sacks, K.S. (2003). *Understanding Emerson*. Princeton: Princeton University Press; Richardson, R.D. (1995). *Emerson*. Berkeley, CA: University of California Press; Lysaken, J.T. (2008). *Emerson and Self-Culture*. Bloomington, IN: Indiana University Press.
3. Beebe, W. (1944, 1988). *The Book of Naturalists*. Princeton: Princeton University Press.
4. Aristotle (1968). *De Anima* (translated by D. Hamylin). Oxford: Oxford University Press: 64.
5. Jefferson, T. (1787, 1982). *Notes on the State of Virginia*. New York: Norton.
6. Becker, C.L. (1932). *The Heavenly City of the Eighteenth Century Philosophers*. New Haven: Yale University Press.
7. Nussbaum, M.C. (1997). *Cultivating Humanity*. Cambridge, MA: Harvard University Press; Nussbaum, M.C. (2004). *Hiding from Humanity*. Princeton: Princeton University Press; Israel, J.I. (2001). *Radical Enlightenment*. Oxford: Oxford University Press.
8. Israel, J.I. (2010). A Revolution of the Mind: Radical Enlightenment and the Intellectual Origins of Modern Democracy. Princeton: Princeton University Press.
9. Miller, P. (1965). *The Life of the Mind in America from the Revolution to the Civil War*. New York: Harcourt Brace and Jovanovich; Miller, C.A. (1988). *Jefferson and Nature*. Baltimore: Johns Hopkins University Press.
10. Thompson, P.B. and Hilde, T.C. (2000). *The Agrarian Roots of Pragmatism*. Nashville: Vanderbilt University Press. See also Pratt, S.L. (2002) *Native Pragmatism*. Bloomington: Indiana University Press.
11. Barrow, M.V. (2009). *Nature's Ghosts*. Chicago: University of Chicago Press.
12. Humboldt, A. Von (1859, 1993). *Cosmos*. New York: George Bell and Sons; Gould, S.J. (1989). *Church, Humboldt and Darwin: The Tension and Harmony of Art and Science* (edited by Franklin Kelly). Washington DC: National Gallery of Art. For more on the connection between Jefferson and Humboldt see Gould, S.J. (2002). *The Structure of Evolutionary Theory*. Cambridge, MA: Harvard University Press, and then especially, Walls, L.D. (2009). *The Passage to Cosmos: Alexander Von Humboldt and the Shaping of America*. Chicago: University of Chicago Press.
13. Miller, C.A. (1988). *Jefferson and Nature*. Baltimore: Johns Hopkins University Press.
14. See for instance Kant, I. (1787, 1965). *Critique of Pure Reason* (translated by L.W. Beck). New York: St. Martin's Press; Kant, I. (1792, 1951). *Critique of Judgment*. New York: Haffner Press; Coleridge, S.T. (1840, 1956). *Confessions*

of an Inquiring Spirit. Stanford: Stanford University Press; Coleridge, S.T. (1956). *Selected Poems.* New York: Appleton-Century Crofts.

15. Two useful scholary works on Kant: Friedman, M. (1992). *Kant the Exact Sciences.* Cambridge, MA: Harvard University Press; and Cassirer, E. (1918, 1981). *Kant's Life and Thought.* New Haven: Yale University Press.
16. Kitcher, P. (1990). *Kant's Transcendental Psychology.* Oxford: Oxford University Press.
17. Abela, P. (2002). *Kant's Empirical Realism.* Oxford: Oxford University Press.
18. Hanna, R. (2001). *Kant and the Foundations of Analytic Philosophy.* Oxford: Oxford University Press.
19. Peirce, C.S. (1899, 1992). *Reasoning and the Logic of Things.* Cambridge, MA: Harvard University Press. See also Lewis, C.L. (1929, 1956). *Mind and the World Order.* New York: Dover Press; Lewis, C.L. (1946, 1971). *An Analysis of Knowledge and Valuation.* LaSalle, IL: Open Court; Apel, K.O. (1981). *Charles S. Peirce: From Pragmatism to Pragmaticism* (translated by J.M. Krois). Amherst: University of Massachusetts Press; and Reschler, N. (1992). *A System of Pragmatic Idealism.* Princeton: Princeton University Press.
20. Emerson, R.W. (1855, 1883). *Nature, Addresses and Lectures.* Cambridge, MA: The Riverside Press: 51.
21. Lovejoy, A.O. (1936, 1978). *The Great Chain of Being.* Cambridge, MA: Harvard University Press; Lovejoy, A.O. (1955). *Essays in the History of Ideas.* New York: George Braziller Inc.
22. Coleridge, S.T. (1840, 1956). *Confessions of an Inquiring Spirit.* Stanford: Stanford University Press; Coleridge, S.T. (1956). *Selected Poems.* New York: Appleton-Century Crofts; Shelley, M. (1817, 1976). *Frankenstein.* New York: Pyram Pub; Schlegel, F. (1800, 1991). *Philosophical Fragments.* Minneapolis: University of Minnesota Press; Humboldt, A. Von (1859, 1993). *Cosmos.* New York: George Bell and Sons.
23. Schiller, F. (1795, 1980). *On the Aesthetic Education of Man.* New York: Frederick Ungar, Publ.
24. Schlegel F. (1800, 1991). *Philosophical Fragments.* Minneapolis: University of Minnesota Press.
25. Novak, B. (1980). *Nature and Culture.* Oxford: Oxford University Press.
26. See further Novak, B. (2007) *Voyages of the Self.* Oxford: Oxford University Press; Walls, L.D. (2009). *The Passage to Cosmos: Alexander Von Humboldt and the Shaping of America.* Chicago: University of Chicago Press; Walls, L.D. (1995). *Seeing New Worlds: Henry David Thoreau and 19th Century Natural Science.* Madison: University of Wisconsin Press; Walls, L.D. (2003). *Emerson's Life in Science: The Culture of Truth.* Ithaca: Cornell University Press.
27. Perry, R.B. (1912). Present Philosophical Tendencies, a Critical Survey of Naturalism, Idealism, Pragmatism, and Realism Together with a Synopsis of the Philosophy of William James. New York: Longmans, Green; Miller, P. (1965). The Life of the Mind in America from the Revolution to the Civil War. New York: Harcourt Brace and Jovanovich; Miller, C.A. (1988). Jefferson and Nature. Baltimore: Johns Hopkins University Press; Walls, L.D. (2009) The Passage to Cosmos: Alexander Von Humboldt and the Shaping of America. Chicago: University of Chicago Press.
28. Novak, B. (2007). *Voyages of the Self.* Oxford: Oxford University Press.

29. Goethe, J.W. von (1790, 2009). *The Metamorphosis of Plants*. Cambridge, MA: MIT Press.
30. Ibid.
31. Goethe, J.W. von (1786–1788, 1999). *The Flight to Italy*. Oxford: Oxford University Press.
32. See, for instance, Peirce, C.S. (1899, 1992). *Reasoning and the Logic of Things* (edited by K.L. Ketner and H. Putnam). Cambridge, MA: Harvard University Press; Dewey, J. (1920, 1948). *Reconstruction in Philosophy*. Boston: Beacon Press; Dewey, J. (1910). *How We Think*. New York: DC Heath and Publ.; Dewey, J. (1929, 1960). *The Quest for Certainty*. New York: Capricorn Books; Dewey, J. (1925, 1989). *Experience and Nature*. La Salle, IL: Open Court; Dewey, J. (1938, 1973). *Experience and Education*. New York: Collier.
33. Cronon, W. (1983, 1991). *Changes in the Land*. New York: Hill and Wang; Merchant, C. (1989). *Ecological Revolutions*. Chapel Hill: University of North Carolina Press.
34. Worster, D. (1977, 1991). *Nature's Economy*. Cambridge: Cambridge University Press; Worster, D. (2008). *A Passion for Nature: The Life of John Muir*. Oxford: Oxford University Press.
35. Nash, R. (1967). *Wilderness and the American Mind*. New Haven: Yale University Press; Oelschlaeger, M. (1991). *The Idea of the Wilderness*. New Haven: Yale University Press.
36. Dewey, J. (1916). *Essays in Experimental Logic*. Chicago: University of Chicago Press; Dewey, J. (1920, 1948). *Reconstruction in Philosophy*. Boston: Beacon Press.
37. Heidegger, M. (1962, 1977). *The Question Concerning Technology*. New York: Harper and Row: 125.
38. Carson, R. (1962, 1987). *Silent Spring*. Boston: Houghton Mifflin Company.
39. Dewey, J. (1925, 1989). *Experience and Nature*. La Salle, IL: Open Court.
40. Sagoff, M. (1988). *The Economy of the Earth*. Cambridge: Cambridge University Press; Oelschlaeger, M. (1991). *The Idea of the Wilderness*. New Haven: Yale University Press.
41. Wilshire, B. (2000) *The Primal Roots of American Philosophy*. University Park: Penn State University Press; Sarokin, D. and Schulkin, J. (1994). Co-evolution of rights and environmental justice. *The Environmentalist*, 14: 121–129; Dewey, J. (1939, 1963). *Freedom and Culture*. New York: Capricorn Books; Dewey, J. (1935, 1963). *Liberalism and Social Action*. New York: Capricorn Books; Dewey, J. (1931). *Philosophy and Civilization*. New York: Minton, Balch and Co.
42. Sagoff, M. (1988). *The Economy of the Earth*. Cambridge: Cambridge University Press; Nash, R. (1967). *Wilderness and the American Mind*. New Haven: Yale University Press; Pratt, S.L. (2002). *Native Pragmatism*. Bloomington: Indiana University Press.
43. Dewey, J. (1931). *Philosophy and Civilization*. New York: Minton, Balch and Co.; Lewis, C.L. (1946, 1971). *An Analysis of Knowledge and Valuation*. LaSalle, IL: Open Court.
44. Moreno, J.D. (1995). *Deciding Together*. Oxford: Oxford University Press; Shook, J.R. (2003, editor). *Pragmatic Naturalism and Realism*. Amherst, NY: Prometheus Press; Godfrey-Smith, P. (2002). Dewey on naturalism,

realism and science. *Philosophy of Science*, 69: S1–S11; Hickman, L.A. (2007). *Pragmatism as Post-Postmodernism*. New York: Fordham University Press.
45. Pratt, S.L. (2002). *Native Pragmatism*. Bloomington: Indiana University Press; Wilshire, B. (2000). *The Primal Roots of American Philosophy*. University Park: Penn State University Press.
46. Thoreau, H.D. (1971). *Great Short Works*. New York: Harper and Row: 4.
47. Ibid. 13; Whitehead, A.N. (1938, 1967). *Modes of Thought*. New York: Free Press.
48. As discussed in Brooks, V.W. (1957). *The Flowering of New England*. New York: E.P. Dutton and Co.
49. Santayana, G. (1967). *Animal Faith and Spiritual Life* (edited by J. Lachs). New York: Appleton, Century, Crofts.
50. Scheler, M. (1928, 1976). *Man's Place in Nature*. New York: Noonday Press.
51. Kahneman, D., Slovic, P., and Tversky, A. (eds) (1982). *Judgment under Uncertainty: Heuristics and Biases*. New York: Cambridge University Press.
52. Margolis, J. (2002). Reinventing Pragmatism: American Philosophy at the End of the Twentieth Century. Ithaca: Cornell University Press.
53. Whitehead, A.N. (1929, 1958). *The Function of Reason*. Boston: Beacon Press.
54. Bowler, P.J. (1983). *The Eclipse of Darwinism*. Baltimore: Johns Hopkins University Press; Bowler, P.J. (1988). *The Non-Darwinian Revolution*. Baltimore: Johns Hopkins University Press.
55. Menand, L. (2001). *The Metaphysical Club*. New York: Farrar, Straus and Giroux; Menand, L. (2002). *American Studies*. New York: Farrar, Straus and Giroux; Menand, L. (1997). *Pragmatism: A Reader*. New York: Random House; Schneider, H.W. (1946, 1963). *A History of American Philosophy*. NY: Columbia University Press.
56. Huxley, T.H. (1863). *Man's Place in Nature*. London: Macmillan; Wallace A.F. (2002) *A Selection of Writings from the Field*. Baltimore: Johns Hopkins University Press.
57. Slotten, R.A. (2004). *The Life of Alfred Russel Wallace*. New York: Columbia University Press; and see also Gould, S.J. (1977). *Ontogeny and Phylogeny*. Cambridge, MA: Harvard University Press; Gould, S.J. (1989). *Church, Humboldt and Darwin: The Tension and Harmony of Art and Science* (edited by Franklin Kelly). Washington, DC: National Gallery of Art; Gould, S.J. (2002). *The Structure of Evolutionary Theory*. Cambridge, MA: Harvard University Press.
58. Goethe, J.W. von (1790, 2009). *The Metamorphosis of Plants*. Cambridge, MA: MIT Press; Humbolt, W. Von (1836, 1971). *Linguistic and Intellectual Development*. Philadelphia: University of Pennsylvania Press.
59. Humboldt, A. Von (1859, 1993). *Cosmos*. New York: George Bell and Sons.
60. Muir, J. (1912, 1962). *The Yosemite*. New York: Doubleday.
61. For more, see the excellent discussion in Gould, S.J. (1989). *Church, Humboldt and Darwin: The Tension and Harmony of Art and Science* (edited by Franklin Kelly). Washington DC: National Gallery of Art; and in Gould, S.J. (2002). *The Structure of Evolutionary Theory*. Cambridge, MA: Harvard University Press.
62. Whitman, W. (1855, 1965). *Leaves of Grass*. New York: Norton.
63. Sellars, R.W. (1922). *Evolutionary Naturalism*. Chicago: Open Court.

64. Walls, L.D. (2003). *Emerson's Life in Science*. Ithaca: Cornell University Press.
65. Krikorian, Y.H. (1944, 1949). *Naturalism and the Human Spirit*. New York: Columbia University Press.
66. Humboldt, A. Von (1859, 1993). *Cosmos*. New York: George Bell and Sons.
67. Meinecke, F. (1957, 1977). *The Age of German Liberation*. Berkeley, CA: University of California Press.
68. Perry, R.B. (1935). *The Thought and Character of William James*, vols 1 and 2. Boston: Little Brown and Co. For a discussion of Agassiz, see Lurie E. (1988). *Louis Agassiz: A Life in Science*. Baltimore, MD: Johns Hopkins University Press.
69. Feinstein, H.M. (1984). *Becoming William James*. Ithaca: Cornell University Press.
70. Lamarck, J.B. (1809, 1984). *Zoological Philosophy*. Chicago: University of Chicago Press; Buckhardt, R.W. (1977, 1995). *Spirit of System: Lamarck and Evolutionary Biology*. Cambridge: Harvard University Press; Keverne, E.B., and Curley, J.P. (2008). Epigenetics, brain evolution, and behavior. *Neuroendocrinology*, 29: 398–412.
71. Schneider, H.W. (1946, 1963). *A History of American Philosophy*. NY: Columbia University Press.
72. Madden, E.H. (1963, 1990). *Chauncey Wright and the Foundations of Pragmatism*. Seattle, WA: University of Washington Press; Madden, E.H. (1964). *Chauncey Wright*. New York: Washington Square Press.
73. See also Wiener, P.P. (1949). *Evolution and the Founders of Pragmatism*. Cambridge, MA: Harvard University Press; Fisch, M.H. (1986). Evolution in American philosophy. In: *Peirce, Semiotic and Pragmatism* (edited by K.L. Ketner and J.W. Kloesel). Bloomington: Indiana University Press.
74. Wright, C. (1878). Letter to Charles Darwin: February 24, 1875. In: *Letters of Chauncey Wright, with Some Account of His life by James Bradley Thayer*. Cambridge, MA: John Wilson & Son: 330–338.
75. Wright, C. (1958). *Philosophical Writings*. New York: Liberal Arts Press; Madden, E.H. (1963, 1990). *Chauncey Wright and the Foundations of Pragmatism*. Seattle, WA: University of Washington Press; Schneider, H.W. (1946, 1963). *A History of American Philosophy*. NY: Columbia University Press; Schneider, H.W. (1964). *Sources of Contemporary Philosophical Realism in America*. New York: Bobbs-Merrill; Miller, P. (1965). *The Life of the Mind in America from the Revolution to the Civil War*. New York: Harcourt Brace and Jovanovich; Miller, C.A. (1988). *Jefferson and Nature*. Baltimore: Johns Hopkins University Press; Wiener, P.P. (1949) *Evolution and the Founders of Pragmatism*. Cambridge, MA: Harvard University Press.
76. Wright, C. (1878). Letter to Charles Darwin: August 29, 1872. In: *Letters of Chauncey Wright, with Some Account of His Life by James Bradley Thayer*. Cambridge, MA: 240–246.
77. Wright, C. (1877). The evolution of self-consciousness. In: *Chauncey Wright, Philosophical Discussions*. New York: Henry Holt and Co.: 199–266.
78. For specific examples, see Sumner, W.G. (1963). *Social Darwinism: Selected Essays*. Englewood, NJ: Prentice Hall; Hofstadter, R. (1963). *Anti-Intellectualism in American Life*. New York: Knopf.
79. Progressivism is a rich historical mine still waiting to be fully plumbed. For a modern view, see Breisach, E.A. (1993). *American Progressive History*.

Chicago: University of Chicago Press; and for contemporary information, see Becker, C.L. (1932). *The Heavenly City of the Eighteenth Century Philosophers*. New Haven: Yale University Press; Becker, C.L. (1915, 1967). *Beginning of the American People*. Ithaca: Cornell University Press; Beard, C.A. (1913, 1935). *An Economic Interpretation of the Constitution of the United States*. New York: Free Press.

80. Merck, F. (1963). Manifest Destiny and Mission in American History. New York: Vintage Press.
81. Peirce, C.S. (2000). *Writings of C.S. Peirce*, vols 1–6. Bloomington: Indiana University Press; Peirce, C.S. (2009). *Writings of C.S. Peirce*, vol. 8. Bloomington: Indiana University Press; Schiller, F.C.S. (1966). *Humanistic Pragmatism: The Philosophy of F.C.S. Schiller* (edited by R. Abel). New York: Free Press.
82. Sanger, M. (1926). The Function of Sterilization, *Birth Control Review*, October.
83. Menand, L. (2001). *The Metaphysical Club*. New York: Farrar, Straus and Giroux; Menand, L. (2002). *American Studies*. New York: Farrar, Straus and Giroux; Menand, L. (1997). *Pragmatism: A Reader*. New York: Random House.
84. Norrell, R.J. (2009). *Up From History: The Life of Booker T. Washington*. Cambridge, MA: Harvard University Press.
85. Hollinger, R., and Depew, D. (1999, editors). *Pragmatism: From Progressivism to Post Modernism*. Santa Barbara, CA: Praeger.
86. Haeckel, E. (1900, 1992). *The Riddle of the Universe*. Buffalo, NY: Prometheus Press; Haeckel E. (1866). *General Morphology of Organisms*. Berlin; Sumner, W.G. (1963). *Social Darwinism: Selected Essays*. Englewood, NJ: Prentice Hall.
87. Randall, J. H. (1977). *Philosophy after Darwin*. New York: Columbia University Press.
88. Woodbridge, F.J.E. (1940). *An Essay on Nature*. New York: Columbia University Press; Nagel, E. (1954). *Sovereign Reason*. New York: The Free Press; Kim, J. (2003). The American origins of philosophical naturalism. *Philosophy in America at the Turn of the Century*: 83–98.
89. See for instance Santayana, G. (1923, 1955). *Skepticism and Animal Faith*. New York: Dover Press; Santayana, G. (1932, 1967). *Character and Opinion in the United States*. New York: Norton; Reck, A.J. (1972). *Speculative Philosophy*. Albuquerque: University of New Mexico Press.
90. Dewey, J. (1910, 1965). *The Influence of Darwin on Philosophy*. Bloomington: Indiana University Press: 12.
91. Daniels, N. (1974, 1989). *Thomas Reid's Inquiry*. Palo Alto: Stanford University Press.
92. Whitehead, A.N. (1927, 1953). *Symbolism*. New York: Macmillan Company; Stengers, I. (2011). *Thinking with Whitehead: A Free and Wild Creation of Concepts* (translated by M. Chase). Cambridge, MA: Harvard University Press.
93. Reid, T. (1785, 1969). *Essays on the Intellectual Powers of Man*. Cambridge, MA: MIT Press.
94. Gibson, J.J. (1966). *The Senses Considered as Perceptual Systems*. New York: Houghton-Mifflin; Gibson, K.R., and Ingold, T. (1993, editors). *Tools, Language and Cognition in Human Evolution*. Cambridge: Cambridge University Press;

Anderson, D.R. (1997). John E. Smith and the Heart of Experience. In: *The Recovery of Philosophy in America: Essays in Honor of John Edwin Smith* (edited by T.P. Kasulis and R.C. Neville). Albany: SUNY Press: 115–130.

95. Schneider, H.W. (1946, 1963). *A History of American Philosophy.* NY: Columbia University Press; Schneider, H.W. (1964). *Sources of Contemporary Philosophical Realism in America.* New York: Bobbs Merrill.
96. Ibid.
97. Campbell, J. (1995). Understanding John Dewey: Nature and Cooperative Intelligence. Chicago: Open Court; Garrison, J. (edited by 2008). Reconstructing Democracy, Recontextualizing Dewey. Albany: SUNY Press.
98. Dewey, J. (1925, 1989). *Experience and Nature.* La Salle, IL: Open Court: 14.
99. Dewey, J. (1938, 1973). *Experience and Education.* New York: Collier: 38.
100. See, for instance, Dewey, J., and Tufts, J.H. (1910). *Ethics.* New York: Henry Holt; Mead, G.H. (1932, 1980). *The Philosophy of the Present.* Chicago: University of Chicago Press; Mead, G.H. (1928, 1972). *The Philosophy of the Act.* Chicago: University of Chicago Press; Mead, G.H. (1934, 1972). *Mind, Self and Society.* Chicago: University of Chicago Press; Addams, J. (1911). *Democracy and Social Ethics.* London: Macmillan; Addams, J. (1915). *Newer Ideals of Peace.* London: Macmillan; Addams, J. (1916, 2002). *The Long Road of Woman's Memory* (ed. C.H. Seigfried). Urbana: University of Illinois Press.
101. Rucker, D. (1969). *The Chicago Pragmatists.* Minneapolis: University of Minnesota Press; Shook, J.R. (2000). *Dewey's Empirical Theory of Knowledge and Reality.* Nashville: Vanderbilt University Press; Shook, J.R. (1998). *Pragmatism: An Annotated Bibliography.* Amsterdam: Rodipi.
102. Addams, J. (1911). *Democracy and Social Ethics.* London: Macmillan; Addams, J. (1915). *Newer Ideals of Peace.* London: Macmillan.
103. Seigfried, C.H. (1996). Pragmatism and Feminism: Reweaving the Social Fabric. Chicago: University of Chicago Press; Kloppenberg, J.T. (1986). Uncertain Victory: Social Democracy and Progressivism in European and American Thought, 1870–1920. New York: Oxford University Press; Kloppenberg, J.T. and Wightman Fox, R. (eds) (1995). A Companion to American Thought. Cambridge: Blackwell Publishers.
104. Hamington, M. (2009). Feminist prophetic pragmatism. *Journal of Speculative Philosophy,* 25: 83–92.
105. Du Bois, W.E.B. (1903, 1982). *The Souls of Black Folk.* New York: Penguin; Du Bois, W.E.B. (1940, 1968). *Dusk of Dawn.* London: Transactions Publishers.
106. Schwartz, J. (2000). *Fighting Poverty with Virtue.* Bloomington: Indiana University Press.
107. James, W. (1910, 1968). The Moral Equivalent of War. In: *The Writings of William James* (edited by John J. McDermott). New York: Modern Library: 660–670.
108. Seigfried, C.H. (1996). *Pragmatism and Feminism: Reweaving the Social Fabric.* Chicago: University of Chicago Press; Rorty, R. (1982). *Consequences of Pragmatism.* Minneapolis: University of Minnesota Press; Green, J.M. (2004). Building a cosmopolitan world future through pragmatist mutual hospitality. In: *Pragmatism and the Problem of Race* (edited by D. Koch and B. Lawson). Bloomington: Indiana University Press; Green, J.M.

(2009). *Pragmatism and Social Hope*. New York: Columbia University Press; Bernstein, R.J. (2010). *The Pragmatic Turn*. Malden, MA: Polity Press.

109. Eames, S. M. (1977). *Pragmatic Naturalism*. Carbondale. IL: Southern Ill. University Press; Wiener, P.P. (1949). *Evolution and the Founders of Pragmatism*. Cambridge, MA: Harvard University Press; Lovejoy, A.O. (1936, 1978). *The Great Chain of Being*. Cambridge, MA: Harvard University Press; Sellars, R.W. (1922). *Evolutionary Naturalism*. Chicago: Open Court Press.
110. Good examples of this practice in play are: Godfrey-Smith, P. (2002). Dewey on naturalism, realism and science. *Philosophy of Science*, 69: S1–S11; Godfrey-Smith, P. (1998). *Complexity and Function of Mind in Nature*. Cambridge: Cambridge University Press; Schulkin, J. (2009). *Cognitive Adaptation: A Pragmatist Perspective*. Cambridge: Cambridge University Press; Weissman, D. (1989). *Hypothesis and the Spiral of Reflection*. Albany, NY: SUNY Press.
111. Santayana, G. (1967). *Animal Faith and Spiritual Life* (edited by J. Lachs). New York: Appleton, Century, Crofts.
112. Woodbridge, F.J.E. (1940). *An Essay on Nature*. New York: Columbia University Press: 60.
113. Whitehead, A.N. (1933, 1961). *Adventures of Ideas*. New York: Free Press; Collingwood, R.G. (1945, 1976). *The Idea of Nature*. Oxford: Oxford University Press; Northrop, F.S.C. (1962). *Man, Nature and God*. New York: Simon and Schuster; Greene, M. (1995). *A Philosophical Testament*. LaSalle, IL: Open Court Press.
114. Grene, M. (1995). *A Philosophical Testament*. LaSalle, IL: Open Court Press.
115. Merleau-Ponty, M. (1968, 1970). *Themes*. Evanston, IL: Northwestern University Press.
116. Quine, W.V.O. (1953, 1961). *From a Logical Point of View*. New York: Harper Torchbooks; Quine, W.V.O. (1969). Epistemology naturalized. In: *Ontological Relativity and Other Essays*. New York: Columbia University Press; Quine, W.V.O. (1974). *The Roots of Reference*. La Salle, IL: Open Court.
117. Peirce, C.S. (1868, 1992). Questions concerning certain faculties claimed for man. In: *The Essential Peirce*, vol. I. Bloomington: Indiana University Press.
118. Hempel, C.G. (1965). *Aspects of Scientific Explanation*. New York: Free Press.
119. Kornblith, H. (1993). *Inductive Inference and its Natural Ground*. Cambridge, MA: MIT Press; Kornblith, H. (1987, editor). *Naturalizing Epistemology*. Cambridge, MA: MIT Press; Kornblith, H. (1994). Naturalism: both metaphysical and epistemelogical, *Midwest Studies in Philosophy*, 19: 39–52; Godfrey-Smith, P. (2002). Dewey on naturalism, realism and science. *Philosophy of Science*, 69: S1–S11; Godfrey-Smith, P. (2008). Induction, samples, and kinds. In: *Carving Nature at its Joints: Topic in Contemporary Philosophy, Volume 8*. Cambridge, MA: MIT Press.
120. Quine, W.V.O. (1969). Epistemology naturalized. In: *Ontological Relativity and Other Essays*. New York: Columbia University Press; Quine, W.V.O. (1974). *The Roots of Reference*. La Salle, IL: Open Court; Heelan, P.A. (1983). *Space Perception and the Philosophy of Science*. Berkeley: University of California Press; Heelan, P.A., and Schulkin, J. (1998). Hermeneutical philosophy and pragmatism: a philosophy of the science. *Synthese*, 115: 269–302.

121. Goodman, N. (1955, 1978). *Fact, Fiction and Forecast.* New York: Bobbs-Merrill Co.; Gigerenzer, G. (2000). *Adaptive Thinking.* Oxford: Oxford University Press; Gigerenzer, G. (2007). *Gut Feelings.* New York: Viking Press.
122. Smith, J.E. (1970). *Themes in American Philosophy.* New York: Harper and Row; Smith, J.E. (1978). *Purpose and Thought.* New Haven: Yale University Press; Quine, W.V.O. (1953, 1961). *From a Logical Point of View.* New York: Harper Torchbooks; Quine, W.V.O. (1969). Epistemology naturalized. In: *Ontological Relativity and Other Essays.* New York: Columbia University Press; Quine, W.V.O. (1974). *The Roots of Reference.* La Salle, IL: Open Court.
123. Dewey, J. (1920, 1948). *Reconstruction in Philosophy.* Boston: Beacon Press; Dewey, J. (1910). *How We Think.* New York: DC Heath; Dewey, J. (1929, 1960). *The Quest for Certainty.* New York: Capricorn Books; Dewey, J. (1925, 1989). *Experience and Nature.* La Salle, IL: Open Court; Dewey, J. (1938, 1973). *Experience and Education.* New York: Collier; Peirce, C.S. (1878). Deduction, induction and hypothesis. *Popular Science Monthly,* 13: 470–482; Peirce, C.S. (1883). *Studies in Logic, by Members of the Johns Hopkins University.* Boston: Little Brown; Peirce, C.S. (1892). The architecture of theories. *The Monist,* 1: 61–76.
124. Hempel, C.G. (1965). *Aspects of Scientific Explanation.* New York: Free Press; Levi, I. (1967). *Gambling with Truth.* Cambridge, MA: MIT Press; Levi, I. (2004). Beware of syllogism: reasoning and conjecture according to Peirce. In: *The Cambridge Companion to Peirce.* Cambridge: Cambridge University Press; Gigerenzer, G. (2000). *Adaptive Thinking.* Oxford: Oxford University Press; Gigerenzer, G., and Selten, R. (2001). *Bounded Rationality.* Cambridge, MA: MIT Press.
125. Dewey, J. (1916). *Essays in Experimental Logic.* Chicago: University of Chicago Press; Dewey, J. (1920, 1948). *Reconstruction in Philosophy.* Boston, MA: Beacon Press; Dewey, J. (1910). *How We Think.* New York: DC Heath; Quine, W.V.O. (1953, 1961). *From a Logical Point of View.* New York: Harper Torchbooks; Quine, W.V.O. (1969). Epistemology naturalized. In: *Ontological Relativity and Other Essays.* New York: Columbia University Press; Quine, W.V.O. (1974). *The Roots of Reference.* La Salle, IL: Open Court.
126. Donald, M. (1991). *Origins of Modern Man.* Cambridge, MA: Harvard University Press; Donald, M. (2004). Hominid enculturation and cognitive evolution. In: *The Development of the Mediated Mind* (edited by J.M. Luraciello, J.A. Hudson, R. Fivush and P.J. Bauer). Mawash, NJ: Erlbaum Press. See also Clark, A. (1997). *Being There.* Cambridge, MA: MIT Press; Clark, R.W. (1983). *Benjamin Franklin.* New York: DaCapo Press; Noe, A. (2004). *Action in Perception.* Cambridge, MA: MIT Press; Johnson, M. (1987, 1990). *The Body in the Mind.* Chicago: University of Chicago Press; Johnson, M. (2007). *The Meaning of the Body.* Chicago: University of Chicago Press; Schulkin, J. (2009). *Cognitive Adaptation: A Pragmatist Perspective.* Cambridge: Cambridge University Press.
127. Goldman A.I. (1999). *Knowledge in a Social World.* Oxford: Clarendon Press; Mead, G.H. (1932, 1980). *The Philosophy of the Present.* Chicago: University of Chicago Press; Mead, G.H. (1928, 1972). *The Philosophy of the Act.* Chicago: University of Chicago Press; Mead, G.H. (1934, 1972). *Mind, Self and Society.* Chicago: University of Chicago Press.

128. Dewey, J. (1938). *Logic: The Theory of Inquiry*. New York: Holt, Rinehart; Dewey, J. (1929, 1962). *Individualism, Old and New*. New York: Capricorn Books; Dewey, J. (1939, 1963). *Freedom and Culture*. New York: Capricorn Books; Dewey, J. (1929, 1966). *A Theory of Valuation*. Chicago: University of Chicago Press.
129. Simon, H.A. (1982). *Models of Bounded Rationality*. Cambridge, MA: MIT Press.
130. Dewey, J. (1929, 1960). *The Quest for Certainty*. New York: Capricorn Books: 3.
131. Dewey, J. (1922, 1957). *Human Nature and Conduct*. New York: Random House; Dewey, J. (1929, 1966). *A Theory of Valuation*. Chicago: University of Chicago Press; Dewey, J. (1934, 1970). *A Common Faith*. New Haven: Yale University Press; Dewey, J., and Tufts, J.H. (1910). *Ethics*. New York: Henry Holt; Fromm, E. (1973). *The Anatomy of Human Destructiveness*. New York: Holt, Rinehart and Winston.
132. For instance, Niebuhr, R. (1952). *The Irony of American History*. New York: Charles Scribner and Sons; Lovejoy, A.O. (1936, 1978). *The Great Chain of Being*. Cambridge, MA: Harvard University Press; Lovejoy, A.O. (1955). *Essays in the History of Ideas*. New York: George Braziller Inc.
133. Dewey, J. (1929, 1960). *The Quest for Certainty*. New York: Capricorn Books: 168.
134. Diggins, J.P. (1991). *The Promise of Pragmatism*. Chicago: University of Chicago Press.
135. Kloppenberg, J.T. (1986). Uncertain Victory: Social Democracy and Progressivism in European and American Thought, 1870–1920. New York: Oxford University Press; Kloppenberg, J.T. and Wightman Fox, R. (1995, editors). A Companion to American Thought. Cambridge: Blackwell Publishers; Kloppenberg, J.T. (1998). The Virtues of Liberalism. New York: Oxford University Press; Kloppenberg, J.T. (2010). James's pragmatism and American culture, 1907–2007. In: 100 Years of Pragmatism: William James's Revolutionary Philosophy (edited by John Stuhr). Bloomington: Indiana University Press.
136. See, for example, Sleeper, R.W. (1986). *The Necessity of Pragmatism: John Dewey's Coneption of Philosophy*. New Haven: Yale University Press; Eldridge, M. (1998). *Transforming Experience: John Dewey's Cultural Instrumentalism*. Nashville: Vanderbilt University Press; Hickman, L.A. (1980, 1992). *John Dewey's Pragmatic Technology*. Bloomington: Indiana University Press; Hickman, L.A. (1998). Dewey's theory of inquiry. In: *Reading Dewey* (edited by Larry Hickman). Bloomington: Indiana University Press; Shook, J.R. (2000). *Dewey's Empirical Theory of Knowledge and Reality*. Nashville: Vanderbilt University Press.
137. Godfrey-Smith, P. (1998). *Complexity and Function of Mind in Nature*. Cambridge: Cambridge University Press.
138. Tomasello, M., and Call, J. (1997) cover this ground thoroughly in *Primate Cognition*. Oxford: Oxford University Press; see also Tomasello, M. (2009). *Why We Cooperate*. Cambridge, MA: MIT Press; Rozin, P. (1976). The evolution of intelligence and access to the cognitive unconscious. In: *Progress in Psychobiology and Physiological Psychology* (edited by J. Sprague and A.N. Epstein). New York: Academic Press; Rozin, P. (1998). Evolution

and development of brains and cultures. In: *Brain and Mind: Evolutionary Perspectives* (edited by M.S. Gazzaniga and J.S. Altman). Strasbourg, France: Human Frontiers Sciences Program; Marler, P.R., and Hamilton, W.J. (1966). *Mechanisms of Animal Behavior.* New York: Wiley and Sons.

139. E.g., Dretkse, F. (1995). *Naturalizing the Mind.* Cambridge, MA: MIT Press; Gigerenzer, G. (2000). *Adaptive Thinking.* Oxford: Oxford University Press.
140. Boyd, R. (1999). Homeostasis, species and higher taxa. In: *Species: New Interdisciplinary Essays* (edited by R.A. Wilson). Cambridge, MA: MIT Press.
141. Stroud, B. (1996). The charm of naturalism. *Proceedings and Addresses of the American Philosophical Association,* 70: 43–45; Williams, M. (1996). *Unnatural Doubts.* Princeton: Princeton University Press; Williamson, T. (2000). *Knowledge and Its Limits.* Oxford: Oxford University Press; Nagel, E. (1954). *Sovereign Reason.* New York: The Free Press.
142. Peirce, C.S. (1868, 1992). Questions concerning certain faculties claimed for man. In: *The Essential Peirce,* vol. 1. Bloomington: Indiana University Press; Hull, D.L. (1988). *Science as a Process.* Chicago: University of Chicago Press.
143. White, M. (1963). *Toward Reunion in Philosophy.* Cambridge, MA: Harvard University Press; White, M. (1973). *Pragmatism and the American Mind.* Oxford: Oxford University Press; Morris, C. (1970). *The Pragmatic Movement in American Philosophy.* NY: George Braziller; Smith, J.E. (1985). Experience in Peirce, James and Dewey. *Monist,* 68: 538–554; Smith, J.E. (1970). *Themes in American Philosophy.* New York: Harper and Row; Smith, J.E. (1978) *Purpose and Thought.* New Haven: Yale University Press.
144. Mead, G.H. (1934, 1972). *Mind, Self and Society.* Chicago: University of Chicago Press; Putnam, H. (2000). *The Collapse of the Fact/Value Distinction.* Cambridge, MA: Harvard University Press; Putnam, H. (1995). *Pragmatism: An Open Question.* Malden, MA: Blackwell; Weissman, D. (2000). *A Social Ontology.* New Haven: Yale University Press; Neville, R.C. (1974). *The Cosmology of Freedom.* New Haven: Yale University Press.
145. Good overviews of the different pragmatist traditions can be found in Mounce, H.O. (1997). *The Two Pragmatisms.* London: Routledge; Hollinger, R., and Depew, D. (1999, editors). *Pragmatism: From Progressivism to Post Modernism.* Santa Barbara, CA: Praeger. See also Shook, J.R. (2000). *Dewey's Empirical Theory of Knowledge and Reality.* Nashville: Vanderbilt University Press; Shook, J.R. (2003, editor). *Pragmatic Naturalism and Realism.* Amherst, NY: Prometheus Press; Shook, J.R. (1998). *Pragmatism: An Annotated Bibliography.* Amsterdam: Rodopi; Shook, J.R. (2000–2006, editor). *Pragmatism in American Thought* series. Bristol, UK: Thoemmes; Shook, J.R., and Margolis, J. (2006). *A Companion to Pragmatism.* Oxford: Blackwell Publishing; Stuhr, J. (1997). *Geneological Pragmatism: Philosophy, Experience, and Community.* Albany: SUNY Press; Stuhr, J. (2002). *Pragmatism, Postmodernism, and the Future of Philosophy.* New York: Routledge; Stuhr, J. (2003). *Experience and Criticism: John Dewey's Reconstruction in Philosophy.* Nashville: Vanderbilt University Press; Stuhr, J. (1987). *Classical American Philosophy.* Oxford: Oxford University Press; Bernstein, R.J. (2010). *The Pragmatic Turn.* Malden, MA: Polity Press.
146. Schulkin, J. (2009). *Cognitive Adaptation: A Pragmatist Perspective.* Cambridge: Cambridge University Press.

147. Moreno, J.D. (1995). *Deciding Together.* Oxford: Oxford University Press; Moreno, J.D. (1999, 2003). Bioethics is a naturalism. In: *Pragmatic Bioethics* (edited by G. McGee). Cambridge, MA: MIT Press; Moreno, J.D. (2005). *Is There an Ethicist in the House? On the Cutting Edge of Bioethics.* Bloomington: Indiana University Press; Edel, A. (1944, 1949). Naturalism and ethical theory. In: *Naturalism and the Human Spirit* (edited by Y.H. Krikorian). New York: Columbia University Press; Edel, A. (1955, 1965). *Ethical Judgment: The Use of Science in Ethics.* New York: Free Press; Edel, A. (2001). *Ethical Theory and Social Change: The Evolution of John Dewey's Ethics, 1908–1932.* New Brunswick, NJ: Transaction Publishers.
148. Hickman, L.A. (1980, 1992). *John Dewey's Pragmatic Technology.* Bloomington: Indiana University Press: Hickman, L.A. (2002). *Philosophical Tools for Technological Culture.* Bloomington: Indiana University Press; Sleeper, R.W. (1986). *The Necessity of Pragmatism: John Dewey's Conception of Philosophy.* New Haven: Yale University Press.

2 C.S. Peirce: A Warranted Perspective on the Culture of Inquiry

1. Cohen, I.B. (1980, editor). *Benjamin Peirce: Father of Pure Mathematics in America.* New York: Arno Press: 89.
2. Ibid.
3. Brent, J. (1993). *Charles Sanders Peirce.* Bloomington: Indiana University Press.
4. Fisch, M.H. (1986). Peirce and Leibniz. In: *Peirce, Semiotic and Pragmatism* (edited by K.L. Ketner and J.W. Kloesel). Bloomington: Indiana University Press.
5. Pearl, R. (1925). Vital Statistics of the National Academy of Sciences II: Elections of Young Men. *Proceedings of the National Academy, vol. II*: 757–760.
6. Potter, M. (2000). *Reason's Nearest Kin.* Oxford: Oxford University Press.
7. See Volume IV, "The Simplest Mathematics," of Peirce, C.S. (1932, 1933). *Collected Papers of Charles Sanders Peirce* (edited by C. Hartshorne and P. Weiss). Cambridge, MA: Harvard University Press.
8. Eisele, C. (1957). The scientist-philosopher C.S. Peirce at the Smithsonian. *Journal of the History of Ideas,* 18: 537–547; Eisele, C. (1979). *Studies in the Scientific Philosophy of C.S. Peirce.* Paris: Mouton Publ.; Peirce, C.S. (2000). *Writings of C.S. Peirce,* vols 1–6. Bloomington: Indiana University Press.
9. Peirce, C.S. (1878). Deduction, induction and hypothesis. *Popular Science Monthly,* 13: 470–482.
10. Hanson, N.R. (1958, 1972). *Patterns of Discovery.* Cambridge: Cambridge University Press; Hanson, N.R. (1971). *Observation and Explanation.* New York: Harper Press; Misak, C.J. (1991). *Truth and the End of Inquiry: A Peirceian Account of Truth.* Oxford: Clarendon Press; Misak, C. (2004). *The Cambridge Companion to Peirce.* Cambridge: Cambridge University Press; Anderson, D.R. (1987). *Creativity and the Philosophy of C.S. Peirce.* New York: Kleuwer.
11. Peirce, C.S. (1878). Deduction, induction and hypothesis. *Popular Science Monthly,* 13: 470–482; Houser, N., Eller, J.R., Lewis, A.C., De Tienne, A.,

Clark, C.L., and Bront Davis, D. (1998). *The Essential Peirce, Volume 2 (1893–1913)*. Bloomington: Indiana University Press; Houser, N., Roberts, D.D. and Van Evra, J. (eds) (1997). *Studies in the Logic of Charles Sanders Peirce*. Indianapolis: University of Indiana Press: 107.

12. Hanson, N.R. (1958, 1972). *Patterns of Discovery*. Cambridge: Cambridge University Press; Anderson, D.R. (1995). *Science of Discovery*. West Lafayette: Purdue University Press; Heelan P.A., and Schulkin J. (1998). Hermeneutical philosophy and pragmatism: a philosophy of the science. *Synthese*, 115: 269–302.
13. Dewey, J. (1910). *How We Think*. New York: DC Heath and Publ; Dewey, J. (1929, 1960). *The Quest for Certainty*. New York: Capricorn Books; Dewey, J. (1925, 1989). *Experience and Nature*. La Salle, IL: Open Court; Gigerenzer, G. (2000). *Adaptive Thinking*. Oxford: Oxford University Press.
14. Peirce, C.S. (1878). Deduction, induction and hypothesis. *Popular Science Monthly*, 13: 470–482; Hanson, N.R. (1958, 1972). *Patterns of Discovery*. Cambridge: Cambridge University Press; Hanson, N.R. (1971). *Observation and Explanation*. New York: Harper Press.
15. Dennett, D. (1987). *The Intentional Stance*. Cambridge, MA: MIT Press.
16. Hanson, N.R. (1958, 1972). *Patterns of Discovery*. Cambridge: Cambridge University Press; Anderson, D.R. (1995). *Science of Discovery*. West Lafayette: Purdue University Press; Heelan, P.A., and Schulkin, J. (1998). Hermeneutical philosophy and pragmatism: A philosophy of the science. *Synthese*, 115: 269–302.
17. Turing, A. (2004). *The Essential Turing*. Oxford: Clarendon Press.
18. Peirce, C.S. (1893, 1992). The fixation of belief. In: *The Essential Peirce*, vol. I. (edited by N. Houser and C. Kloesel). Bloomington: Indiana University Press: 115.
19. Buchler, J. (1939, 1966). *Charles Peirce's Empiricism*. New York: Octagon Books; Hanna, R. (2001). *Kant and the Foundations of Analytic Philosophy*. Oxford: Oxford University Press; Hookway, C. (2000). *Truth, Rationality, and Pragmatism: Themes from Peirce*. Oxford: Clarendon Press; Rosenthal, S.B. (2007). *C.I. Lewis in Focus: The Pulse of Pragmatism*. Bloomington: Indiana University Press. Anderson, D.R. (1995). *Strands of System: The Philosophy of Charles Peirce*. West Lafayette: Purdue University Press.
20. Peirce C.S. *Collected Papers* (1932, 1933). Vols 1–4 (edited by C. Hartshorne and P. Weiss), Volumes VII and VIII (edited by A.W. Burks). Cambridge, MA: Harvard University Press.
21. Cohen, M.R. (1923). *Chance, Logic and Love*. New York: Harcourt.
22. Hollinger, D.A. (1975). *Morris Cohen and the Scientific Ideal*. Cambridge, MA: MIT Press.
23. For more on triads, see Esposito, J.L. (1980). *Evolutionary Metaphysics: The Development of Peirce's Theory of Categories*. Athens, OH: Ohio State University Press; Hausman, C. (1993). *The Evolutionary Philosophy of Charles S. Peirce*. Cambridge: Cambridge University Press; Colapietro, V. (1996, editor). *Peirce's Philosophical Perpectives*. New York: Fordham University Press.
24. Peirce, C.S. (1887, 1988). The triad in metaphysics. In: *The Writings of C.S. Peirce*, vol. 6. Bloomington: Indiana University Press: 181.
25. Wittgenstein, L. (1953, 1968). *Philosophical Investigations*. New York: Macmillan Publishing; Goodman, R.B. (1998). Wittgenstein and pragmatism.

Paralas, 4: 91–105; Hildebrand, D.L. (1996). Genuine doubt and the community in Peirce's theory of inquiry. *Southwest Philosophy Review*, 12: 33–43; Hildebrand, D.L. (2000). Putnam, pragmatism and Dewey. *Transactions of CS Peirce Society*, 36: 109–123.

26. Hookway, C. (1984). Naturalism, fallibilism, and evolutionary epistemology. In: *Minds, Machines, and Evolution* (edited by C. Hookway). Cambridge: Cambridge University Press; Hookway, C. (2000). *Truth, Rationality, and Pragmatism: Themes from Peirce*. Oxford: Clarendon Press; Haack, S. (1998). *Manifesto of a Passionate Moderate*. Chicago: University of Chicago Press; Fisch, M.H. (1986). Peirce and Leibniz. In: *Peirce, Semiotic and Pragmatism* (edited by K.L. Ketner and J.W. Kloesel). Bloomington: Indiana University Press.
27. Hanson, N.R. (1958, 1972). *Patterns of Discovery*. Cambridge: Cambridge University Press; Hanson, N.R. (1971). *Observation and Explanation*. New York: Harper Press.
28. Whitehead, A.N. (1929, 1978). *Process and Reality*. New York: Free Press; Whitehead, A.N. (1927, 1953) *Symbolism*. New York: Macmillan Company.
29. Quine, W.V.O. (1969). Epistemology naturalized. In: *Ontological Relativity and Other Essays*. New York: Columbia University Press; Quine, W.V.O. (1974). *The Roots of Reference*. La Salle, IL: Open Court; Smith, J.E. (1985). Experience in Peirce, James and Dewey. *Monist*, 68: 538–554; Smith, J.E. (1970). *Themes in American Philosophy*. New York: Harper and Row; Smith, J.E. (1978) *Purpose and Thought*. New Haven: Yale University Press.
30. Peirce, C.S. (1866). On a method for search for categories. In: *The Essential Peirce*, vol. 1. Bloomington: Indiana University Press.
31. Peirce, C.S. (1878). Deduction, induction and hypothesis. *Popular Science Monthly*, 13: 470–482; Peirce, C.S. (1883). *Studies in Logic, by Members of the Johns Hopkins University*. Boston: Little Brown; Peirce, C.S. (1892). The architecture of theories. *The Monist*, 1: 61–76; Hacking, I. (1999). *The Taming of Chance*. Cambridge: Cambridge University Press; Gigerenzer, G. (2000). *Adaptive Thinking*. Oxford: Oxford University Press.
32. Weiss, P. (1965). Charles S. Peirce: philosopher. In: *Perspectives on Peirce* (edited by R.J. Bernstein). New Haven: Yale University Press; Fisch, M.H. (1986). Peirce and Leibniz. In: *Peirce, Semiotic and Pragmatism* (edited by K.L. Ketner and J.W. Kloesel). Bloomington: Indiana University Press; Brent, J. (1993). *Charles Sanders Peirce*. Bloomington: Indiana University Press.
33. Eisele, C. (1957). The scientist-philosopher C.S. Peirce at the Smithsonian. *Journal of the History of Ideas*, 18: 537–547.
34. Peirce, C.S. (1883). Determinations of gravity at Allegheny, Edensburgh and York, PA in 1879 and 1890. In: *Writings of C.S. Peirce*, vol. 5. Bloomington: Indiana University Press: 9.
35. Fisch, M.A. (1986). Peirce and Leibniz. In *Peirce, Semiotic and Pragmatism* (edited by K.L. Ketner and J.W. Kloesel). Bloomington: Indiana University Press.
36. See Nathan Houser's beautiful introduction to vol. 8 (2009). *Writings of C.S. Peirce*. Bloomington, IN: Indiana University Press.
37. See Cadwallader, T.C. (1975). Peirce as an experimental psychologist. *Transactions of the Charles S. Peirce Society*, 11: 167–186.
38. Fisch, M.H. (1986). Peirce at the Johns Hopkins University. In: *Peirce, Semiotic and Pragmatism* (edited by K.L. Ketner and J.W. Kloesel). Bloomington:

Indiana University Press; Fisch, M.H. (1986). Peirce and Leibniz. In: *Peirce, Semiotic and Pragmatism* (edited by K.L. Ketner and J.W. Kloesel). Bloomington: Indiana University Press.
39. Peirce, C.S., and Jastrow, J. (1885). On small differences of sensation. *Memoirs of the National Academy of Sciences*, 3: 75–83; Peirce, C.S. (2000). *Writings of C.S. Peirce*, vol. 6. Bloomington: Indiana University Press.
40. Furomoto, L. (1992). Joining separate spheres – Christine Ladd-Franklin. *American Psychologists*, 47: 175–181.
41. See, for instance, Ladd-Franklin, C. (1887). A method for the experimental determination of the horopter. *The American Journal of Psychology*, 1: 99–111; Ladd-Franklin, C. (1889). On some characteristics of symbolic logic. *The American Journal of Psychology*, 2: 543–567; Ladd-Franklin, C. (1927). Visible radiation from excited nerve fiber: the reddish blue arcs and the reddish blue glow of the retina. *Science*, 66: 239–241.
42. See Peirce, C.S. (2009). *Writings of C.S. Peirce*, vol. 8. Bloomington: Indiana University Press: xcvi.
43. Peirce, C.S., and Jastrow, J. (1885). On small differences of sensation. *Memoirs of the National Academy of Sciences*, 3: 75–83.
44. Fisch, M.H. (1986). Peirce at the Johns Hopkins University. In: *Peirce, Semiotic and Pragmatism* (edited by K.L. Ketner and J.W. Kloesel). Bloomington: Indiana University Press; Fisch, M.H. (1986). Peirce and Leibniz. In *Peirce, Semiotic and Pragmatism* (edited by K.L. Ketner and J.W. Kloesel). Bloomington: Indiana University Press.
45. Peirce, C.S. (1883). *Studies in Logic, by Members of the Johns Hopkins University*. Boston: Little Brown.
46. Peirce, C.S. (1880). Logic, chapter 1. In: *Writings of C.S. Peirce*, vol. 4. Bloomington: Indiana University Press.
47. See Peirce, C.S. (1877). The fixation of belief. *Popular Science Monthly*, 12: 1–15.
48. Fisch, M.A. (1954). Alexander Bain and the geneology of pragmatism. *Journal of the History of Ideas*, 15; and see also Fisch, M.H. (1986). Evolution in American philosophy. In: *Peirce, Semiotic and Pragmatism* (edited by K.L. Ketner and J.W. Kloesel). Bloomington: Indiana University Press.
49. Rozin, P. (1976). The evolution of intelligence and access to the cognitive unconscious. In: *Progress in Psychobiology and Physiological Psychology* (edited by J. Sprague and A.N. Epstein). New York: Academic Press; Rozin, P. (1998). Evolution and development of brains and cultures. In: *Brain and Mind: Evolutionary Perspectives* (edited by M.S. Gazzaniga and J.S. Altman). Strasbourg, France: Human Frontiers Sciences Program.
50. Gallistel, C.R. (1980). *The Organization of Action: A New Synthesis*. Hillsdale, NJ: Lawrence Erlbaum; Schutz, A. (1932, 1967). *The Phenomenology of the Social World* (transl. G. Walsh and F. Lehnert). Evanston, IL: Northwestern University Press.
51. Peirce, C.S. (1868, 1992). Questions concerning certain faculties claimed for man. In: *The Essential Peirce*, vol. 1. Bloomington: Indiana University Press.
52. Parrott, G.W., and Schulkin, J. (1993). Neuropsychology and the cognitive nature of emotions. *Cognition and Emotion*, 7: 43–59; Dewey, J. (1896). The reflex arc concept in psychology. *Psychological Review*, 3: 357–370.
53. Brent, J. (1993). *Charles Sanders Peirce*. Bloomington: Indiana University Press.

54. Weiss, P. (1965). Charles S. Peirce: philosopher. In: *Perspectives on Peirce* (edited by R.J. Bernstein). New Haven: Yale University Press.
55. Brent, J. (1993). *Charles Sanders Peirce*. Bloomington: Indiana University Press; Perry, R.B. (1935). *The Thought and Character of William James*. Boston: Little, Brown and Co.; Perry, R.B. (1938, 1958). *In the Spirit of William James*. Bloomington: Indiana Univ. Press.
56. Brent, J. (1993). *Charles Sanders Peirce*. Bloomington: Indiana University Press.
57. Fisch, M.H. (1986). Peirce at the Johns Hopkins University. In: *Peirce, Semiotic and Pragmatism* (edited by K.L. Ketner and J.W. Kloesel). Bloomington: Indiana University Press; Fisch, M.H. (1986). Peirce and Leibniz. In: *Peirce, Semiotic and Pragmatism* (edited by K.L. Ketner and J.W. Kloesel). Bloomington: Indiana University Press.
58. Peirce, C.S., and Victoria, Lady Welby (1903–1912, 1977). *Semiotic and Significs: The Correspondence between Charles S. Peirce and Victoria, Lady Welby* (edited by C.S. Hardwick). Bloomington: Indiana University Press.
59. Ibid. 2.
60. Ibid. 108.
61. Ibid. 65.
62. Ibid. 21.
63. Clark, A. (1997). *Being There*. Cambridge, MA: MIT Press.
64. See, for instance, Marler, P.R., and Hamilton, W.J. (1966). *Mechanisms of Animal Behavior.* New York: Wiley and Sons; Smith, W.J. (1977). *The Behavior of Communicating: An Ethological Approach*. Cambridge, MA: Harvard University Press; Houser, N., and Kloesel, C. (1992, 1998) *The Essential Peirce, Volume 1 (1867–1893)*. Bloomington: Indiana University Press; Houser, N., Eller, J.R., Lewis, A.C., De Tienne, A., Clark, C.L., and Bront Davis, D. (1998). *The Essential Peirce, Volume 2 (1893–1913)*. Bloomington: Indiana University Press.
65. Smith, W.J. (1977). *The Behavior of Communicating: An Ethological Approach*. Cambridge, MA: Harvard University Press.
66. Ibid.
67. Rozin, P. (1976). The Evolution of intelligence and access to the cognitive unconscious. In: *Progress in Psychobiology and Physiological Psychology* (edited by J. Sprague and A.N. Epstein). New York: Academic Press; Rozin, P. (1998). Evolution and development of brains and cultures. In: *Brain and Mind: Evolutionary Perspectives* (edited by M.S. Gazzaniga and J.S. Altman). Strasbourg, France: Human Frontiers Sciences Program; Gigerenzer, G. (2000). *Adaptive Thinking*. Oxford: Oxford University Press.
68. Brent, J. (1993). *Charles Sanders Peirce*. Bloomington: Indiana University Press.
69. Misak, C.J. (1991). *Truth and the End of Inquiry: A Peirceian Account of Truth*. Oxford: Clarendon Press; Misak, C. (2004). *The Cambridge Companion to Peirce*. Cambridge: Cambridge University Press; Smith, J.E. (1985). Experience in Peirce, James and Dewey. *Monist*, 68: 538–554; Smith, J.E. (1970). *Themes in American Philosophy.* New York: Harper and Row; Smith, J.E. (1978). *Purpose and Thought*. New Haven: Yale University Press.
70. Menand, L. (2001). *The Metaphysical Club*. New York: Farrar, Straus and Giroux; Menand, L. (2002). *American Studies*. New York: Farrar, Straus and Giroux.

71. Ketner, K.L. (1998). *His Glassy Essence: An Autobiography of C.S. Peirce.* Nashville: Vanderbilt University Press.
72. Peirce, C.S. (1892). The architecture of theories. *The Monist,* 1: 61–76.
73. See *The Essential Peirce,* vol. 1 (edited by N. Houser and C. Kloesel). Bloomington: Indiana University Press.
74. For example, see Todes, D.P. (1989). *Darwin without Malthus.* Oxford: Oxford University Press; Shapin, S. (2008). *The Scientific Life.* Chicago: University of Chicago Press; Gliboff, S. (2008). *H.G. Bronn, Ernst Haeckel, and the Origins of German Darwinism.* Cambridge, MA: MIT Press; Shapin, S. (1995). *A Social History of Truth.* Chicago: University of Chicago Press; Shapin, S. (1996). *The Scientific Revolution.* Chicago: University of Chicago Press; Shapin, S., and Schaffer, S. (1985). *Leviathan and the Air-Pump.* Princeton: Princeton University Press; Dear, P. (1995). *Discipline and Experience.* Chicago: University of Chicago Press; Dear, P. (2006). *The Intelligibility of Nature.* Chicago: University of Chicago Press; Harrington, A. (1987). *Medicine, Mind and the Double Brain.* Princeton: Princeton University Press; Collingwood, R.G. (1945, 1976). *The Idea of Nature.* Oxford: Oxford University Press.
75. Examples of a Peirceian orientation may be found in Clark, A. (1997). *Being There.* Cambridge, MA: MIT Press; Donald, M. (2004). Hominid encultura- tion and cognitive evolution. In: *The Development of the Mediated Mind* (edited by J.M. Luraciello et al.). Mawash, NJ: Erlbaum Press; Noe, A. (2004). *Action in Perception.* Cambridge, MA: MIT Press; Wheeler, M., and Clark, A. (2008). Culture, embodiment, and genes: unravelling the triple helix. *Philosophical Transactions of the Royal Society B,* 373: 3563–3575; and Wheeler, M., and Clark, A. (1999). Genic representation: reconciling content and causal complexity. *British Journal for the Philosophy of Science,* 50: 103–135.
76. Hanson, N.R. (1958, 1972). *Patterns of Discovery.* Cambridge: Cambridge University Press; Heelan, P.A., and Schulkin, J. (1998). Hermeneutical philosophy and pragmatism: A philosophy of the science. *Synthese,* 115:269–302.
77. Descartes, R. (1637, 1993). *Discourse on Method.* Indianapolis: Hackett Publishing Co.; Descartes, R. (1649, 1989). *The Passions of the Soul* (translated by S. Voss). Cambridge, MA: Hackett Publ. Co.
78. Rozin, P. (1976). The evolution of intelligence and access to the cognitive unconscious. In: *Progress in Psychobiology and Physiological Psychology* (edited by J. Sprague and A.N. Epstein). New York: Academic Press; Rozin, P. (1998). Evolution and development of brains and cultures. In: *Brain and Mind: Evolutionary Perspectives* (edited by M.S. Gazzaniga and J.S. Altman). Strasbourg, France: Human Frontiers Sciences Program.
79. Gigerenzer, G. (2000). *Adaptive Thinking.* Oxford: Oxford University Press; Gigerenzer, G. (2007). *Gut Feelings.* New York: Viking Press; Gigerenzer, G. (2003). Why does framing influence judgment? *Journal of General Internal Medicine,* 18: 906–961.
80. Hacking, I. (1964). *Logic of Statistical Inference.* Cambridge: Cambridge University Press; Hacking, I. (1975). *The Emergence of Probability.* Cambridge: Cambridge University Press; Hacking, I. (1999). *The Taming of Chance.* Cambridge: Cambridge University Press.
81. Gigerenzer, G., Gaismaiser W., Milcke, E.K., Schwartz, L.M., and Woloshin, S. (2008). Helping doctors and patients make sense of health statistics. *Psychological Science in the Public Interest,* 8: 53–96.

82. Hobart, M.E., and Schiffman, Z.S. (1998). *Information Ages*. Baltimore: Johns Hopkins University Press.
83. Gibson, J.J. (1966). *The Senses Considered as Perceptual Systems*. New York: Houghton-Mifflin; Gigerenzer, G. (2000). *Adaptive Thinking*. Oxford: Oxford University Press.
84. Goldstein, D.G., and Gigerenzer, G. (2002). Models of ecological rationality: the recognition heuristic. *Psychological Review*, 109: 75–90.
85. Leslie, A., Gelman, R., and Gallistel, C.R. (2008). The generative basis of natural number concepts. *Trends in Cognitive Science*, 12: 213–218; Carey, S. (2009). *On the Origins of Concepts*. Oxford: Oxford University Press; Peirce, C.S. (1899, 1992). *Reasoning and the Logic of Things* (edited by K.L. Ketner and H. Putnam). Cambridge, MA: Harvard University Press; Peirce, C.S. (2000). *Writings of C.S. Peirce*, vols 1–6. Bloomington: Indiana University Press; Peirce, C.S. (2009). *Writings of C.S. Peirce*, vol. 8. Bloomington: Indiana University Press.
86. Lakoff, G., and Núñez, R.E. (2000). *Where Mathematics Comes From*. New York: Basic Books; Dehaene, S. (2007). *The Number Sense*. Oxford: Oxford University Press.
87. Lakoff, G., and Núñez, R.E. (2000). *Where Mathematics Comes From*. New York: Basic Books: 352.
88. Whitehead, A.N. (1919, 1982). *An Enquiry Concerning the Principles of Natural Knowledge*. New York: Dover Press.
89. Kline, M. (1959). *Mathematics and the Physical World*. New York: Dover Press; Izard, V., Sann, C., Spelke, E.S., and Steri, A. (2009). Newborn infants perceive abstract numbers. *Proceedings of the National Academy of Sciences*, 106: 10382–10385.
90. Plato. (1985). *Meno*. Oxford: Aris and Phillips.
91. Lakoff, G., and Núñez, R.E. (2000). *Where Mathematics Comes From*. New York: Basic Books: 5.
92. Gallistel, C.R. (1993). *The Organization of Learning*. Cambridge, MA: MIT Press; Gallistel, C.R., Gelman, R.M., and Cordes, I.S. (2006). The Cultural and Evolutionary History of the Real Numbers. In: *Culture and Evolution* (edited by S. Levinson and P. Jaisson). Cambridge, MA: MIT Press.
93. Clark, A. (1997). *Being There*. Cambridge, MA: MIT Press.
94. Deheane, S., Izard, V., Pica, P., and Spelke, E. (2006). Core knowledge of geometry in an Amazonian Indigene group. *Science*, 311: 381–384.
95. Desmurget, M., Reilly, K.T., Richard, N., Szathmari, A., Mottolese, C., and Sirigu, A. (2009). Movement intention after parietal cortex stimulation in humans. *Science*, 324: 811–813.
96. Knops, A., Thirion, B., Hubbard, E.M, Michel, V., and Dehaene, S. (2009). Recruitment of an area involved in eye movements during mental arithmetic. *Science*, 324: 1583–1585; Piazza, M., and Izard, V. (2009). How humans count. *The Neuroscientist*, 15: 261–273.
97. For more on this, see: Heidegger, M. (1927, 1962). *Being and Time*. New York: Harper and Row; Heidegger, M. (1962, 1977). *The Question Concerning Technology*. New York: Harper and Row; Heidegger, M. (1934–1935, 1942, 1996). *Holderlin's Hymn "The Ister."* Bloomington: Indiana University Press; Heelan, P.A. (1983). *Space Perception and the Philosophy of Science*. Berkeley: University of California Press; Heelan, P.A. (2001). The Lifeworld and

Scientific Interpretation. In: *Handbook of Phenomenology and Medicine*. Waco, TX: Baylor University.

98. Lakoff, G., and Johnson, M. (1999). *Philosophy in the Flesh: The Embodied Mind and its Challenge to Western Thought*. New York: Basic Books; Johnson, M. (1987, 1990). *The Body in the Mind*. Chicago: University of Chicago Press; Johnson, M. (2007). *The Meaning of the Body*. Chicago: University of Chicago Press; Pinker, S. (1994). *The Language Instinct*. New York: William Morrow and Co.; Levinson, S. (2006). Cognition at the heart of human interaction. *Discourse Studies*, 8: 85–93; Levinson, S. (2003). *Space in Language and Cognition*. Cambridge: Cambridge University Press.
99. See, for instance, Kantian a pragmatist such as Habermas, J. (1967, 1988). *On the Logic of the Social Sciences* (translated by S.W. Nicholson and J.A. Stark). Cambridge, MA: MIT Press. See also Merton, P.K. (1973). *The Sociology of Science*. Chicago: University of Chicago Press; Weber, M. (1904, 1949). *The Methodology of the Social Sciences* (translated by E. Shils and H. Finch). New York: Free Press; Weber, M. (1905, 1958). *The Protestant Ethic and the Spirit of Capitalism* (translated by T. Parkson). New York: Scribners; Durkheim, E. (1974). *Sociology and Philosophy* (translated by D.F. Pocock). New York: Free Press.
100. Atran, S. (1990, 1996). *Cognitive Foundations of Natural History*. New York: Cambridge University Press.
101. Sabini, J., and Schulkin, J. (1994). Biological realism and social constructivism. *Journal for the Theory of Social Behavior*, 224: 207–217; Sehon, S. (2005). *Teleological Realism: Mind, Agency and Explanation*. Cambridge, MA: MIT Press; Kukla, A. (1998). *Studies in Scientific Realism*. Oxford: Oxford University Press.
102. Peirce, C.S. (2000). *Writings of C.S. Peirce*, vols 1–6. Bloomington: Indiana University Press; Peirce, C.S. (2009). *Writings of C.S. Peirce*, vol. 8. Bloomington: Indiana University Press; Wittgenstein, L. (1953, 1968). *Philosophical Investigations*. New York: Macmillan Publishing; Heelan, P.A., and Schulkin, J. (1998). Hermeneutical philosophy and pragmatism: A philosophy of the science. *Synthese*, 115: 269–302; Hanson, N.R. (1958, 1972). *Patterns of Discovery*. Cambridge: Cambridge University Press; Hanson, N.R. (1971). *Observation and Explanation*. New York: Harper Press.
103. Diderot, D. (1755, 1964). The encyclopedia. In: *Rameau's Nephew and Other Works*. New York: The Library of Liberal Arts; Kagan, J. (2009). *The Three Cultures: Natural Sciences, Social Sciences and the Humanities in the 21st Century*. Cambridge: Cambridge University Press; Snow, C.P. (1959, 1998). *The Two Cultures*. Cambridge: Cambridge University Press.
104. Diderot, D. (1755, 1964). The encyclopedia. In: *Rameau's Nephew and Other Works*. New York: The Library of Liberal Arts.
105. Friedman, M. (1998). *The Consequences of Logical Positivism*. Cambridge: Cambridge University Press; Friedman, M. (2001). *Dynamics of Reason*. Stanford: Stanford University Press.
106. Aristotle (1962). *The Politics* (translated by T. Sinclair). New York: Penguin Books; Aristotle (1968). *De Anima* (translated by D. Hamylin). Oxford: Oxford University Press.
107. Snow, C.P. (1961). *Science and Government*. Cambridge, MA: Harvard University Press; Kagan, J. (2009). *The Three Cultures: Natural Sciences, Social*

Sciences and the Humanities in the 21st Century. Cambridge: Cambridge University Press; Schulkin, J. (2009). *Cognitive Adaptation: A Pragmatist Perspective*. Cambridge: Cambridge University Press.

108. Gadamer, H.G. (1981, 1986). *Reason in the Age of Science* (translated by F.G. Lawrence). Cambridge: MIT Press; Sullivan, W.M. (1986). *Reconstructing Public Philosophy*. Berkeley: University of California Press.
109. Humboldt, A. Von (1859, 1993). *Cosmos*. New York: George Bell and Sons; Schiller, F.C.S. (1966). *Humanistic Pragmatism: The Philosophy of F.C.S. Schiller* (edited by R. Abel). New York: Free Press; Schiller, F. (1795, 1980). *On the Aesthetic Education of Man*. New York: Frederick Ungar, Publ.; Zammito, J.H. (2002). *Kant, Herder and the Birth of Anthropology*. Chicago: University of Chicago Press; Beiser, F.C. (2002). *German Idealism*. Cambridge: Harvard University Press.
110. Hull, D.L. (1988). *Science as a Process*. Chicago: University of Chicago Press.
111. Hacking, I. (1975). *The Emergence of Probability*. Cambridge: Cambridge University Press; Hacking, I. (1999). *The Taming of Chance*. Cambridge: Cambridge University Press.
112. Friedman, M. (2001). *Dynamics of Reason*. Stanford: Stanford University Press.
113. Hull, D.L. (1988). *Science as a Process*. Chicago: University of Chicago Press; Kitcher, P. (1993). *The Advancement of Science*. Oxford: Oxford University Press; Rosenberg, A. (1985). *The Structure of Biological Science*. Cambridge: Cambridge University Press; Rosenberg, A. (1994). *Instrumental Biology or the Disunity of Science*. Chicago: University of Chicago Press.
114. Dewey, J. (1916). *Essays in Experimental Logic*. Chicago: University of Chicago Press.
115. Peirce, C.S. (1878). Deduction, induction and hypothesis. *Popular Science Monthly*, 13: 470–482; Hanson, N.R. (1958, 1972). *Patterns of Discovery*. Cambridge: Cambridge University Press; Hanson, N.R. (1971). *Observation and Explanation*. New York: Harper Press; Heelan, P.A., and Schulkin, J. (1998). Hermeneutical philosophy and pragmatism: A philosophy of science. *Synthese*, 115: 269–302.
116. Hacking, I. (1964). *Logic of Statistical Inference*. Cambridge: Cambridge University Press; Hacking, I. (1975). *The Emergence of Probability*. Cambridge: Cambridge University Press; Hacking, I. (1999). *The Taming of Chance*. Cambridge: Cambridge University Press.
117. See, in particular, a beautiful example by Kitcher, P. (1993). *The Advancement of Science*. Oxford: Oxford University Press. Also see: Laudan, L. (1977). *Progress and Its Problems*. Berkeley: University of California Press; Laudan, L. (1997, 2003). Progress or rationality? The prospects for normative naturalism. In: *The Philosophy of Science* (edited by D. Papineau). Oxford: Oxford University Press.
118. Galison, P. (1988). History, philosophy and the central metaphor. *Science in Context*, 3: 197–212; Shapin, S. (2008). *The Scientific Life*. Chicago: University of Chicago Press; Shapin, S. (1995). *A Social History of Truth*. Chicago: University of Chicago Press; Shapin, S. (1996). *The Scientific Revolution*. Chicago: University of Chicago Press; Dear, P. (1995). *Discipline and Experience*. Chicago: University of Chicago Press; Dear, P. (2006). *The Intelligibility of Nature*. Chicago: University of Chicago Press.

119. Kahneman, D., Slovic, P., and Tversky, A. (eds) (1982). *Judgment under Uncertainty: Heuristics and Biases*. New York: Cambridge University Press; Gigerenzer, G. (2000). *Adaptive Thinking*. Oxford: Oxford University Press; Gigerenzer, G. (2007). *Gut Feelings*. New York: Viking Press; Gigerenzer, G. (2003). Why does framing influence judgment? *Journal of General Internal Medicine*, 18: 906–961.

3 Evolutionary Origins: Oriented to Kinds

1. Peirce, C.S. (1899, 1992). *Reasoning and the Logic of Things*. Cambridge, MA: Harvard University Press; Carnap, R. (1928, 1969). *The Logical Structure of the World*. Berkeley: University of California Press; Hanson, N.R. (1958, 1972). *Patterns of Discovery*. Cambridge: Cambridge University Press; Hanson, N.R. (1971). *Observation and Explanation*. New York: Harper Press.
2. Dewey, J. (1896). The reflex arc concept in psychology. *Psychological Review*, 3: 357–70; Quine, W.V.O. (1969). Epistemology naturalized. In: *Ontological Relativity and Other Essays*. New York: Columbia University Press; Quine, W.V.O. (1974). *The Roots of Reference*. La Salle, IL: Open Court; Reid, T. (1785, 1969). *Essays on the Intellectual Powers of Man*. Cambridge, MA: MIT Press.
3. Quine, W.V.O. (1969). Natural kinds. In: *Ontological Relativity and Other Essays*. New York: Columbia University Press.
4. Quine, W.V.O. (1974). *The Roots of Reference*. La Salle, IL: Open Court.
5. Farber, P.L. (2000). *Finding Order in Nature*. Baltimore: Johns Hopkins University Press; Grant, E. (2007). *A History of Natural Philosophy*. Cambridge: Cambridge University Press.
6. Keil, F.C. (1989). *Concepts, Kinds and Cognitive Development*. Cambridge, MA: MIT Press; Keil, F. (1979). *Semantic and Conceptual Development: An Ontological Perspective*. Cambridge, MA: Harvard University Press; Carey, S. (1985, 1987). *Conceptual Change in Childhood*. Cambridge, MA: MIT Press; Carey, S. (2004). Bootstrapping and the origins of concepts. *Daedalus*, Winter 133: 59–68; Carey, S. (2009). *On the Origins of Concepts*. Oxford: Oxford University Press; Atran, A., Medin, D.L., and Ross, N.O. (2005). The cultural mind. *Psychological Review*, 112: 744–776.
7. Keil, F.C. (1989). *Concepts, Kinds and Cognitive Development*. Cambridge, MA: MIT Press; Keil, F. (1979). *Semantic and Conceptual Development: An Ontological Perspective*. Cambridge, MA: Harvard University Press; Carey, S. (1985, 1987). *Conceptual Change in Childhood*. Cambridge, MA: MIT Press; Carey, S. (2004). Bootstrapping and the origins of concepts. *Daedalus*, Winter 133: 59–68; Carey, S. (2009). *On the Origins of Concepts*. Oxford: Oxford University Press.
8. Shook, J.R. (2003, editor). *Pragmatic Naturalism and Realism*. Amherst, NY: Prometheus.
9. Mayr, E. (1942, 1982). *Systemics and the Origin of Species*. New York: Columbia University Press; Mayr, E. (1963). *Animal Species and Evolution*. Cambridge, MA: Harvard University Press; Rozin, P. (1976). The evolution of intelligence and access to the cognitive unconscious. In: *Progress in Psychobiology and Physiological Psychology* (edited by J. Sprague and A.N. Epstein). New York: Academic Press.

10. Lyell, C. (1830, 2010). *Principles of Geology*. New York: Nabu Press.
11. Malthus, T.R. (1798, 1970). *An Essay on the Principle of Population*. Baltimore: Penguin Books.
12. Dobzhansky, T.C. (1962). *Mankind Evolving*. New Haven: Yale University Press; Simpson, G.G. (1949). *The Meaning of Evolution*. New Haven: Yale University Press.
13. Mellars, P. (2006). Why did modern human populations disperse from Africa ca. 60,000 years ago? *Proceedings of the National Acacdemy of Science*, 103: 9381–9386; Boyd, R., and Richerson, P. (2005). *Not by Genes Alone: How Culture Transformed Human Evolution*. Chicago: University of Chicago Press.
14. Lahr, M.M. and Foley, R. (2004). Human evolution writ small. *Nature*, 431: 1043–1044.
15. Darwin, C. (1868). *The Variation of Animals and Plants under Domestication*, 2 vols. London: John Murray.
16. Hauser, M.D., Chomsky, N. and Tecumseh Fitch, W. (2002). The faculty of language: what is it, who has it, and how did it evolve? *Science*, 298: 1569–1576.
17. Corballis, M.C. (2002). *From Hand to Mouth*. Princeton: Princeton University Press; Dunbar, R.I.M. (1996). *Grooming, Gossip and the Evolution of Language*. Cambridge, MA: Harvard University Press.
18. Mithen, S. (1996). *The Prehistory of the Mind: The Cognitive Origins of Art and Science*. London: Thames and Hudson, Ltd.
19. Ibid.
20. Mithen, S. (1996). *The Prehistory of the Mind. The Cognitive Origins of Art and Science*. London: Thames and Hudson, Ltd; and see also Mellars, P. (2006). Why did modern human populations disperse from Africa ca. 60,000 years ago? *Proceedings of the National Academy of Science*, 103: 9381–9386.
21. Lahr, M.M., and Foley, R. (1998). Towards a theory of modern human origins: Geography, demography, and diversity in recent human evolution. *Yearbook of Physical Anthropology*, 41: 137–176.
22. Grant, E. (2007). *A History of Natural Philosophy*. Cambridge: Cambridge University Press; Collingwood, R.G. (1945, 1976). *The Idea of Nature*. Oxford: Oxford University Press; Schulkin, J. (1996). *The Delicate Balance*. Lanham, MD: University Press of America.
23. Whitehead, A.N. (1938, 1967). *Modes of Thought*. New York: Free Press.
24. Oelschlaeger, M. (1991). *The Idea of the Wilderness*. New Haven: Yale University Press.
25. Emerson, R.W. (1855, 1883). *Nature, Addresses and Lectures*. Cambridge, MA: The Riverside Press; Whitehead, A.N. (1933, 1961). *Adventures of Ideas*. New York: Free Press.
26. Neville, R.C. (2010). *Pragmatism, Metaphysics, Comparison and Realism*, unpublished manuscript. Smith, J.E. (1978). *Purpose and Thought*. New Haven: Yale University Press.
27. Whitehead, A.N. (1929, 1978). *Process and Reality*. New York: Free Press: xiii.
28. Whitehead, A.N. (1938, 1967). *Modes of Thought*. New York: Free Press; Pratt, S.L. (2002). *Native Pragmatism: Rethinking the Roots of American Philosophy*. Bloomington: Indiana University Press.
29. Maclean, I. (2002). *Logic, Signs and Nature in the Renaissance*. Cambridge: Cambridge University Press.

30. Whitehead, A.N. (1927, 1953). *Symbolism*. New York: Macmillan Company: 106. For interesting scholarship with regard to Whitehead on action and intensity to detail focuses on adaptation and real events see, for instance: Neville, R.C. (1992). *The Highroad around Modernism*. Albany: SUNY Press; Jones, J.A. (1998). *Intensity: An Essay in Whiteheadian Ontology*. Nashville: Vanderbilt University Press.
31. Whitehead, A.N. (1927, 1953). *Symbolism*. New York: Macmillan Company: 91.
32. Dewey, J. (1896). The reflex arc concept in psychology. *Psychological Review*, 3: 357–370; Mead, G.H. (1936). *Movements of Thought in the Nineteenth Century* (edited by C.W. Morris). Chicago: University of Chicago Press.
33. Nash, R. (1967). *Wilderness and the American Mind*. New Haven: Yale University Press; Nicholls, S. (2009). *Paradise Found*. Chicago; University of Chicago Press.
34. Thoreau, H.D. (1971). *Great Short Works*. New York: Harper and Row; Worster, D. (1977, 1991). *Nature's Economy*. Cambridge: Cambridge University Press; Midgley, M. (1979, 1995). *Beast and Man*. London: Routledge.
35. Bury, J.B. (1933, 1960). *The Idea of Progress*. New York: Dover Press; Huxley, T.H. (1863). *Man's Place in Nature*. London: Macmillan.
36. Goldsmith, R.B. (1940, 1982). *The Material Basis of Evolution*. New Haven, CT: Yale University Press; Gould, S.J., and Eldridge, N. (1977). Punctuated equilibria: the tempo and mode of evolution reconsidered. *Paleobiology*, 3: 115–151.
37. Humbolt W. Von (1836, 1971). *Linguistic and Intellectual Development*. Philadelphia: University of Pennsylvania Press.
38. Dobzhansky, T.C. (1962). *Mankind Evolving*. New Haven: Yale University Press.
39. Grant, E. (2007). *A History of Natural Philosophy*. Cambridge: Cambridge University Press.
40. Particularly well described in Gibson, K.R., and Ingold, T. (eds) (1993). *Tools, Language and Cognition in Human Evolution*. Cambridge: Cambridge University Press.
41. See, for instance, Donald, M. (1991). *Origins of Modern Man*. Cambridge, MA: Harvard University Press; Rozin, P. (1976). The evolution of intelligence and access to the cognitive unconscious. In: *Progress in Psychobiology and Physiological Psychology* (edited by J. Sprague and A.N. Epstein). New York: Academic Press; Rozin, P. (1998). Evolution and development of brains and cultures. In: *Brain and Mind: Evolutionary Perspectives* (edited by M.S. Gazzaniga and J.S. Altman). Strasbourg, France: Human Frontiers Sciences Program; Mithen, S. (1996). *The Prehistory of the Mind. The Cognitive Origins of Art and Science*. London: Thames and Hudson, Ltd; Hauser, M.D. (1997). *The Evolution of Communication*. Cambridge, MA: MIT Press.
42. Humbolt W. Von (1836, 1971). *Linguistic and Intellectual Development*. Philadelphia: University of Pennsylvania Press; Chomsky, N. (1972). *Language and Mind*. New York: Harcourt Brace Jovanovich; Pinker, S. (1994). *The Language Instinct*. New York: William Morrow and Co.
43. Mellars, P. (2006). Why did modern human populations disperse from Africa ca. 60,000 years ago? *Proceedings of the National Academy of Science*, 103: 9381–9386.

44. Dewey, J. (1925, 1989). *Experience and Nature.* La Salle, IL: Open Court; Sterelny, K. (2000). *The Evolution of Agency and Other Essays.* Cambridge: Cambridge University Press; Sterelny, K. (2003). *Thought in a Hostile World.* New York: Blackwell; Sterelny, K. (2004). Genes, memes and human history. *Mind and Language,* 19: 249–257.
45. Gallistel, C.R. (1993). *The Organization of Learning.* Cambridge, MA: MIT Press.
46. Sabini, J. and Schulkin, J. (1994). Biological realism and social constructivism. *Journal for the Theory of Social Behavior,* 224: 207–217; Moreno, J.D. (1995). *Deciding Together.* Oxford: Oxford University Press.
47. Mead, G.H. (1934, 1972). *Mind, Self and Society.* Chicago: University of Chicago Press.
48. Gallistel, C.R. (1980). *The Organization of Action: A New Synthesis.* Hillsdale, NJ: Lawrence Erlbaum.
49. Jackson, J.H. (1884, 1958). Evolution and dissolution of the nervous system. In: *Collected Works of John Hughlings Jackson,* vol. II (edited by J. Taylor). London: Staples Press; James, W. (1890, 1952). *The Principles of Psychology,* vols 1 and 2. New York: Henry Holt.
50. Schulkin, J. (2004). *Bodily Sensibility: Intelligent Action.* Oxford: Oxford University Press; Schulkin, J. (2007). *Effort: A Behavioral Neuroscience Perspective on the Will.* Mahwah, NJ: Erlbaum Press; Barton, R.A. (2004). Binocularity and brain evolution in primates. *Proceedings of the National Academy of Sciences,* 101: 10113–10115.
51. Diamond, J. (1998). *Guns, Germs and Steel.* New York: Norton; Rozin, P. (1976). The evolution of intelligence and access to the cognitive unconscious. In: *Progress in Psychobiology and Physiological Psychology* (edited by J. Sprague and A.N. Epstein). New York: Academic Press.
52. Mithen, S. (1996). *The Prehistory of the Mind: The Cognitive Origins of Art and Science.* London: Thames and Hudson, Ltd.
53. Foley, R. (2006). The emergence of culture in the context of hominid evolutionary patterns. In: *Evolution and Culture* (edited by S.C. Levinson and P. Jaisson). Cambridge, MA: MIT Press; Gibson, K.R., and Ingold, T. (1993, editors). *Tools, Language and Cognition in Human Evolution.* Cambridge: Cambridge University Press; Mellars, P. (2006). Why did modern human populations disperse from Africa ca. 60,000 years ago? *Proceedings of the National Academy of Sciences,* 103: 9381–9386.
54. Jackson, J.H. (1884, 1958). Evolution and dissolution of the nervous system. In: *Selected Writings of John Hughlings Jackson,* vol. II (edited by J. Taylor). London: Staples Press.
55. Decety, J., and Jackson, P.W. (2006). A social neuroscience perspective on empathy. *Current Directions in Psychological Science,* 15: 54–58; Barton, R.A. (2004). Binocularity and brain evolution in primates. *Proceedings of the National Academy of Sciences,* 101: 10113–10115.
56. Gibson, K.R., and Ingold, T. (eds) (1993). *Tools, Language and Cognition in Human Evolution.* Cambridge: Cambridge University Press.
57. Rizzolatti, G., and Luppino, G. (2001). The cortical motor system. *Neuron,* 31: 889–901.
58. Martin, A. (2007). The representation of object concepts in the brain. *Annual Review of Psychology,* 58: 25–45; Ullman, M.T. (2001). A

neurocognitive perspective on language: the declarative procedural model. *Nature Neuroscience*, 9: 266–286; Lieberman, P. (2000). *Human Language and Our Reptilian Brain*. Cambridge, MA; Harvard University Press.
59. Martin, A. (2007). The representation of object concepts in the brain. *Annual Review of Psychology*, 58: 25–45; Martin, A., and Weisber, J. (2003). Neural foundations for understanding social and mechanical concepts. *Cognitive Neuropsychology*, 20: 575–587; Martin, A., Wiggs, C.L., Ungerleider, L.G., and Haxby, J.V. (1996). Neural correlates of category specific knowledge. *Nature*, 379: 649–652.
60. Gibson, K.R., and Ingold, T. (eds) (1993). *Tools, Language and Cognition in Human Evolution*. Cambridge: Cambridge University Press; Martin, A. (2007). The representation of object concepts in the brain. *Annual Review of Psychology*, 58: 25–45.
61. Barton, R.A. (2004). Binocularity and brain evolution in primates. *Proceedings of the National Academy of Sciences*, 101: 10113–10115; Schulkin, J. (2007). *Effort: A Behavioral Neuroscience Perspective on the Will*. Mahwah, NJ: Erlbaum Press; Reader, S.M., and Laland, K.N. (2002). Social intelligence, innovation, and enhanced brain size in primates. *Proceedings of the National Academy of Sciences*, 99: 4436–4441; Decety, J., and Jackson, P.W. (2006). A social neuroscience perspective on empathy. *Current Directions in Psychological Science*, 15: 54–58; Rizzolatti, G., and Luppino, G. (2001). The cortical motor system. *Neuron*, 31: 889–901.
62. Aristotle. (1968). *De Anima* (translated by D. Hamylin). Oxford: Oxford University Press; Levinson, S. (2003). *Space in Language and Cognition*. Cambridge: Cambridge University Press; Keil, F.C., and Wilson, R.A. (2000). *Explanation and Cognition*. Cambridge, MA: MIT Press; Carey, S. (2009). *On the Origins of Concepts*. Oxford: Oxford University Press.
63. Dewey, J. (1896). The reflex arc concept in psychology. *Psychological Review*, 3: 357–370; Lakoff, G., and Johnson, M. (1999). *Philosophy in the Flesh: The Embodied Mind and its Challenge to Western Thought*. New York: Basic Books; Gallistel, C.R. (1993). *The Organization of Learning*. Cambridge, MA: MIT Press; Schulkin, J. (2000). *Roots of Social Sensibility and Neural Function*. Cambridge, MA: MIT Press.
64. Cassirer, E. (1944, 1978). *An Essay on Man*. New Haven: Yale University Press; Cassirer, E. (1951). *The Philosophy of the Enlightenment*. Princeton: Princeton University Press; Cassirer, E. (1957). *The Philosophy of Symbolic Forms*, vol. III. New Haven: Yale University Press.
65. Keil, F.C. (1989). *Concepts, Kinds and Cognitive Development*. Cambridge, MA: MIT Press; Keil, F.C. (1979). *Semantic and Conceptual Development: An Ontological Perspective*. Cambridge, MA: Harvard University Press.
66. Keil, F.C. (1979). *Semantic and Conceptual Development: An Ontological Perspective*. Cambridge, MA: Harvard University Press.
67. Cosmides, L. and Tooby, J. (1992). Cognitive adaptations for social exchange. In: *The Adapted Mind* (edited by J. Barkow, L. Cosmides, and J. Tooby). New York: Oxford University Press; Atran, S., Medin, D., Ross, N., Lynch, E., Coley, J., Ucan Ek', E., and Vapnarsky, V. (1999). Folkecology and commons management in the Maya Lowlands. *Proceedings of the National Academy of Sciences*, 96: 7598–7603.
68. Gibson, J.J. (1966). *The Senses Considered as Perceptual Systems*. New York: Houghton-Mifflin; Clark, A. (1997). *Being There*. Cambridge, MA: MIT Press.

For drawing links between Gibson and Dewey and for a defense of the logic of inquiry and perception in Dewey, see Burke, T. (1994). *Dewey's New Logic: A Reply to Russell*. Chicago: University of Chicago Press.

69. Atran, S., Medin, D.L., and Ross, N.O. (2005). The cultural mind. *Psychological Review*, 112: 744–776.
70. Aristotle. (1968). *De Anima* (translated by D. Hamylin). Oxford: Oxford University Press: 18.
71. Ibid. 65.
72. Power, M., and Schulkin, J. (2008). Anticipatory physiological regulation in feeding biology: Cephalic phase responses. *Appetite*, 50: 194–206.
73. Santayana, G. (1923, 1955). *Skepticism and Animal Faith*. New York: Dover Press; Santayana, G. (1932, 1967). *Character and Opinion in the United States*. New York: Norton; Santayana, G. (1967). *Animal Faith and Spiritual Life* (edited by J. Lachs). New York: Appleton, Century, Crofts; Santayana, G. (1988). Some gleanings from *The Life of Reason*. *Journal of Philosophy*, 51: 46–49; Weissman, D. (1989). *Hypothesis and the Spiral of Reflection*. Albany, NY: SUNY Press.
74. Lakoff, G., and Johnson, M. (1999). *Philosophy in the Flesh: The Embodied Mind and its Challenge to Western Thought*. New York: Basic Books.
75. See, for instance: Barsalou, L.W. (2003). Abstraction in perceptual symbol systems. *Philosophical Transactions of the Royal Society B*, 358: 1177–1187; Levinson, S. (2006). Cognition at the heart of human interaction. *Discourse Studies*, 8: 85–93; Levinson, S. (2003). *Space in Language and Cognition*. Cambridge: Cambridge University Press; Martin, A. (2007). The representation of object concepts in the brain. *Annual Review of Psychology*, 58: 25–45.
76. Locke, J. (1690, 1959). *An Essay Concerning Human Understanding*. New York: Dover Press; Quine, W.V.O. (1969). *Ontological Relativity and Other Essays*. New York: Columbia University Press.
77. Rorty, R. (1979). *Philosophy and the Mirror of Nature*. Chicago: University of Chicago Press.
78. Aristotle (1968). *De Anima*. Oxford: Oxford University Press: 57.
79. Locke, J. (1690, 1959). *An Essay Concerning Human Understanding*. New York: Dover Press; Landau, B., and Gleitman, L.R. (1985). *Language and Experience*. Cambridge, MA: Harvard University Press.
80. Leibniz, G. (1764, 1996). *New Essays on Human Understanding*. Cambridge: Cambridge University Press.
81. Ibid. 93.
82. Daniels, N. (1974, 1989). *Thomas Reid's Inquiry*. Palo Alto: Stanford University Press.
83. Hume, D. (1748, 1962). *An Inquiry Concerning Human Understanding*. New York: Washington Square Press.
84. Reid, T. (1785, 1969). *Essays on the Intellectual Powers of Man*. Cambridge, MA: MIT Press: 31.
85. Ibid. 189.
86. Gigerenzer, G. (2000). *Adaptive Thinking*. Oxford: Oxford University Press; Gigerenzer, G. (2007). *Gut Feelings*. New York: Viking Press.
87. Scerri, E.R. (2002). *The Periodic Table*. Oxford: Oxford University Press.
88. Schulkin J. (2009). *Cognitive Adaptation: A Pragmatist Perspective*. Cambridge: Cambridge University Press.

89. Kant, I. (1787, 1965). *Critique of Pure Reason* (translated by L.W. Beck). New York: Bobbs-Merrill: 41.
90. Friedman, M. (1992). *Kant and the Exact Sciences*. Cambridge, MA: Harvard University Press. For linking Peirce to neo-Kantian philosophers, see: Nordmann A. (2010). Critical realism, critical idealism, critical commonsensism. In: *The Kantian Legacy in Nineteenth-Century Science* (edited by M. Friedman and A. Nordmann). Cambridge, MA: MIT Press.
91. Kant, I. (1787, 1965). *Critique of Pure Reason* (translated by L.W. Beck). New York: Bobbs-Merrill: 248.
92. Ibid. 616.
93. Kuhn, T.S. (2000). *The Road Since Structure*. Chicago: University of Chicago Press.
94. See, for instance: Cassirer, E. (1957). *The Philosophy of Symbolic Forms*, vol. III. New Haven: Yale University Press; Carnap, R. (1928, 1969). *The Logical Structure of the World*. Berkeley: University of California Press; Lewis, C.L. (1929, 1956). *Mind and the World Order*. New York: Dover Press; Lewis, C.L. (1946, 1971). *An Analysis of Knowledge and Valuation*. LaSalle, IL: Open Court; Sellars, W. (1956, 1997). *Empiricism and the Philosophy of Mind*. Cambridge, MA: Harvard University Press; Sellars, W. (1962). *Science, Perception, and Reality*. New York: Routledge; Sellars, W. (1968). *Science and Metaphysics*. New York: Humanities Press; Brandom, R.B. (2009). *Reason in Philosophy*. Cambridge, MA: Harvard University Press.
95. Duns Scotus (1962). *Philosophical Writings of Duns Scotus*. Indianapolis: Bobbs-Merrill.
96. Houser, N., and Kloesel, C. (1992, 1998). *The Essential Peirce, Volume 1 (1867–1893)*. Bloomington: Indiana University Press: 107.
97. Ibid. 30.
98. See also Kagan, J. (2002). *Surprise, Uncertainty and Mental Structure*. Cambridge, MA: Harvard University Press; Loewenstein, G. (1994). The psychology of curiosity. *Psychological Bulletin*, 116: 75–98.
99. Peirce, C.S. (1992, 1998). The nature of meaning. In: *The Essential Peirce*, vol. 2 (edited by N. Houser and C. Kloesel). Bloomington: Indiana University Press: 216.
100. Perry, R.B. (1935). *The Thought and Character of William James*. Boston, MA: Little Brown and Co.; Perry, R.B. (1938, 1958). *In the Spirit of William James*. Bloomington: Indiana University Press; Perry, R.B. (1912). *Present Philosophical Tendencies, a Critical Survey of Naturalism, Idealism, Pragmatism, and Realism together with a Synopsis of the Philosophy of William James*. New York: Longmans, Green; Fisch, M.H. (1986). Evolution in American philosophy. In: *Peirce, Semiotic and Pragmatism* (edited by K.L. Ketner and J.W. Kloesel). Bloomington: Indiana University Press.
101. Dear, P. (2006). *The Intelligibility of Nature*. Chicago: University of Chicago Press.
102. Dewey, J. (1925, 1989). *Experience and Nature*. La Salle, IL: Open Court: 14.
103. Kitcher, P. (1993). *The Advancement of Science*. Oxford: Oxford University Press: 131.
104. Lakoff, G., and Johnson, M. (1999). *Philosophy in the Flesh: The Embodied Mind and its Challenge to Western Thought*. New York: Basic Books; Gallistel, C.R. (1980). *The Organization of Action: A New Synthesis*. Hillsdale, NJ:

Lawrence Erlbaum; Dewey, J. (1896). The reflex arc concept in psychology. *Psychological Review,* 3: 357–370; Dewey, J. (1902, 1974). *The Child and the Curriculum and the School and Society.* Chicago: University of Chicago Press.

105. Quine W.V.O. (1951). Two dogmas of empiricism. *The Philosophical Review,* 60: 20–43.
106. Ibid. 46.
107. Goodman, N. (1955, 1978). *Fact, Fiction and Forecast.* New York: Bobbs-Merrill Co.
108. Hempel, C.G. (1965). *Aspects of Scientific Explanation.* New York: Free Press.
109. Wittgenstein, L. (1953, 1968). *Philosophical Investigations.* New York: Macmillan Publishing; Goodman, N. (1955, 1978). *Fact, Fiction and Forecast.* New York: Bobbs-Merrill Co.; Goodman, R.B. (1990). *American Philosophy and the Romantic Tradition.* Cambridge: Cambridge University Press; Goodman, R.B. (1990). *American Philosophy and the Romantic Tradition.* Cambridge: Cambridge University Press; Goodman, R.B. (1998). Wittgenstein and pragmatism. *Paralas,* 4: 91–105; Goodman, R.B. (2002). *Wittgenstein and William James.* Cambridge: Cambridge University Press.
110. Quine, W.V.O. (1969). *Ontological Relativity and Other Essays.* New York: Columbia University Press: 123.
111. Ibid. 125.
112. Kant, I. (1787, 1965). *Critique of Pure Reason* (translated by L.W. Beck). New York: St. Martin's Press; Lewis, C.L. (1929, 1956). *Mind and the World Order.* New York: Dover Press; Lewis, C.L. (1946, 1971). *An Analysis of Knowledge and Valuation.* LaSalle, IL: Open Court.
113. Hanson, N.R. (1958, 1972). *Patterns of Discovery.* Cambridge: Cambridge University Press; Hanson, N.R. (1971). *Observation and Explanation.* New York: Harper Press.
114. Dewey, J. (1929, 1966). *A Theory of Valuation.* Chicago: University of Chicago Press; Dewey, J., and Tufts, J.H. (1910). *Ethics.* New York: Henry Holt; Lewis, C.L. (1946, 1971). *An Analysis of Knowledge and Valuation.* LaSalle, IL: Open Court; Neville, R.C. (1974). *The Cosmology of Freedom.* New Haven: Yale University Press; Putnam, H. (2000). *The Collapse of the Fact/Value Distinction.* Cambridge, MA: Harvard University Press; Putnam, H. (1995). *Pragmatism: An Open Question.* Malden, MA: Blackwell.
115. Hanson, N.R. (1958, 1972). *Patterns of Discovery.* Cambridge: Cambridge University Press: 16.
116. Carey, S. (1985, 1987). *Conceptual Change in Childhood.* Cambridge, MA: MIT Press; Carey, S. (2004). Bootstrapping and the origins of concepts. *Daedalus,* Winter 133: 59–68; Carey, S. (2009). *On the Origins of Concepts.* Oxford: Oxford University Press.
117. Dewey, J. (1938, 1973). *Experience and Education.* New York: Collier; Dewey, J. (1934, 1958). *Art as Experience.* New York: Capricorn Books.
118. Hanson, N.R. (1958, 1972). *Patterns of Discovery.* Cambridge: Cambridge University Press: 59.
119. Ibid. 72.
120. Wittgenstein, L. (1953, 1968). *Philosophical Investigations.* New York: Macmillan Publishing.

121. Sellars, W. (1956, 1997). *Empiricism and the Philosophy of Mind*. Cambridge, MA: Harvard University Press: 3.
122. Kripke, S. (1980). *Naming and Necessity*. Cambridge, MA: Harvard University Press; Mill, J.S. (1843, 1873). *A System of Logic*. London: Longmans, Green, Reader and Dyer.
123. Dewey, J. (1925, 1989). *Experience and Nature*. La Salle, IL: Open Court.
124. Wittgenstein, L. (1953, 1968). *Philosophical Investigations*. New York: Macmillan Publishing.
125. Ibid. 116.
126. Hacking, I. (1964). *Logic of Statistical Inference*. Cambridge: Cambridge University Press; Hacking, I. (1975). *The Emergence of Probability*. Cambridge: Cambridge University Press; Hacking, I. (1999). *The Taming of Chance*. Cambridge: Cambridge University Press.
127. Humbolt, W. Von (1836, 1971). *Linguistic and Intellectual Development*. Philadelphia: University of Pennsylvania Press; Chomsky, N. (1965). *Aspects of the Theory of Syntax*. Cambridge, MA: MIT Press; Chomsky, N. (1972). *Language and Mind*. New York: Harcourt Brace Jovanovich; Pinker, S. (1994). *The Language Instinct*. New York: William Morrow and Co.
128. Wittgenstein, L. (1953, 1968). *Philosophical Investigations*. New York: Macmillan Publishing: 194.
129. Kagan, J. (1984). *The Nature of the Child*. New York: Basic Books; Kagan, J. (2002). *Surprise, Uncertainty and Mental Structure*. Cambridge, MA: Harvard University Press.
130. Boyd, R. (1999). Homeostasis, species and higher taxa. In: *Species: New Interdisciplinary Essays* (edited by R.A. Wilson). Cambridge, MA: MIT Press.
131. Goodman, N. (1955, 1978). *Fact, Fiction and Forecast*. New York: Bobbs-Merrill Co.
132. Elder, C.L. (2009). Biological species are natural kinds. *Southern Journal of Philosophy*, 46: 339–362.
133. Boyd, R. (1999). Homeostasis, species and higher taxa. In: *Species: New Interdisciplinary Essays* (edited by R.A. Wilson). Cambridge, MA: MIT Press.
134. Sterling, P. (2004). Principles of allostasis. In: *Allostasis, Homeostasis and the Costs of Physiological Adaptation* (edited by J. Schulkin). Cambridge: Cambridge University Press.
135. Dupre, J. (1981). Natural kinds and biological taxa. *The Philosophical Review*, 90: 66–90.
136. Sellars, R.W. (1922). *Evolutionary Naturalism*. Chicago: Open Court Press; Kitcher, P. (1992). The naturalists return. *Philosophical Review*, 101: 3–51; Hacking, I. (1999). *The Taming of Chance*. Cambridge: Cambridge University Press.
137. Dupre, J. (1981). Natural kinds and biological taxa. *The Philosophical Review*, 90: 66–90; Wilson, E.O. (1992). *The Diversity of Life*. Cambridge: Harvard University Press; Wilson, E.O. (1994). *Naturalist*. Washington, DC: Island Press; Wilson, R.A. (2004). *Boundaries of the Mind*. Cambridge: Cambridge University Press; Wilson, R.A. (2005). *Genes and the Agents of Life*. Cambridge: Cambridge University Press.
138. Keil, F.C. (1989). *Concepts, Kinds and Cognitive Development*. Cambridge, MA: MIT Press; Keil, F.C. (1979). *Semantic and Conceptual Development: An Ontological Perspective*. Cambridge, MA: Harvard University Press; Craver,

C.F. (2009). Mechanisms and natural kinds. *Philosophical Psychology,* 22: 575–594; Wilson, R.A. (2005). *Genes and the Agents of Life.* Cambridge: Cambridge University Press.

139. Elder, C.L. (2009). Biological species are natural kinds. *Southern Journal of Philosophy,* 46: 339–362.
140. Dupre, J. (1981). Natural kinds and biological taxa. *The Philosophical Review,* 90: 66–90; Kornblith, H. (1993). *Inductive Inference and its Natural Ground.* Cambridge, MA: MIT Press; Schulkin, J. (2009). *Cognitive Adaptation: A Pragmatist Perspective.* Cambridge: Cambridge University Press.
141. Dupre, J. (1981). Natural kinds and biological taxa. *The Philosophical Review,* 90: 66–90.
142. Dewey, J. (1910, 1965). *The Influence of Darwin on Philosophy.* Bloomington: Indiana University Press.
143. Dupre, J. (1981). Natural kinds and biological taxa. *The Philosophical Review,* 90: 66–90.
144. Mayr, E. (1991). *One Long Argument.* Cambridge, MA: Harvard University Press; Mayr, E. (1942, 1982). *Systemics and the Origin of Species.* New York: Columbia University Press; Mayr, E. (1963). *Animal Species and Evolution.* Cambridge: Harvard University Press; Hull, D.L. (1988). *Science as a Process.* Chicago: University of Chicago Press; Boyd, R., and Richerson, P. (2005). *Not By Genes Alone: How Culture Transformed Human Evolution.* Chicago: University of Chicago Press; Wilson, R.A. (2005). *Genes and the Agents of Life.* Cambridge: Cambridge University Press.
145. Murphy, G.L. (2002). *The Big Book of Concepts.* Cambridge, MA: MIT Press.
146. Carey, S. (1985, 1987). *Conceptual Change in Childhood.* Cambridge, MA: MIT Press; Carey, S. (2004). Bootstrapping and the origins of concepts. *Daedalus,* Winter 133: 59–68; Carey, S. (2009). *On the Origins of Concepts.* Oxford: Oxford University Press.
147. Keil, F.C. (1989). *Concepts, Kinds and Cognitive Development.* Cambridge, MA: MIT Press; Keil, F.C. (1979). *Semantic and Conceptual Development: An Ontological Perspective.* Cambridge, MA: Harvard University Press; Carey, S. (1985, 1987). *Conceptual Change in Childhood.* Cambridge, MA: MIT Press; Carey, S. (2004). Bootstrapping and the origins of concepts. *Daedalus,* Winter 133: 59–68; Carey, S. (2009). *On the Origins of Concepts.* Oxford: Oxford University Press.
148. Barsalou, L.W. (2003). Abstraction in perceptual symbol systems. *Philosophical Transactions of the Royal Society B,* 358: 1177–1187; Martin, A. (2007). The representation of object concepts in the brain. *Annual Review of Psychology,* 58: 25–45.
149. Schulkin, J. (2009). *Cognitive Adaptation: A Pragmatist Perspective.* Cambridge: Cambridge University Press;
150. Pylshyn, Z.W. (2007). *Things and Places.* Cambridge, MA: MIT Press; Gigerenzer, G. (2000). *Adaptive Thinking.* Oxford: Oxford University Press.
151. Carey, S. (1985, 1987). *Conceptual Change in Childhood.* Cambridge, MA: MIT Press; Carey, S. (2004). Bootstrapping and the origins of concepts. *Daedalus,* Winter 133: 59–68; Carey, S. (2009). *On the Origins of Concepts.* Oxford: Oxford University Press.
152. Kitcher, P. (1993). *The Advancement of Science.* Oxford: Oxford University Press.

153. Carey, S. (1985, 1987). *Conceptual Change in Childhood*. Cambridge, MA: MIT Press; Carey, S. (2004). Bootstrapping and the origins of concepts. *Daedalus*, Winter 133: 59–68; Carey, S. (2009). *On the Origins of Concepts*. Oxford: Oxford University Press.
154. Buffon, G.L. (1749–1767). *Histoire naturelle generale et particuliere*, 15 vols. Paris: Imprimerie Royale.
155. Moyal, A. (2001). *Platypus*. Baltimore: Johns Hopkins University Press.
156. See especially Lovejoy, A.O. (1936, 1978). *The Great Chain of Being*. Cambridge, MA: Harvard University Press; Lovejoy, A.O. (1955). *Essays in the History of Ideas*. New York: George Braziller Inc.
157. Martin, A. (2007). The representation of object concepts in the brain. *Annual Review of Psychology*, 58: 25–45; Barsalou, L.W. (2003). Abstraction in perceptual symbol systems. *Philosophical Transactions of the Royal Society B*, 358: 1177–1187; Johnson, M. (1987, 1990). *The Body in the Mind*. Chicago: University of Chicago Press; Heelan, P.A., and Schulkin, J. (1998). Hermeneutical philosophy and pragmatism: A philosophy of the science. *Synthese*, 115: 269–302.
158. Santayana, G. (1923, 1955). *Scepticism and Animal Faith*. New York: Dover Press: 34. See also Santayana, G. (1967). *Animal Faith and Spiritual Life* (edited by J. Lachs). New York: Appleton, Century, Crofts: 453.
159. Moyal, A. (2000). *Platypus*. Baltimore: Johns Hopkins University Press.
160. Darwin, C. (1859, 1958). *The Origin of Species*. New York: Mentor Books; Darwin, C. (1868). *The Variation of Animals and Plants under Domestication*, 2 vols. London: John Murray; Darwin, C. (1871, 1874). *Descent of Man*. New York: Rand McNally and Company; Darwin, C. (1872, 1965). *The Expression of Emotions in Man and Animals*. Chicago: University of Chicago Press; Mayr, E. (1991). *One Long Argument*. Cambridge, MA: Harvard University Press; Mayr, E. (1942, 1982). *Systemics and the Origin of Species*. New York: Columbia University Press; Mayr, E. (1963). *Animal Species and Evolution*. Cambridge, MA: Harvard University Press.
161. Dunbar, R.I.M. (1996). *Grooming, Gossip and the Evolution of Language*. Cambridge, MA: Harvard University Press; Barton, R.A. (2004). Binocularity and brain evolution in primates. *Proceedings of the National Academy of Sciences*, 101: 10113–10115; Barton, R.A. (2006). Primate brain evolution: integrating comparative neurophysiological and ethological data. *Evolutionary Anthropology*, 15: 224–236.
162. Levinson, S. (2006). Cognition at the heart of human interaction. *Discourse Studies*, 8: 85–93; Levinson, S. (2003). *Space in Language and Cognition*. Cambridge: Cambridge University Press; Pratt, S.L. (2002). *Native Pragmatism: Rethinking the Roots of American Philosophy*. Bloomington: Indiana University Press.
163. Pinker, S. (1994). *The Language Instinct*. New York: William Morrow and Co; Aiello, L.C., and Dunbar, R.I.B. (1993). Neocortex size, group size and the evolution of language. *Current Anthropology*, 34: 184–192; Dunbar, R.I.M. (1996). *Grooming, Gossip and the Evolution of Language*. Cambridge, MA: Harvard University Press; Dunbar, R.I.M. (1992). Neocortex size as a constraint on group size in primates. *Journal of Human Evolution*, 22: 469–493.

164. Heelan, P.A. (1983). *Space Perception and the Philosophy of Science*. Berkeley: University of Calif. Press; Heelan, P.A. (2001). The lifeworld and scientific interpretation. In: *Handbook of Phenomenology and Medicine*. Waco, TX: Baylor University; Heelan, P.A., and Schulkin, J. (1998). Hermeneutical philosophy and pragmatism: a philosophy of the science. *Synthese*, 115: 269–302.
165. Barton, R.A. (2004). Binocularity and brain evolution in primates. *Proceedings of the National Academy of Sciences*, 101: 10113–10115; Barton, R.A. (2006). Primate brain evolution: Integrating comparative neurophysiological and ethological data. *Evolutionary Anthropology*, 15: 224–236.

4 Cephalic Capabilities and Medical Decision Making: Endlessly Imperfect

1. Starr, P. (1984). *The Social Transformation*. New York: Basic Books.
2. Oshingsky, D.M. (2005). *Polio: An American Story*. Oxford: Oxford University Press.
3. Ibid.
4. Rozin, P. (1976). the evolution of intelligence and access to the cognitive unconscious. In: *Progress in Psychobiology and Physiological Psychology* (edited by J. Sprague and A.N. Epstein). New York: Academic Press; Rozin, P. (1998). Evolution and development of brains and cultures. In: *Brain and Mind: Evolutionary Perspectives* (edited by M.S. Gazzaniga and J.S. Altman). Strasbourg, France: Human Frontiers Sciences Program; Richter, C.P. (1943). *Total Self-Regulatory Functions in Animals and Man*. New York: Harvey Lecture Series.
5. Clark, A. (1997). *Being There*. Cambridge, MA: MIT Press; Clark, R.W. (1983). *Benjamin Franklin*. New York: DaCapo Press; Donald, M. (1991). *Origins of Modern Man*. Cambridge, MA: Harvard University Press; Donald, M. (2004). Hominid enculturation and cognitive evolution. In: *The Development of the Mediated Mind* (edited by J.M. Luraciello et al.). Mawash, NJ: Erlbaum Press.
6. Dunbar, R.I.M. (1996). *Grooming, Gossip and the Evolution of Language*. Cambridge, MA: Harvard University Press; Dunbar, R.I.M. (1992). Neocortex size as a constraint on group size in primates. *Journal of Human Evolution*, 22: 469–493; Dunbar, R.I.M., and Shultz, S. (2007). Understanding primate evolution. *Philosophical Transactions of the Royal Society*, 362: 649–658.
7. Aristotle (1968). *De Anima* (translated by D. Hamylin). Oxford: Oxford University Press; Ray, J. (1660). *Catalogus plantarum circa Cantabrigiam Nascentium*. Cambridge: Field; Linnaeus, C. (1735). *Systema Naturae*. Leiden: Haak.
8. Atran, S. (1990, 1996). *Cognitive Foundations of Natural History*. New York: Cambridge University Press.
9. Whorton, J.C. (2002). *Nature Cures: The History of Alternative Medicine in America*. Oxford: Oxford University Press; Cooper, D.E. (2002, 2007). *The Measure of Things*. Oxford: Clarendon Press.
10. Ulett, G.A., Han, J., and Han, S. (1998). Traditional and evidence-based acupuncture: history, mechanisms, and present status. *Southern Medical Journal*, 91: 1115–1127.

11. Atran, A., Medin, D.L., and Ross, N.O. (2005). The cultural mind. *Psychological Review*, 112: 744–766; Carey, S. (2009). *On the Origins of Concepts*. Oxford: Oxford University Press.
12. Farber, P.L. (2000). *Finding Order in Nature*. Baltimore: Johns Hopkins University Press; Grant, E. (2007). *A History of Natural Philosophy*. Cambridge: Cambridge University Press.
13. Kagan, J. (1984). *The Nature of the Child*. New York: Basic Books.
14. Foley, R.A. (2001). The evolutionary consequences of increased carnivory in hominids. In: *Meat-Eating and Human Evolution* (edited by C.B. Stanford and H.T. Bunn). New York: Oxford University Press; Foley, R.A. and Lee, P.D. (1991). Ecology and energetics of encephalization in hominid evolution. *Philosophical Transactions of the Royal Society, Series B*, 334: 223–232.
15. Galileo (1610, 1957). The starry messenger. In: *Discoveries and Opinions of Galileo* (edited by S. Drake). New York: Doubleday; Galileo, G. (1957). *Discoveries and Opinions*. New York: Anchor Books.
16. Grant, E. (2007). *A History of Natural Philosophy*. Cambridge: Cambridge University Press.
17. Carey, S. (2009). *On the Origins of Concepts*. Oxford: Oxford University Press.
18. Kagan, J. (1984). *The Nature of the Child*. New York: Basic Books; Kagan, J. (2002) *Surprise, Uncertainty and Mental Structure*. Cambridge, MA: Harvard University Press; Tomasello, M. (2009). *Why We Cooperate*. Cambridge, MA: MIT Press.
19. Lakoff, G., and Johnson, M. (1999). *Philosophy in the Flesh: The Embodied Mind and its Challenge to Western Thought*. New York: Basic Books; Schulkin, J. (2009). *Cognitive Adaptation*. Cambridge: Cambridge University Press; Weissman, D. (1989). *Hypothesis and the Spiral of Reflection*. Albany: SUNY Press; Weissman, D. (2000). *A Social Ontology*. New Haven: Yale University Press.
20. Kahneman, D. (2011). *Thinking Fast and Slow*. New York: Farrar, Straus and Giroux; Gigerenzer, G. (2000). *Adaptive Thinking*. Oxford: Oxford University Press; Gigerenzer, G. (2007). *Gut Feelings*. New York: Viking Press; Gigerenzer, G. (2003). Why does framing influence judgment? *Journal of General Internal Medicine*, 18: 906–961; Baron, J. (1988, 2008). *Thinking and Deciding*. Cambridge: Cambridge University Press.
21. Carey, S. (1985, 1987). *Conceptual Change in Childhood*. Cambridge, MA: MIT Press.
22. Langer, S.K. (1937). *Philosophy in a New Key*. Cambridge, MA: Harvard University Press; Langer, S.K. (1962). *Philosophical Sketches*. New York: Mentor Books; Langer S.K. (1972). *Mind: An Essay on Human Feeling*. Baltimore: Johns Hopkins University Press.
23. Dewey, J. (1938). *Logic: The Theory of Inquiry*. New York: Holt, Rinehart.
24. Popper, K.R. (1934, 1968). *The Logic of Scientific Discovery*. New York: Harper Torch Books; Popper, K.R. (1962, 1971). *Conjectures and Refutations*. New York: Harper and Row.
25. Dewey, J. (1910). *How We Think*. New York: DC Heath and Publ.: 39.
26. Ibid.
27. Ibid. 57.
28. Schulkin, J. (2004). *Bodily Sensibility*. Oxford: Oxford University Press; Johnson, M. (1987, 1990). *The Body in the Mind*. Chicago: University of

Chicago Press; Johnson, M. (1993). *Moral Imagination*. Chicago: University of Chicago Press; Johnson, M. (2007). *The Meaning of the Body*. Chicago: University of Chicago Press; Merleau-Ponty, M. (1942, 1967). *The Structure of Behavior*. Boston: Beacon Press; Gallagher, S. (2005). *How the Body Shapes the Mind*. Oxford: Oxford University Press; Varela, F., Thompson, E., and Rosch, E. (1991). *The Embodied Mind*. Cambridge: Cambridge University Press.

29. Schulkin, J. (2009). *Cognitive Adaptation: A Pragmatist Perspective*. Cambridge: Cambridge University Press.
30. Gibson, K.R., and Ingold, T. (1993, editors). *Tools, Language and Cognition in Human Evolution*. Cambridge: Cambridge University Press.
31. Mithen, S. (1996). *The Prehistory of the Mind: The Cognitive Origins of Art and Science*. London: Thames and Hudson, Ltd.
32. For biographical information about Hippocrates, see Pinault, J.R. (1992). *Hippocratic Lives and Legends*. Leiden: Brill Academic Publishers.
33. Hippocrates. (1983). *Hippocratic Writings*. Harmondsworth: Penguin Books.
34. Galen (1968). *On the Usefulness of the Parts of the Body*, vol. 1. Ithaca: Cornell University Press.
35. Schultz, S. (2002). William Harvey and the circulation of the blood: the birth of a scientific revolution and modern physiology. *News in Psychological Sciences*, 17: 175–180.
36. Singer, C. (1957). *A Short History of Anatomy and Physiology from the Greeks to Harvey*. New York: Dover Press.
37. Kagan, J. (1994). *Galen's Prophecy*. New York: Basic Books.
38. Ibid.
39. Schmidt, L., and Schulkin, J. (1999). *Extreme Fear, Shyness and Social Behavior*. Oxford: Oxford University Press.
40. Lakoff, G., and Johnson, M. (1999). *Philosophy in the Flesh: The Embodied Mind and its Challenge to Western Thought*. New York: Basic Books; Atran, S. (1990, 1996). *Cognitive Foundations of Natural History*. New York: Cambridge University Press.
41. Atran, S. (1990, 1996). *Cognitive Foundations of Natural History*. New York: Cambridge University Press.
42. Bynum, W. (2008). *The History of Medicine*. Oxford: Oxford University Press.
43. DeWulf, M. (1907, 1956). *Scholastic Philosophy*. New York: Dover Press; Fakhry, M. (1970, 1983). *A History of Islamic Philosophy*. New York: Columbia University Press.
44. Maclean, I. (2002). *Logic, Signs and Nature in the Renaissance*. Cambridge: Cambridge University Press.
45. French, R. (2008). *Medicine Before Science*. Cambridge: Cambridge University Press.
46. Kristeller, P.O. (1955, 1961). *Renaissance Thought*. New York: Harper and Row.
47. Cassirer, E. (1926, 1963). *The Individual and the Cosmos in Renaissance Philosophy* (translated by M. Domandi). New York: Harper and Row; Cassirer, E. (1951). *The Philosophy of the Enlightenment*. Princeton: Princeton University Press; Cassirer, E. (1957). *The Philosophy of Symbolic Forms*, vol. III. New Haven: Yale University Press; Langer, S.K. (1937). *Philosophy in a New Key*. Cambridge, MA: Harvard University Press; Langer, S.K. (1962). *Philosophical Sketches*. New York: Mentor Books; Burckhardt, J. (1929, 1958). *The Civilization of the Renaissance in Italy*. New York: Colophon Books.

48. Cassirer, E. (1926, 1963). *The Individual and the Cosmos in Renaissance Philosophy* (translated by M. Domandi). New York: Harper and Row; Cassirer, E. (1951). *The Philosophy of the Enlightenment*. Princeton: Princeton University Press; Cassirer, E. (1957). *The Philosophy of Symbolic Forms*, vol. III. New Haven: Yale University Press.
49. Machiavelli, M. (1525, 1988). *Florentine Histories* (translated by L.F. Banfield and H.C. Mansfield, Jr). Princeton: Princeton University Press.
50. Nietzsche, F. (1878, 1984). *Human, All Too Human* (translated by M. Faber and S. Lehmann). Lincoln: University of Nebraska Press.
51. Chan, W.J. (1963). *A Source Book in Chinese Philosophy*. Princeton: Princeton University Press.
52. Haskins, C.H. (1923, 1957). *The Rise of the Universities*. Ithaca: Cornell University Press; Haskins, C.H. (1957, 1966). *The Renaissance of the 12th Century*. New York: Meridian Press.
53. Boas, F. (1911, 1968). *The Mind of Primitive Man*. New York: Free Press.
54. Singer, C. (1957). *A Short History of Anatomy and Physiology from the Greeks to Harvey*. New York: Dover Press.
55. Maclean, I. (2002). *Logic, Signs and Nature in the Renaissance*. Cambridge: Cambridge University Press.
56. Locke, J. (1690, 1959). *An Essay Concerning Human Understanding*. New York: Dover Press.
57. Hall, M.B. (1965). *Robert Boyle on Natural Philosophy: An Essay With Selections From His Writings*. Bloomington: Indiana University Press.
58. Bacon, F. (1620, 1887). *Novum Organum*. New York: P.F. Collier and Sons.
59. Schultz, S. (2002). William Harvey and the circulation of the blood: the birth of a scientific revolution and modern physiology. *News in Psychological Sciences*, 17: 175–180.
60. Harvey, W. (1651, 1965). Anatomical exercises on the generation of animals. In: *Works of William Harvey*. New York: Johnson Reprint Corp.; Shapin, S., and Schaffer, S. (1985). *Leviathan and the Air-Pump*. Princeton: Princeton University Press.
61. Maclean, I. (2002). *Logic, Signs and Nature in the Renaissance*. Cambridge: Cambridge University Press; Singer, C. (1957). *A Short History of Anatomy and Physiology from the Greeks to Harvey*. New York: Dover Press; French, R. (2008). *Medicine before Science*. Cambridge: Cambridge University Press.
62. Shapin, S., and Schaffer, S. (1985). *Leviathan and the Air-Pump*. Princeton: Princeton University Press.
63. Shapin, S. (2008). *The Scientific Life*. Chicago: University of Chicago Press; Shapin, S. (1995). *A Social History of Truth*. Chicago: University of Chicago Press; Shapin, S. (1996). *The Scientific Revolution*. Chicago: University of Chicago Press.
64. Shapin, S. (2008). *The Scientific Life*. Chicago: University of Chicago Press. Also, see his new book: Shapin, S. (2010). *Never Pure*. Baltimore: Johns Hopkins University Press.
65. Dewey, J. (1910). *How We Think*. New York: DC Heath and Publ.
66. Mettrie, J.O. de La (1748, 1912). *Man a Machine*. LaSalle, IL: Open Court Press.
67. Descartes, R. (1637, 1993). *Discourse on Method*. Indianapolis: Hackett Publishing Co; Descartes, R. (1649, 1989). *The Passions of the Soul* (translated by S. Voss). Cambridge, MA: Hackett Publishing Co.

68. Heelan, P. (1983). *Space Perception and the Philosophy of Science Time*. Berkeley: University of California Press.
69. Hacking, I. (1999). *The Taming of Chance*. Cambridge: Cambridge University Press: 1.
70. Ibid. 55.
71. Ibid.; Porter, T.M. (2004). *Karl Pearson: The Scientific Life in a Statistical Age*. Princeton: Princeton University Press; Porter, R. (1998). *The Greatest Benefit to Mankind*. New York: Norton.
72. Wirth, J.B. (1989). *Richter and Magendie*. The First International Cyberconference on the Psychobiology of Curt P. Richter.
73. Bernard, C. (1865, 1957). *An Introduction to the Study of Experimental Medicine*. New York: Dover Publications.
74. Bergson, H. (1919, 1946). *The Creative Mind* (translated by M. Andison). New York: Citadel Press; Bergson, H. (1911, 1998). *Creative Evolution*. New York: Dover Press.
75. Bernard, C. (1865, 1957). *An Introduction to the Study of Experimental Medicine*. New York: Dover Publications.
76. Todes, D.P. (1989). *Darwin without Malthus*. Oxford: Oxford University Press; Todes, D.P. (1997). Pavlov's physiology factory. *History of Science Society*, 88: 205–246.
77. Pavlov, I.P. (1897, 1902). *The Work of the Digestive Glands*. London: Charles Griffin; Pavlov, I.P. (1928, 1963). *Lectures on Conditioned Reflexes*. New York: International Publishing Co.
78. Powley, T.L. (1977). The ventralmedial hypothalamic syndrome, satiety and cephalic phase. *Psychological Review*, 84: 89–126; Smith, G.P. (2000). Pavlov and integrative physiology. *American Journal of Physiology*, 279: R743–R755.
79. Todes, D.P. (1997). Pavlov's physiology factory. *History of Science Society*, 88: 205–246.
80. Power, M., and Schulkin, J. (2008). Anticipatory physiological regulation in feeding biology: Cephalic phase responses. *Appetite*, 50: 194–206.
81. Todes, D.P. (1997). Pavlov's physiology factory. *History of Science Society*, 88: 205–246.
82. Gallistel, C.R. (1980). *The Organization of Action: A New Synthesis*. Hillsdale, NJ: Lawrence Erlbaum.
83. Dewey, J. (1896). The reflex arc concept in psychology. *Psychological Review*, 3: 357–370.
84. Todes, D.P. (2002). *Pavlov's Physiology Factory*. Baltimore: Johns Hopkins University Press.
85. Todes, D.P. (1997). Pavlov's physiology factory. *History of Science Society*, 88: 205–246.
86. Dewey, J. (1896). The reflex arc concept in psychology. *Psychological Review*, 3: 357–370.
87. Richter, C.P. (1943). *Total Self-Regulatory Functions in Animals and Man*. New York: Harvey Lecture Series.
88. Power, M., and Schulkin, J. (2008). Anticipatory physiological regulation in feeding biology: Cephalic phase responses. *Appetite*, 50: 194–206; Smith, G.P. (2000). Pavlov and integrative physiology. *American Journal of Physiology*, 279: R743–R755; Todes, D.P. (1997). Pavlov's physiology factory. *History of Science Society*, 88: 205–246.

89. Power, M., and Schulkin, J. (2008). Anticipatory physiological regulation in feeding biology: Cephalic phase responses. *Appetite*, 50: 194–206; Richter, C.P. (1943). *Total Self-Regulatory Functions in Animals and Man*. New York: Harvey Lecture Series.
90. Powley, T.L. (1977). The ventralmedial hypothalamic syndrome, satiety and cephalic phase. *Psychological Review*, 84: 89–126.
91. Richter, C.P. (1943). *Total Self-Regulatory Functions in Animals and Man*. New York: Harvey Lecture Series.
92. Power, M., and Schulkin, J. (2008). Anticipatory physiological regulation in feeding biology: cephalic phase responses. *Appetite*, 50: 194–206.
93. Bliss, M. (1999). *William Osler: A Life in Medicine*. Oxford: Oxford University Press; Bliss, M. (2003). *Harvey Cushing: A Life in Surgery*. Oxford: Oxford University Press.
94. Bernard, C. (1865, 1957). *An Introduction to the Study of Experimental Medicine*. New York: Dover Publications.
95. Schulkin, J. (2003). *Rethinking Homeostasis*. Cambridge, MA: MIT Press.
96. Strand, F.L. (1999). *Neuropeptides: Regulators of Physiological Processes*. Cambridge, MA: MIT Press.
97. Powley, T.L. (1977). The ventralmedial hypothalamic syndrome, satiety and cephalic phase. *Psychological Review*, 84: 89–126.
98. Gabbe, S.G. (1996). The alphabet of academic medicine. *Obstetrics & Gynecology*, 88: 479–481.
99. Power, M., and Schulkin, J. (2009). *The Evolution of Obesity*. Baltimore: Johns Hopkins University Press.
100. Craig, W. (1918). Appetites and aversions as constituents of instinct. *Biological Bulletin*, 34: 91–107; Tinbergen, N. (1951, 1969). *The Study of Instinct*. Oxford: Oxford University Press.
101. Bliss, M. (2011). *The Making of Modern Medicine: Turning Points in the Treatment of Disease*. Chicago: The University of Chicago Press.
102. Dear, P. (1995). *Discipline and Experience*. Chicago: University of Chicago Press.
103. For specific examples of this shift, see: Koyre, A. (1961). *Renaissance Thought: The Classic, Scholastic, and Humanist Strains*. New York: Harper Torchbooks; Koyre, A. (1968). *Metaphysics and Measurement: Essays in Scientific Revolution*. Cambridge, MA: Harvard University Press; Kuhn, T.S. (1962). *The Structure of Scientific Revolution*. Chicago: University of Chicago Press; Kuhn, T.S. (2000). *The Road Since Structure*. Chicago: University of Chicago Press.
104. Hobbes, T. (1651, 1949). *De Cive*. New York: Appleton-Century Crafts.
105. Scarry, E. (1985). *The Body in Pain*. Oxford: Oxford University Press.
106. Bernard, C. (1865, 1957). *An Introduction to the Study of Experimental Medicine*. New York: Dover Publications; Canquilhem, G. (1991). *The Normal and the Pathological*. New York: Zone Books.
107. Lawrence C. (1998). Medical minds: surgical bodies. In: *Science Incarnate* (edited by C. Lawrence and S. Shapin). Chicago: University of Chicago Press.
108. Shapin, S. (2008). *The Scientific Life*. Chicago: University of Chicago Press; Shapin, S. (1995). *A Social History of Truth*. Chicago: University of Chicago Press; Shapin, S. (1996). *The Scientific Revolution*. Chicago: University of Chicago Press.

109. Jackson, J.H. (1884, 1958). Evolution and dissolution of the nervous system. In: *Selected Writings of John Hughlings Jackson*, vol. II (edited by J. Taylor). London: Staples Press.
110. Duhem, P. (1915, 1991). *German Science*. La Salle, IL: Open Court Press.
111. Otis, L. (2007). *Muller's Lab*. Oxford: Oxford University Press.
112. Helmhotlz, H. von (1995). *Science and Culture* (edited by D. Cahan). Chicago: University of Chicago Press; Helmholtz, H. Von (1867, 1963). *Handbook of Physiological Optics*. New York: Dover Press.
113. Osler, W. (1892). *The Principles and Practices of Medicine*. New York: D. Appleton and Company.
114. Bliss, M. (1999). *William Osler: A Life in Medicine*. Oxford: Oxford University Press; Bliss, M. (2003). *Harvey Cushing: A Life in Surgery*. Oxford: Oxford University Press; Osler, W. (1923). *The Evolution of Modern Medicine*. New York: D. Appleton and Company.
115. Ibid.
116. Longo, L.D. (1980). Obstetrics and gynecology. In: *The Education of American Physicians* (edited by R.L. Numbers). Berkeley: University of California Press.
117. Schulkin, J. (2005). *Curt Richter: A Life in the Laboratory*. Baltimore: Johns Hopkins University Press.
118. Flexner, S. (1941). *William Welsh and the Heroic Age of American Medicine*. New York: Viking Press.
119. Flexner, S. (1910). *Medical Education in the United States and Canada: A Report to the Carnegie Foundation for the Advancement of Teaching. Bulletin Number Four*. Boston, MA: Updyke.
120. Schulkin, J. (2000). Decision sciences and evidence-based medicine. Two intellectual movements to support clinical decision making. *Academic Medicine*, 75: 816–818; Schulkin, J. (2008). *Medical Decisions, Estrogen, and Aging*. New York: Springer; Sullivan, W.M. (1986). *Reconstructing Public Philosophy*. Berkeley: University of California Press.
121. Dewey, J. (1910). *How We Think*. New York: DC Heath and Publ.
122. Dewey, J. (1929, 1966). *A Theory of Valuation*. Chicago: University of Chicago Press; Dewey, J. (1929, 1962). *Individualism, Old and New*. New York: Capricorn Books; Dewey, J. (1929, 1960). *The Quest for Certainty*. New York: Capricorn Books.
123. Dewey, J. (1925, 1989). *Experience and Nature*. La Salle, IL: Open Court; Langer, S.K. (1937). *Philosophy in a New Key*. Cambridge, MA: Harvard University Press; Langer, S.K. (1962). *Philosophical Sketches*. New York: Mentor Books; Langer, S.K. (1972). *Mind: An Essay on Human Feeling*. Baltimore: Johns Hopkins University Press; Schulkin, J. (2004). *Bodily Sensibility: Intelligent Action*. Oxford: Oxford University Press.
124. Fulton, J.F. (1946). *Harvey Cushing: A Biography*. Springfield, IL: Charles Thomas.
125. Ibid.; Bliss, M. (2003). *Harvey Cushing: A Life in Surgery*. Oxford: Oxford University Press.
126. Rowland, L.P. (2009). *The Legacy of Tracy J. Putnam and H. Houson Merrit: Modern Neurology in the United States*. Oxford: Oxford University Press; Bliss, M. (2003). *Harvey Cushing: A Life in Surgery*. Oxford: Oxford University Press; Brodel, M. (1946). *Three Unpublished Drawings of the Anatomy of the Human Ear*. Philadelphia: W.B. Sanders Co.

127. Crosby, R.W., and Cody, J. (1991). *Max Brodel: The Man Who Put Art into Medicine*. New York: Springer Verlag; Hendricks, M. (2000). Applying art to medicine. *Johns Hopkins Magazine*, 12: 20–25.
128. Dewey, J. (1934, 1958). *Art as Experience*. New York: Capricorn Books.
129. Ibid. 38.
130. Dewey, J. (1920, 1948). *Reconstruction in Philosophy*. Boston: Beacon Press; Alexander, T.M. (1987). *John Dewey's Theory of Art, Experience, and Nature: The Horizons of Feeling*. Albany, NY: SUNY Press; Alexander, T.M. (1996). The fourth world of American philosophy: the philosophical significance of Native American culture. *Transactions of the Charles S. Peirce Society*, 32; Shusterman, R. (1997). *Practicing Philosophy: Pragmatism and the Philosophical Life*. New York: Routledge.
131. Brodel, M. (1946). *Three Unpublished Drawings of the Anatomy of the Human Ear*. Philadelphia: W.B. Sanders Co.
132. Dewey, J. (1934, 1958). *Art as Experience*. New York: Capricorn Books: 124.
133. Hickman, L.A. (1980, 1992). *John Dewey's Pragmatic Technology*. Bloomington: Indiana University Press; Hickman, L.A. (2002). *Philosophical Tools for Technological Culture*. Bloomington: Indiana University Press; Hickman, L.A. (2007). *Pragmatism as Post-Postmodernism*. New York: Fordham University Press.
134. Heelan, P.A., and Schulkin, J. (1998). Hermeneutical philosophy and pragmatism: A philosophy of science. *Synthese*, 115: 269–302.
135. Schumpeter, J.A. (1934). *Theory of Economic Development*. Cambridge, MA: Harvard University Press.
136. Elster, J. (1979, 1988). *Ulysses and the Sirens*. Cambridge: Cambridge University Press; Elster, J. (1983). *Explaining Technical Change*. Cambridge: Cambridge University Press.
137. Dewey, J. (1925, 1989). *Experience and Nature*. La Salle, IL: Open Court; Langer, S.K. (1937). *Philosophy in a New Key*. Cambridge, MA: Harvard University Press; Langer, S.K. (1962). *Philosophical Sketches*. New York: Mentor Books.
138. Hacking, I. (1964). *Logic of Statistical Inference*. Cambridge: Cambridge University Press; Hacking, I. (1975). *The Emergence of Probability*. Cambridge: Cambridge University Press; Hacking, I. (1999). *The Taming of Chance*. Cambridge: Cambridge University Press.
139. Porter, T.M. (2004). *Karl Pearson: The Scientific Life in a Statistical Age*. Princeton: Princeton University Press.
140. Kuhn, T.S. (1962). *The Structure of Scientific Revolution*. Chicago: University of Chicago Press; Kuhn, T.S. (2000). *The Road since Structure*. Chicago: University of Chicago Press; Gillispie, C.C. (1960). *The Edge of Objectivity*. Princeton: Princeton University Press; Galison, P. (1988). History, philosophy and the central metaphor. *Science in Context*, 3: 197–212.
141. Keynes, J.M. (1921, 1957). *A Treatise on Probability*. New York: Harper and Row.
142. Hacking, I. (1999). *The Taming of Chance*. Cambridge: Cambridge University Press.
143. Gigerenzer, G. (2000). *Adaptive Thinking*. Oxford: Oxford University Press; Toulmin, S. (1977). *Human Understanding*. Princeton: Princeton University Press; Toulmin, S. (2001). *Return to Reason*. Cambridge, MA: Harvard University Press.

144. Moreno, J.D. (1995). *Deciding Together.* Oxford: Oxford University Press; Smith, A. (1759, 1882). *The Theory of Moral Sentiments.* Indianapolis: Liberty Classics; Wilson, J.Q. (1993). *The Moral Sense.* New York: Free Press; Schulkin, J. (2004). *Bodily Sensibility: Intelligent Action.* Oxford: Oxford University Press.
145. Kahneman, D., Slovic, P., and Tversky, A. (1982, editors). *Judgment under Uncertainty: Heuristics and Biases.* New York: Cambridge University Press; Gigerenzer, G. (2000). *Adaptive Thinking.* Oxford: Oxford University Press.
146. Gigerenzer, G. (2000). *Adaptive Thinking.* Oxford: Oxford University Press.
147. Baron, J. (1988, 2008). *Thinking and Deciding.* Cambridge: Cambridge University Press.
148. Schulkin, J. (2008). *Medical Decisions, Estrogen, and Aging.* New York: Springer.
149. Kahneman, D., Slovic, P., and Tversky, A. (1982, editors). *Judgment under Uncertainty: Heuristics and Biases.* New York: Cambridge University Press.
150. Dewey, J. (1939, 1963). *Freedom and Culture.* New York: Capricorn Books; Neville, R.C. (1974). *The Cosmology of Freedom.* New Haven: Yale University Press; Moreno, J.D. (1995). *Deciding Together.* Oxford: Oxford University Press; Moreno, J.D. (1999, 2003). Bioethics is a naturalism. In: *Pragmatic Bioethics* (edited by G. McGee). Cambridge, MA: MIT Press.
151. Dewey, J. (1939, 1963). *Freedom and Culture.* New York: Capricorn Books; Levinson, S. (2006). Cognition at the heart of human interaction. *Discourse Studies*, 8: 85–93.
152. Moreno, J.D. (1999, 2003). Bioethics is a naturalism. In *Pragmatic Bioethics* (edited by G. McGee). Cambridge, MA: MIT Press: 5.
153. Peirce, C.S. (1899, 1992). *Reasoning and the Logic of Things* (edited by K.L. Ketner and H. Putnam). Cambridge, MA: Harvard University Press; Gigerenzer, G. (2000). *Adaptive Thinking.* Oxford: Oxford University Press.
154. Kahneman, D., Slovic, P., and Tversky, A. (eds) (1982). *Judgment under Uncertainty: Heuristics and Biases.* New York: Cambridge University Press; Baron, J. (1988, 2008). *Thinking and Deciding.* Cambridge: Cambridge University Press.
155. Kahneman, D., Slovic, P., and Tversky, A. (eds) (1982). *Judgment under Uncertainty: Heuristics and Biases.* New York: Cambridge University Press; Loewenstein, G. (1996). Out of control: visceral influences on behavior. *Organizational Behavior and Human Decision Processes*, 65: 272–292; Loewenstein, G. (1994). The psychology of curiosity. *Psychological Bulletin*, 116: 75–98.
156. Simon, H.A. (1982). *Models of Bounded Rationality*, vol. 1. Cambridge, MA: MIT Press; Gigerenzer, G., Todd, R.M., and ABC Research Group (1999). *Simple Heuristics That Make Us Smart.* New York: Oxford University Press.
157. Chomsky, N. (1965). *Aspects of the Theory of Syntax.* Cambridge, MA: MIT Press; Chomsky, N. (1972). *Language and Mind.* New York: Harcourt Brace Jovanovich.
158. Baron, J. (1988, 2008). *Thinking and Deciding.* Cambridge: Cambridge University Press; Elstein, A.S. (2004). On the origins and development of evidence-based medicine and medical decision-making. *Inflammation Research*, 53: 8184–8189; Schulkin, J. (1996). *The Delicate Balance.* Lanham, MD: University Press of America.

159. Elstein, A.S. (2004). On the origins and development of evidence-based medicine and medical decision-making. *Inflammation Research*, 53: 8184–8189.
160. Cook, D.J., Murrow, C.D., and Haynes, R.B. (1997). Systematic reviews: synthesis of the best evidence in clinical medicine. *Annals of Internal Medicine*, 126: 377–391.
161. Farquhar, C.M., Kofa, E., Power, M.L., Zinberg, S., and Schulkin, J. (2002). Clinical practice guidelines as educational tools for obstetrician-gynecologists. *Journal of Reproductive Medicine*, 47: 897–902; Schulkin, J. (2000). Decision sciences and evidence-based medicine. Two intellectual movements to support clinical decision making. *Academic Medicine*, 75: 816–818; Schulkin, J. (2008). *Medical Decisions, Estrogen, and Aging*. New York: Springer.
162. Elstein, A.S., Christensen, C., Cottrell, J.J., Polson, A., and Ng, M. (1999). Effects of prognosis, perceived benefit and decision style upon decision making in critical care. *Critical Care Medicine*, 27: 58–65.
163. Dewey, J. (1929, 1966). *A Theory of Valuation*. Chicago: University of Chicago Press; Neville, R.C. (1974). *The Cosmology of Freedom*. New Haven: Yale University Press; Weissman, D. (1993). *Truth's Debt to Value*. New Haven, Yale University Press; Seigfried, C.H. (1996). *Pragmatism and Feminism: Reweaving the Social Fabric*. Chicago: University of Chicago Press; Seigfried, C.H. (2002, editor). *Feminist Interpretations of John Dewey*. University Park, PA: Pennsylvania State University Press; Seigfried, C.H. (1998). John Dewey's pragmatist feminism. In: *Reading Dewey* (edited by L. Hickman). Bloomington: Indiana University Press.
164. Schulkin, J. (2008). *Medical Decisions, Estrogen, and Aging*. New York: Springer.
165. Sarokin, D. and Schulkin, J. (1994). Co-evolution of rights and environmental justice. *The Environmentalist*, 14: 121–129.
166. Moreno, J.D. (1995). *Deciding Together*. Oxford: Oxford University Press; Moreno, J.D. (2005). *Is There an Ethicist in the House? On the Cutting Edge of Bioethics*. Bloomington: Indiana University Press; Moreno, J.D. (1999, 2003). Bioethics is a naturalism. In: *Pragmatic Bioethics* (edited by G. McGee). Cambridge, MA: MIT Press.
167. Dewey, J. (1916). *Essays in Experimental Logic*. Chicago: University of Chicago Press; Hickman, L.A. (1998). Dewey's theory of inquiry. In: *Reading Dewey* (edited by L. Hickman). Bloomington: Indiana University Press.
168. Peirce, C.S. (1878). Deduction, induction and hypothesis. *Popular Science Monthly*, 13: 470–482.
169. Kahneman, D., Slovic, P., and Tversky, A. (1982, editors). *Judgment under Uncertainty: Heuristics and Biases*. New York: Cambridge University Press; Gigerenzer, G. (2000). *Adaptive Thinking*. Oxford: Oxford University Press; Baron, J. (1988, 2008). *Thinking and Deciding*. Cambridge: Cambridge University Press.
170. Kant, I. (1788, 1956). *Critique of Practical Reason* (transl. L.W. Beck). New York: Bobbs-Merrill.
171. Dewey, J. (1935, 1963). *Liberalism and Social Action*. New York: Capricorn Books; Rawls, J. (1971). *A Theory of Justice*. Cambridge, MA: Harvard University Press.

172. Schulkin, J. (1992). *The Pursuit of Inquiry*. Albany, NY: SUNY Press.
173. Dewey, J. (1939, 1963). *Freedom and Culture*. New York: Capricorn Press.
174. Pellegrino, E.D., and Thomasma, D.C. (1988). *For the Patient's Good*. Oxford: Oxford Univeristy Press.
175. Veatch, R.M. (1981). *A Theory of Medical Ethics*. Boston: Jones and Bartlett.
176. Moreno, J.D., and Bergerm, S. (2010). *Progress in Bioethics: Science, Policy and Politics*. Cambridge, MA: MIT Press.
177. Moore, D.A., Cain, D.M., Loewenstein, G., and Bazerman, M.H. (2005). *Conflicts of Interest*. Cambridge: Cambridge University Press.
178. Warner, R. (1980). *Morality in Medicine*. Sherman Oaks, CA: Alfred Publishing; Veatch, R.M. (1981). *A Theory of Medical Ethics*. Boston: Jones and Bartlett.
179. Moreno, J.D. (1995). *Deciding Together*. Oxford: Oxford University Press; Moreno, J.D. (2005). *Is There an Ethicist in the House? On the Cutting Edge of Bioethics*. Bloomington: Indiana University Press.
180. Schulkin, J. (1996). *The Delicate Balance*. Lanham, MD: University Press of America.
181. Moreno, J.D. (1995). *Deciding Together*. Oxford: Oxford University Press; Moreno, J.D. (1999, 2003). Bioethics is a naturalism. In: *Pragmatic Bioethics* (edited by G. McGee). Cambridge, MA: MIT Press; Moreno, J.D. (2005). *Is There an Ethicist in the House? On the Cutting Edge of Bioethics*. Bloomington: Indiana University Press.
182. Ludmerer, K.L. (1985). *Learning to Heal*. New York: Basic Books; Ludmerer, K.L. (1999). *Time to Heal*. Oxford: Oxford University Press; Flexner, A. (1910). *Medical Education in the United States and Canada: A Report to the Carnegie Foundation for the Advancement of Teaching. Bulletin Number Four.* Boston: Updyke.
183. Kohler, R. E. (1991). *Partners in Science*. Chicago: University of Chicago Press.
184. Oshinsky, D.M. (2005). *Polio: An American Story*. Oxford: Oxford University Press.

5 Psychobiology: A Jamesian Self-regulatory View

1. Sterling, P. (2004). Principles of allostasis. In: *Allostasis, Homeostasis and the Costs of Physiological Adaptation* (edited by J. Schulkin). Cambridge: Cambridge University Press.
2. Stuhr, J. (2010). *100 Years of Pragmatism: William James's Revolutionary Philosophy*. Bloomington: Indiana University Press; Wilshire, B. (2000). *The Primal Roots of American Philosophy*. Univ. Park: Penn State University Press; Gale, R.M. (1999). *The Philosophy of William James: An Introduction*. Cambridge: Cambridge University Press.
3. For the actual experiments that demonstrate self regulation, see: Richter, C.P. (1943). *Total Self-Regulatory Functions in Animals and Man*. New York: Harvey Lecture Series; Richter, C.P. (1965, 1979). *Biological Clocks in Medicine and Psychiatry*. Springfield, IL: Charles C. Thomas.
4. Matthiessen, F.O. (1947, 1980). *The James Family*. New York: Vintage Press.

5. Goodman R.B. (1990). *American Philosophy and the Romantic Tradition.* Cambridge: Cambridge University Press.
6. Barzun, J. (1983). *A Stroll with William James.* New York: Harper and Row.
7. Dewey, J. (1896). The reflex arc concept in psychology. *Psychological Review,* 3: 357–370.
8. See Schulkin, J. (2007). *Effort: A Behavioral Neuroscience Perspective on the Will.* Mahwah, NJ: Earlbaum Press.
9. Barton, R.A. (2004). Binocularity and brain evolution in primates. *Proceedings of the National Academy of Sciences,* 101: 10113–10115; Barton, R.A. (2006). Primate brain evolution: integrating comparative neurophysiological and ethological data. *Evolutionary Anthropology,* 15: 224–236; Lakoff, G., and Johnson, M. (1999). *Philosophy in the Flesh: The Embodied Mind and its Challenge to Western Thought.* New York: Basic Books; Schulkin, J. (2002). *Bodily Sensibility and Intelligent Action.* Oxford: Oxford University Press.
10. Bjork, D.W. (1983). *The Compromised Scientist: William James in the Development of American Psychology.* New York: Columbia University Press.
11. Seigfried, C.H. (1996). *Pragmatism and Feminism: Reweaving the Social Fabric.* Chicago: University of Chicago Press.
12. Novak, B. (2007). *Voyages of the Self.* Oxford: Oxford University Press.
13. Cadwallader, T.C. (1975). Peirce as an experimental psychologist. *Transactions of the Charles S. Peirce Society,* 11: 167–186; Fisch, M.H. (1986). Evolution in American philosophy. In: *Peirce, Semiotic and Pragmatism* (edited by K.L. Ketner and J.W. Kloesel). Bloomington: Indiana University Press.
14. James W. (1890, 1952). *The Principles of Psychology,* vols 1 and 2. New York: Henry Holt.
15. Harrington, A. (1987). *Medicine, Mind and the Double Brain.* Princeton: Princeton University Press; Critchley, H.D. (2005). Neural mechanisms of autonomic, affective, and cognitive integration. *Journal of Comparative Neurology,* 493: 154–166.
16. Barton, R.A. (2004). Binocularity and brain evolution in primates. *Proceedings of the National Academy of Sciences,* 101: 10113–10115; Barton, R.A. (2006). Primate brain evolution: integrating comparative neurophysiological and ethological data. *Evolutionary Anthropology,* 15: 224–236.
17. James, W. (1890, 1952). *Principles of Psychology.* New York: Henry Holt: vol. 1: 79.
18. Schull, J. (1992). Selection: James's principal principle. In: *Reinterpreting the Legacy of William James* (edited by M.E. Donnelly). Washington, DC: American Psychological Assocation: 139–152.
19. Seigfried, C.H. (1984). The positivist foundation in Williams James' *Principles. Review of Metaphysics,* 38: 579–593.
20. Schulkin, J. (2007). *Effort: A Behavioral Neuroscience Perspective on the Will.* Mahwah, NJ: Erlbaum Press.
21. James, W. (1890, 1952). *The Principles of Psychology.* New York: Dover Press; Prinz, J.J. (2002). *Furnishing the Mind.* Cambridge, MA: MIT Press.
22. Merleau-Ponty, M. (1968, 1970). *Themes.* Evanston, IL: Northwestern University Press; Johnson, M. (1987, 1990). *The Body in the Mind.* Chicago: University of Chicago Press; Johnson, M. (1993). *Moral Imagination.* Chicago: University of Chicago Press; Johnson, M. (2007). *The Meaning of the Body.*

Chicago: University of Chicago Press; Kelly, S.D. (2002). Merleau-Ponty on the Body. *Ratio*, 4: 0034–0006; Damasio, A.R. (1999). *The Feeling of What Happens*. New York: Harcourt Press.

23. Jeannerod, M. (1997). *The Cognitive Neuroscience of Action*. Oxford: Blackwell Publishers; Jeannerod, M. (1999). To act or not to act: perspectives on the representation of action. *Quarterly Journal of Experimental Psychology*, 52: 1–29; Berthoz, A. (2000). *The Brain's Sense of Movement*. Cambridge, MA: Harvard University Press; Schulkin, J. (2007). *Effort: A Behavioral Neuroscience Perspective on the Will*. Mahwah, NJ: Erlbaum Press.
24. Desmurget, M., Reilly, K.T., Richard, N., Szathmari, A., Mottolese, C., and Sirigu, A. (2009). Movement intention after parietal cortex stimulation in humans. *Science*, 324: 811–813.
25. James, W. (1890, 1952). *The Principles of Psychology*. New York: Dover Press: vol. 2, 252.
26. Elster, J. (1979, 1988). *Ulysses and the Sirens*. Cambridge: Cambridge University Press; Elster, J. (1983). *Explaining Technical Change*. Cambridge: Cambridge University Press; Schulkin, J. (2007). *Effort: A Behavioral Neuroscience Perspective on the Will*. Mahwah, NJ: Erlbaum Press.
27. Parrott, G.W., and Schulkin, J. (1993). Neuropsychology and the cognitive nature of emotions. *Cognition and Emotion*, 7: 43–59.
28. Dewey, J. (1967). John Dewey, psychology. In: *John Dewey: The Early Works 1882–1898*. Carbondale, IL: Southern Illinois University Press.
29. Ibid. 309.
30. Gallistel, C.R. (1980). *The Organization of Action: A New Synthesis*. Hillsdale, NJ: Lawrence Erlbaum.
31. Dewey, J. (1896). The reflex arc concept in psychology. *Psychological Review*, 3: 357–370.
32. Berthoz, A., and Petite, J.L. (2006, 2008). *The Physiology and Phenomenology of Action*. Oxford: Oxford University Press; Gallistel, C.R. (1993). *The Organization of Learning*. Cambridge, MA: MIT Press; Jackson, P.L., and Decety, J. (2004). Motor cognition: a new paradigm to self and other interactions. *Current Opinion in Neurobiology*, 14: 259–263; Kimura, D. (1993). *Neuromotor Mechanism in Human Communication*. Oxford: Oxford University Press.
33. Eichenbaum, H., and Cohen, N.J. (2001). *From Conditioning to Conscious Recollection*. Oxford: Oxford University Press.
34. Schulkin, J. (2007). *Effort: A Behavioral Neuroscience Perspective on the Will*. Mahwah, NJ: Erlbaum Press.
35. Spinoza, B. (1668, 1955). *On the Improvement of the Understanding* (transl. by R.H.M. Elwes). New York: Dover Press.
36. James, W. (1890, 1952). *The Principles of Psychology*. New York: Henry Holt: vol. 1, 234.
37. Whitehead, A.N. (1927, 1953). *Symbolism*. New York: Macmillan Company; Whitehead, A.N. (1929, 1958). *The Function of Reason*. Boston: Beacon Press; Whitehead, A.N. (1929, 1978). *Process and Reality*. New York: Free Press; Whitehead, A.N. (1933, 1961). *Adventures of Ideas*. New York: Free Press; Whitehead, A.N. (1938, 1967). *Modes of Thought*. New York: Free Press.
38. James, W. (1890, 1952). *The Principles of Psychology*. New York: Henry Holt: vol. 1, 284.

39. See also, e.g., Helmholtz, H. von (1867, 1963). *Handbook of Physiological Optics*. New York: Dover Press.
40. Quine, W.V.O. (1974). *The Roots of Reference*. La Salle, IL: Open Court; Putnam, H. (1990). *Realism With a Human Face*. Cambridge, MA: Harvard University Press.
41. James, W. (1890, 1952). *The Principles of Psychology*. New York: Henry Holt: vol. 1, 423.
42. Menand, L. (2001). *The Metaphysical Club*. New York: Farrar, Straus and Giroux; Menand, L. (2002). *American Studies*. New York: Farrar, Straus and Giroux; Menand, L. (1997). *Pragmatism: A Reader*. New York: Random House.
43. Rosen, J., and Schulkin, J. (1998). From normal fear to pathological anxiety. *Psychological Review*, 105: 325–350; Gloor, P. (1978). *Inputs and Outputs of the Amygdala: What the Amygdala is Trying to Tell the Rest of the Brain*. New York: Wiley Press; Gloor, P. (1997). *The Temporal Lobe and Limbic System*. New York: Oxford University Press.
44. Freud, S. (1924, 1960). *A General Introduction to Psychoanalysis*. New York: Washington Square Press.
45. Piaget, J. (1971, 1975). *Biology and Knowledge*. Chicago: University of Chicago Press.
46. James, W. (1890, 1952). *The Principles of Psychology*. New York: Henry Holt: vol. 1, 104.
47. Rozin, P. (1976). The evolution of intelligence and access to the cognitive unconscious. In: *Progress in Psychobiology and Physiological Psychology* (edited by J. Sprague and A.N. Epstein). New York: Academic Press; Rozin, P. (1998). Evolution and development of brains and cultures. In: *Brain and Mind: Evolutionary Perspectives* (edited by M.S. Gazzaniga and J.S. Altman). Strasbourg, France: Human Frontiers Sciences Program.
48. James, W. (1912, 1967). *Radical Empiricism* (edited by R.B. Perry). New York: Peter Smith; Perry, R.B. (1935). *The Thought and Character of William James*. Boston: Little Brown and Co.
49. Damasio, A.R. (1994). *Descartes' Error: Emotion, Reason, and the Human Brain*. New York: Grosset/Putnam; Damasio, A.R. (1999). *The Feeling of What Happens*. New York: Harcourt Press; Bechara, A. (2005). Decision-making, impulse control and loss of willpower to resist drugs: a neurocognitive perspective. *Nature Neuroscience*, 8: 1458–1463.
50. James, W. (1890, 1952). *The Principles of Psychology*. New York: Dover Press: vol. 2, 389.
51. Ibid. 390.
52. Gallistel, C.R. (1980). *The Organization of Action: A New Synthesis*. Hillsdale, NJ: Lawrence Erlbaum.
53. Langer, S. (1972). *Mind: An Essay on Human Feeling*. Baltimore: Johns Hopkins University Press.
54. Ibid. 45.
55. Rosen, J., and Schulkin, J. (1998). From normal fear to pathological anxiety. *Psychological Review*, 105: 325–350; LeDoux, J.E. (1996). *The Emotional Brain*. New York: Simon and Schuster.
56. James, W. (1890, 1952). *Principles of Psychology*. New York: Henry Holt: vol. 2, 454.

57. Langer, S.K. (1937). *Philosophy in a New Key*. Cambridge, MA: Harvard University Press; Langer, S.K. (1962). *Philosophical Sketches*. New York: Mentor Books; Langer, S.K. (1972). *Mind: An Essay on Human Feeling*. Baltimore: Johns Hopkins University Press.
58. Johnson, M. (1987, 1990). *The Body in the Mind*. Chicago: University of Chicago Press; Johnson, M. (1993). *Moral Imagination*. Chicago: University of Chicago Press; Johnson, M. (2007). *The Meaning of the Body*. Chicago: University of Chicago Press; Damasio, A.R. (1994). *Descartes' Error: Emotion, Reason, and the Human Brain*. New York: Grosset/Putnam; Schulkin, J. (2004). *Bodily Sensibility and Intelligent Action*. Oxford: Oxford University Press.
59. Merleau-Ponty, M. (1968, 1970). *Themes*. Evanston, IL: Northwestern University Press; Varela, F., Thompson, E., and Rosch, E. (1991). *The Embodied Mind*. Cambridge: Cambridge University Press; Gallagher, S. (2005). *How the Body Shapes the Mind*. Oxford: Oxford University Press.
60. LeDoux, J.E. (1996). *The Emotional Brain*. New York: Simon and Schuster; Kagan, J. (1984). *The Nature of the Child*. New York: Basic Books; Kagan, J. (2002). *Surprise, Uncertainty and Mental Structure*. Cambridge, MA: Harvard University Press.
61. James, W. (1902, 1974). *The Varieties of Religious Experience: A Study in Human Nature*. New York: Collier Books, Macmillan Publishing Co.; James, W. (1896, 1956). *The Will to Believe, Human Immortality*. New York: Dover Press.
62. Menand, L. (2001). *The Metaphysical Club*. New York: Farrar, Straus and Giroux; Menand, L. (2002). *American Studies*. New York: Farrar, Straus and Giroux; Menand, L. (1997). *Pragmatism: A Reader*. New York: Random House.
63. Brinkley, D. (2009). *The Wilderness Warrior: Theodore Roosevelt and the Crusade for America*. New York: HarperCollins.
64. James, W. (1896, 1956). *The Will to Believe, Human Immortality*. New York: Dover Press: 223.
65. Ibid. 245.
66. James, W. (1902, 1974). *The Varieties of Religious Experience: A Study in Human Nature*. New York: Collier Books, Macmillan Publishing Co.: 220.
67. Feinstein, H.M. (1984). *Becoming William James*. Ithaca: Cornell University Press; Barzun, J. (1983). *A Stroll with William James*. New York: Harper and Row.
68. Bordogna, R. (2008). *William James at the Boundaries*. Chicago: University of Chicago Press.
69. James, W. (1890, 1952). *The Principles of Psychology*, vol. 2. New York: Henry Holt.
70. Merleau-Ponty, M. (1968, 1970). *Themes*. Evanston, IL: Northwestern University Press; Mead, G.H. (1934, 1972). *Mind, Self and Society*. Chicago: University of Chicago Press.
71. Edie, J.M. (1987). *William James and Phenomenology*. Bloomington: Indiana University Press; Schutz, A. (1967). *The Phenomenology of the Social World*. Evanston, IL: Northwestern University Press.
72. Merleau-Ponty, M. (1942, 1967). *The Structure of Behavior*. Boston: Beacon Press.
73. Heelan, P.A. (1983). *Space Perception and the Philosophy of Science*. Berkeley: University of California Press; Heelan, P.A. (2001). The lifeworld and

scientific interpretation. In: *Handbook of Phenomenology and Medicine*. Waco, TX: Baylor University Press.

74. Stuhr, J. (1997). *Geneological Pragmatism: Philosophy, Experience, and Community*. Albany: SUNY Press; Stuhr, J. (2002). *Pragmatism, Postmodernism, and the Future of Philosophy*. New York: Routledge; Stuhr, J. (2003). *Experience and Criticism: John Dewey's Reconstruction in Philosophy*. Nashville: Vanderbilt University Press; Stuhr, J. (1987). *Classical American Philosophy*. Oxford: Oxford University Press; Stuhr, J. (1980). Santayana's unnatural naturalism. In: *Two Centuries of Philosophy in America* (edited by P. Caw). Totowa, NJ: Rowman and Littlefield; McDermott, J.J. (2007). *The Drama of Possibility: Experience as Philosophy of Culture*. New York: Fordham University Press.
75. James, W. (1912, 1967) *Radical Empricism* (edited by R.B. Perry). New York: Peter Smith. See also Whitehead, A. N. (1927, 1953). *Symbolism*. New York: Macmillan Company.
76. Bergson, H. (1919, 1946). *The Creative Mind* (translated by M. Andison). New York: Citadel Press; Bergson, H. (1911, 1998). *Creative Evolution*. New York: Dover Press; Whitehead, A.N. (1929, 1958). *The Function of Reason*. Boston: Beacon Press; Whitehead, A.N. (1929, 1978). *Process and Reality*. New York: Free Press; Whitehead, A.N. (1933, 1961). *Adventures of Ideas*. New York: Free Press; Whitehead, A.N. (1938, 1967). *Modes of Thought*. New York: Free Press.
77. Bjork, D.W. (1983). *The Compromised Scientist: William James in the Development of American Psychology*. New York: Columbia University Press.
78. Bergson, H. (1919, 1946). *The Creative Mind* (translated by M. Andison). New York: Citadel Press; Bergson, H. (1911, 1998). *Creative Evolution*. New York: Dover Press: 218–219.
79. Darwin, C. (1872, 1965). *The Expression of Emotions in Man and Animals*. Chicago: University of Chicago Press.
80. Ibid. 228.
81. Craig, W. (1918). Appetites and aversions as constituents of instinct. *Biological Bulletin*, 34: 91–107; Tinbergen, N. (1951, 1969). *The Study of Instinct*. Oxford: Oxford University Press.
82. Midgley, M. (1979,1995). *Beast and Man*. London: Routledge: 51.
83. Darwin, C. (1872, 1965). *The Expression of Emotions in Man and Animals*. Chicago: University of Chicago Press: 256.
84. Darwin, C. (1871, 1874). *Descent of Man*. New York: Rand McNally and Company: 65.
85. Darwin, C. (1872, 1965). *The Expression of Emotions in Man and Animals*. Chicago: University of Chicago Press; Gould, S.J. (2002). *The Structure of Evolutionary Theory*. Cambridge, MA: Harvard University Press.
86. James, W. (1887) Some human instincts. *Popular Science Monthly*, 31: 366.
87. Ibid.
88. James, W. (1890, 1952). *The Principles of Psychology*. New York: Henry Holt.
89. Ibid. 366.
90. Ibid. 390.
91. Langer, S. (1972). *Mind: An Essay on Human Feeling*. Baltimore: Johns Hopkins University Press: 36.
92. Meyer, A. (1915). Objective psychology or psychobiology with subordination of the medically useless contrast of mental and physical. *Journal of the American Medical Association*, 65: 860–863; Meyer, A. (1935). Scope and teaching of psychobiology. *Journal of the American Medical Association*, 10: 93–98.

93. Schulkin J. (2005). *Curt Richter: A Life in the Laboratory.* Baltimore: Johns Hopkins University Press.
94. Meyer, A. (1915). Objective psychology or psychobiology with subordination of the medically useless contrast of mental and physical. *Journal of the American Medical Association*, 65: 861.
95. Meyer A. (1935). Scope and teaching of psychobiology. *Journal of the American Medical Association*, 10: 94.
96. Ibid.
97. Meyer, A. (1915). Objective psychology or psychobiology with subordination of the medically useless contrast of mental and physical. *Journal of the American Medical Association*, 65: 861.
98. Meyer A. (1951). *The Collected Papers of Adolf Meyer*, 4 vols (edited by E.E. Winters). Baltimore: Johns Hopkins University Press; Leys, R. (1984). Meyer, Watson, and the dangers of behaviorism. *Journal of the History of the Behavioral Sciences*, 20: 128–149.
99. Schneider, H.W. (1946, 1963). *A History of American Philosophy.* New York: Columbia University Press.
100. Klerman, G. (1979). The Psychobiology of affective states: the legacy of Adolph Meyer. In: *Research in the Psychobiology of Human Behavior.* Baltimore, MD: Johns Hopkins University Press.
101. Meyer, A. (1931, 1957). *Psychobiology.* Springfield, IL: Charles C. Thomas: 47.
102. Ibid.
103. Dewey, J. (1916). *Essays in Experimental Logic.* Chicago: University of Chicago Press.
104. James, W. (1907, 1969). *Pragmatism.* Cleveland: Meridian Books, The World Publishing Co.: 45.
105. Dewey, J. (1925, 1989). *Experience and Nature.* La Salle, IL: Open Court.
106. Richter, C.P. (1943). *Total Self-Regulatory Functions in Animals and Man.* New York: Harvey Lecture Series.
107. Cadwallader, T.C. (1975). Peirce as an experimental psychologist. *Transactions of the Charles S. Peirce Society*, 11: 167–186; Schulkin, J. (2005). *Curt Richter: A Life in the Laboratory.* Baltimore: Johns Hopkins University Press.
108. Smith, J.E. (1978). *Purpose and Thought.* New Haven: Yale University Press; Peirce, C.S. (1878). Deduction, induction and hypothesis. *Popular Science Monthly*, 13: 470–482; Peirce, C.S. (1899, 1992). *Reasoning and the Logic of Things* (edited by K.L. Ketner and H. Putnam). Cambridge, MA: Harvard University Press.
109. James, W. (1907, 1969). *Pragmatism.* Cleveland: Meridian Books, The World Publishing Co.: 54.
110. Wirth, J.B. (1989). *Richter and Magendie.* The First International Cyberconference on the Psychobiology of Curt P. Richter.
111. James, W. (1890, 1952). *The Principles of Psychology.* New York: Henry Holt.
112. Dewey, J. (1910, 1965). *The Influence of Darwin on Philosophy.* Bloomington: Indiana University Press.
113. Cannon, W.B. (1932, 1963). *The Wisdom of the Body.* New York: W.W. Norton.
114. Dewey, J. (1925, 1989). *Experience and Nature.* La Salle, IL: Open Court; Richter, C.P. (1943). *Total Self-Regulatory Functions in Animals and Man.* New York: Harvey Lecture Series.

115. James, W. (1887) Some human instincts. *Popular Science Monthly*, 31: 160–176; James, W. (1890, 1952). *The Principles of Psychology*. New York: Henry Holt.
116. Darwin, C. (1872, 1965). *The Expression of Emotions in Man and Animals*. Chicago: University of Chicago Press.
117. Meyer, A. (1915). Objective psychology or psychobiology with subordination of the medically useless contrast of mental and physical. *Journal of the American Medical Association*, 65: 860–863.
118. Dewey, J. (1925, 1989). *Experience and Nature*. La Salle, IL: Open Court.
119. Richter, C.P. (1952). Domestication of the Norway rat and its implication for the study of genetics in man. *American Journal of Human Genetics*, 4: 273–285.
120. Richter, C.P. (1943). *Total Self-Regulatory Functions in Animals and Man*. New York: Harvey Lecture Series.
121. Dewey, J. (1925, 1989). *Experience and Nature*. New York: Dover Press.
122. Perry, R.B. (1935). *The Thought and Character of William James*. Boston: Little Brown and Co.
123. Cannon, W.B. (1932, 1963). *The Wisdom of the Body*. New York: W.W. Norton: 24.
124. Cannon, W.B. (1915, 1929, 1963). *Bodily Changes in Pain, Hunger, Fear and Rage*. New York: Harper Torchbooks.
125. Cannon, W.B. (1932, 1963). *The Wisdom of the Body*. New York: W.W. Norton.
126. Richter, C.P. (1941). The biology of drive. *Psychosomatic Medicine*, 3: 105–110.
127. Schulkin, J. (2003). *Rethinking Homeostasis*. Cambridge, MA: MIT Press.
128. See two different views that underlie the cephalic anticipatory mechanisms: Rizzolatti, G., and Luppino, G. (2001). The cortical motor system. *Neuron*, 31: 889–901; Buzsaki, G. (2006). *Rhythms of the Brain*. Oxford: Oxford University Press.
129. Sterling, P. (2004) Principles of allostasis. In: *Allostasis, Homeostasis and the Costs of Physiological Adaptation* (edited by J. Schulkin). Cambridge: Cambridge University Press.
130. Tinbergen, N. (1951, 1969). *The Study of Instinct*. Oxford: Oxford University Press; Hinde, R. A. (1970). *Animal Behavior* (2nd edn). New York: McGraw-Hill; Hinde, R.A. (1968). Critique of energy models of motivation. In: *Motivation* (edited by D. Bindra and J. Stewart). Baltimore: Penguin Books.
131. Sterling, P. (2004). Principles of allostasis. In: *Allostasis, Homeostasis and the Costs of Physiological Adaptation* (edited by J. Schulkin). Cambridge: Cambridge University Press.
132. Jackson, J.H. (1884, 1958). Evolution and dissolution of the nervous system. In: *Collected Works of John Hughlings Jackson*, vol. II (edited by J. Taylor). London: Staples Press; James, W. (1890, 1952). *The Principles of Psychology*. New York: Henry Holt.
133. Pavlov, I.P. (1897, 1902). *The Work of the Digestive Glands* (translated by W.H. Thompson). London: Charles Griffin; Powley, T.L. (1977). The ventralmedial hypothalamic syndrome, satiety and cephalic phase. *Psychological Review*, 84: 89–126.
134. Adolphs, R. (1999). Social cognition and the human brain. *Trends in Cognitive Sciences*, 3: 469–479; Adolphs, R., Denburg, N.L., and Tranel,

D. (2001). The amygdala's role in long-term declarative memory for gist and detail. *Behavioral Neuroscience*, 112: 983–992; Gigerenzer, G. (2000). *Adaptive Thinking*. Oxford: Oxford University Press.

135. Adolphs, R., Denburg, N.L., and Tranel, D. (2001). The amygdala's role in long-term declarative memory for gist and detail. *Behavioral Neuroscience*, 112: 983–992.
136. Gigerenzer, G. (2000). *Adaptive Thinking*. Oxford: Oxford University Press; Gigerenzer, G. (2007). *Gut Feelings*. New York: Viking Press; and see Chapter 4.
137. Adolphs, R., Denburg, N.L., and Tranel, D. (2001). The amygdala's role in long-term declarative memory for gist and detail. *Behavioral Neuroscience*, 112: 983–992; Schulkin, J. (2004). *Bodily Sensibility: Intelligent Action*. Oxford: Oxford University Press.
138. Dewey, J. (1925, 1989). *Experience and Nature*. La Salle, IL: Open Court.
139. See also Merleau-Ponty, M. (1968, 1970). *Themes*. Evanston, IL: Northwestern University Press.

6 Social Smarts, Moral Sentiments, Social Constructs

1. Addams, J. (1911). *Democracy and Social Ethics*. London: Macmillan; Addams, J. (1915). *Newer Ideals of Peace*. London: Macmillan.
2. Mead, G.H. (1928, 1972). *The Philosophy of the Act*. Chicago: University of Chicago Press; Mead, G.H. (1934, 1972). *Mind, Self and Society*, Chicago: University of Chicago Press; Mead, G. H. (1934, 1964). Evolution becomes a general idea. In: *Selected Papers*. Chicago: University of Chicago Press; Mead, G.H. (1936). *Movements of Thought in the Nineteenth Century* (edited by C.W. Morris). Chicago: University of Chicago Press; Joas, H. (1985). *G.H. Mead: A Contemporary Rexamination of his Thought*. Cambridge, MA: MIT Press; Joas, H. (1993). *Pragmatism and Social Theory*. Chicago: University of Chicago Press; Morris, C. (1970). *The Pragmatic Movement in American Philosophy*. New York: George Braziller.
3. Watson, J.B. (1930, 1961). Autobiography. In: *A History of Psychology in Autobiography*. New York: Russell and Russell.
4. Habermas, J. (1967, 1988). *On the Logic of the Social Sciences* (translated by S.W. Nicholson and J.A. Stark). Cambridge, MA: MIT Press; Habermas, J. (1996). *The Liberating Power of Symbols*. Cambridge, MA: MIT Press.
5. Whitehead, A.N. (1929, 1978). *Process and Reality*. New York: Free Press.
6. Dalton, T.C. (2002). *Becoming John Dewey: Dilemmas of a Philosopher and Naturalist*. Bloomington: Indiana University Press.
7. Mead, G.H. (1932, 1980). *The Philosophy of the Present*. Chicago: University of Chicago Press. Bergson, H. (1919, 1946). *The Creative Mind* (translated by M. Andison). New York: Citadel Press; Bergson, H. (1911, 1998). *Creative Evolution*. New York: Dover Press; Whitehead, A.N. (1927, 1953). *Symbolism*. New York: Macmillan Company; Whitehead, A.N. (1929, 1978). *Process and Reality*. New York: Free Press.
8. Mead, G.H. (1932, 1980). *The Philosophy of the Present*. Chicago: University of Chicago Press; Merleau-Ponty, M. (1968, 1970). *Themes*. Evanston, IL: Northwestern University Press.
9. Mead, G.H. (1934). *Mind, Self, and Society: From the Standpoint of a Social Behaviorist*. Chicago: University of Chicago Press: vol. 1, 78.

10. Ibid. 134.
11. Mead, G.H. (1932, 1980). *The Philosophy of the Present*. Chicago: University of Chicago Press; Mead, G.H. (1928, 1972). *The Philosophy of the Act*. Chicago: University of Chicago Press; Mead, G.H. (1934, 1972). *Mind, Self and Society*. Chicago: University of Chicago Press; Mead, G.H. (1934, 1964). Evolution becomes a general idea. In: *Selected Papers*. Chicago: University of Chicago Press; Mead, G.H. (1936). *Movements of Thought in the Nineteenth Century* (edited by C.W. Morris). Chicago: University of Chicago Press.
12. Mead, G.H. (1928, 1972). *The Philosophy of the Act*, vol. 3. Chicago: University of Chicago Press: 25.
13. Franks, D. (2010). *Neurosociology: The Nexus Between Neuroscience and Social Psychology*. New York: Springer.
14. Mead, G.H. (1928, 1972). *The Philosophy of the Act*, vol. 3. Chicago: University of Chicago Press; Dennett, D. (1987). *The Intentional Stance*. Cambridge, MA: MIT Press.
15. Merleau-Ponty, M. (1968, 1970). *Themes*. Evanston, IL: Northwestern University Press.
16. Joas, H. (1985). *G.H. Mead: A Contemporary Reexamination of his Thought*. Cambridge, MA: MIT Press; Joas, H. (1993). *Pragmatism and Social Theory*. Chicago: University of Chicago Press.
17. Mead, G.H. (1932, 1980). *The Philosophy of the Present*. Chicago: University of Chicago Press: 68.
18. Mead, G.H. (1936). *Movements of Thought in the Nineteenth Century* (edited by C.W. Morris). Chicago: University of Chicago Press: 18.
19. Mead, G.H. (1928, 1972). *The Philosophy of the Act*, vol. 3. Chicago: University of Chicago Press: 184.
20. Ibid. 183.
21. Schulkin, J. (2009). *Cognitive Adaptation: A Pragmatist Perspective*. Cambridge: Cambridge University Press; Barton, R.A. (2004). Binocularity and brain evolution in primates. *Proceedings of the National Academy of Sciences*, 101: 10113–10115; Barton, R.A. (2006). Primate brain evolution: integrating comparative neurophysiological and ethological data. *Evolutionary Anthropology*, 15: 224–236; Decety, J., and Jackson, P.W. (2006). A social neuroscience perspective on empathy. *Current Directions in Psychological Science*, 15: 54–58.
22. Jackson, J.H. (1884, 1958). Evolution and dissolution of the nervous system. In: *Selected Writings of John Hughlings Jackson*, vol. II (edited by J. Taylor). London: Staples Press; James, W. (1890, 1952). *The Principles of Psychology*. New York: Henry Holt; Gallistel, C.R. (1980). *The Organization of Action: A New Synthesis*. Hillsdale, NJ: Lawrence Erlbaum.
23. Passingham, R.E. (2008). *What Is Special about the Human Brain?* Oxford: Oxford University Press.
24. Dunbar, R.I.M. (2003). The social brain. *Annual Review of Anthropology*, 32: 163–181; Barton, R.A., Aggleton, J.P., and Grenyer, R. (2003). Evolutionary coherence of the mammalian amygdala. *Procedings of the Royal Society B: Biological Sciences*, 270: 539–543.
25. Panksepp, J. (1998). *Affective Neuroscience: The Foundations of Human and Animal Emotions*. New York: Oxford University Press; Schulkin, J. (2004). *Bodily Sensibility*. Oxford: Oxford University Press.

26. LeDoux, J.E. (1996). *The Emotional Brain*. New York: Simon and Schuster; Aggleton, J. (1992, 2000). *The Amygdala*. Oxford: Oxford University Press.
27. Stephan, H., Frahm, H.D., and Baron, G. (1981). New and revised data on volumes of brain structure in insectivores and primates. *Folia Primatologica*, 35: 1–29.
28. Barger, N., Stefanacci, L., and Semendeferi, K. (2007). A comparative volumetric analysis of the amygdaloid complex and basolateral division in the human and ape brain. *American Journal of Physical Anthropology*, 134: 392–403.
29. Whiten, A., and Byrne, R.W. (1997, editors). *Machiavellian Intelligence II: Extensions and Evaluations*. Cambridge: Cambridge University Press; Mead, G.H. (1934, 1972). *Mind, Self, and Society*, vol. 1. Chicago: University of Chicago Press: 78.
30. Byrne, R.W., and Corp, N. (2004). Neocortex size predicts deception rate in primates. *Proceedings of the Royal Society*, 271: 1693–1699.
31. Jackson, J.H. (1884, 1958). Evolution and dissolution of the nervous system. In: *Selected Writings of John Hughlings Jackson*, vol. II (edited by J. Taylor). London: Staples Press; James, W. (1890, 1952). *The Principles of Psychology*. New York: Henry Holt.
32. Rozin, P. (1976). The Evolution of intelligence and access to the cognitive unconscious. In: *Progress in Psychobiology and Physiological Psychology* (edited by J. Sprague and A.N. Epstein). New York: Academic Press; Gibson, J.J. (1966). *The Senses Considered as Perceptual Systems*. New York: Houghton-Mifflin.
33. Humphrey, N. (1976). The social function of intellect. In: *Growing Points in Ethology* (edited by P.P.G. Bateson and R.A. Hinde). Cambridge: Cambridge University Press: 307–317; Mithen, S. (1996). *The Prehistory of the Mind: The Cognitive Origins of Art and Science*. London: Thames and Hudson, Ltd; Mithen, S. (2006). *The Singing Neanderthal*. Cambridge, MA: Harvard University Press; Levinson, S. (2006). Cognition at the heart of human interaction. *Discourse Studies*, 8: 85–93; Levinson, S. (2003). *Space in Language and Cognition*. Cambridge: Cambridge University Press.
34. Levinson, S. (2006). Cognition at the heart of human interaction. *Discourse Studies*, 8: 85–93; Levinson, S. (2003). *Space in Language and Cognition*. Cambridge: Cambridge University Press.
35. Gibson, K.R., and Ingold, T. (1993, editors). *Tools, Language and Cognition in Human Evolution*. Cambridge: Cambridge University Press.
36. deWaal, F., and Lanting, F. (1997). *Bonobo: The Forgotten Ape*. Berkeley: University of California Press.
37. Schulkin, J. (2004). *Bodily Sensibility: Intelligent Action*. Oxford: Oxford University Press.
38. Tomasello, M., Kruger, A.C., and Ratner, H.H. (1993). Cultural learning. *Behavioral and Brain Sciences*, 16: 495–552.
39. Herrmann, E., Call, J., Hernandez-Lloreda, M.V. et al. (2007). Humans have evolved specialized skills of social cognition. *Science*, 317: 1360–1366.
40. Noordzij, M.L., Newman-Norlund, S.E., de Ruiter, J. P., Hagoort, P., Levinson, S.C., and Toni, I. (2009). Brain mechanisms underlying human communication. *Frontiers in Human Neuroscience*, 3: 1–13.

41. Habermas, J. (1967, 1988). *On the Logic of the Social Sciences* (transl. S.W. Nicholson and J.A. Stark). Cambridge, MA: MIT Press; Apel, K.O. (1981). *Charles S. Peirce: From Pragmatism to Pragmaticism* (translated by J.M. Krois). Amherst: University of Massachusetts Press; Kitcher, P. (2011). *The Ethical Project*. Cambridge, MA: Harvard University Press.
42. Dewey, J. (1920, 1948). *Reconstruction in Philosophy*. Boston: Beacon Press.
43. Dewey, J. (1908, 1970). *Theory of the Moral Life*. New York: Holt, Rinehart and Winston: 124; Jaspers, K. (1913, 1997). *General Psychopathology*. Baltimore: Johns Hopkins University Press.
44. Seigfried. C.H. (1996). *Pragmatism and Feminism: Reweaving the Social Fabric*. Chicago: University of Chicago Press; Seigfried, C.H. (1998). John Dewey's Pragmatist Feminism. In: *Reading Dewey* (edited by L. Hickman). Bloomington: Indiana University Press.
45. Hauser, M.D., Chomsky, N. and Tecumseh Fitch, W. (2002). The faculty of language: What is it, who has it, and how did it evolve? *Science*, 298: 1569–1576; Ray, J. (1660). *Catalogus plantarum circa Cantabrigiam Nascentium*. Cambridge: Field.
46. Lakoff, G., and Johnson, M. (1999). *Philosophy in the Flesh: The Embodied Mind and its Challenge to Western Thought*. New York: Basic Books.
47. Greene, J.D., Sommerville, R.B., Nystrom, L.E., Darley, J.M., and Cohen, J.D. (2001). An fMRI investigation of emotional engagement in moral judgment. *Science*, 293: 2105–2108; Greene, J.D., Morelli, S.A., Lowenberg, K., Nystrom, L.E., and Cohen, J.D. (2008). Cognitive load selectively interferes with utilitarian moral judgement. *Cognition*, 107: 1144–1154.
48. Moll, J., and Schulkin J. (2009). Social attachment and aversion in human moral cognition. *Neuroscience and Biobehavioral Reviews*, 33: 456–465.
49. Sabini, J. and Schulkin, J. (1994). Biological realism and social constructivism. *Journal for the Theory of Social Behavior*, 224: 207–217.
50. Gilligan, C. (1982, 1993). *In a Different Voice: Psychological Theory and Women's Development*. Cambridge, MA: Harvard University Press; Seigfried, C.H. (1996). *Pragmatism and Feminism: Reweaving the Social Fabric*. Chicago: University of Chicago Press; Seigfried, C.H. (ed.) (2002). *Feminist Interpretations of John Dewey*. University Park, PA; Pennsylvania State University Press; Seigfried, C.H. (1998). John Dewey's Pragmatist Feminism. In: *Reading Dewey* (edited by L. Hickman). Bloomington: Indiana University Press.
51. Schulkin, J. (2004). *Bodily Sensibility: Intelligent Action*. Oxford: Oxford University Press.
52. Wilson, J.Q. (1993). *The Moral Sense*. New York: Basic Books.
53. Moll, J., and Schulkin, J. (2009). Social attachment and aversion in human moral cognition. *Neuroscience and Biobehavioral Reviews*, 33: 456–465.
54. Parrott, G.W., and Schulkin, J. (1993). Neuropsychology and the cognitive nature of emotions. *Cognition and Emotion*, 7: 43–59.
55. Moll, J., De Oliveira-Souza, R., and Zahn, R. (2008). The neural basis of moral cognition: Sentiments, concepts, and values. *Annals of the New York Academy of Sciences*, 1124: 161–180.
56. Moll, J., and Schulkin, J. (2009). Social attachment and aversion in human moral cognition. *Neuroscience and Biobehavioral Reviews*, 33: 456–465.

57. Rozin, P. (1976). The evolution of intelligence and access to the cognitive unconscious. In: *Progress in Psychobiology and Physiological Psychology* (edited by J. Sprague and A.N. Epstein). New York: Academic Press.
58. Rozin, P. (1999). The process of moralization. *Psychological Science*, 10: 218–221; Rozin, P. (2005). The meaning of natural: process more important than content. *Psychological Science*, 16: 652–658; Rozin, P., and Fallon, A.E. (1987). A perspective on disgust. *Psychological Review*, 94: 23–41; Rozin, P., Haidt, J., and McCauley, C.R. (1993). Disgust. In: *Handbook of Emotions* (edited by M. Lewis and J.M. Haviland). New York: Guilford.
59. Nussbaum, M.C. (2004). *Hiding from Humanity*. Princeton: Princeton University Press: 83.
60. Gigerenzer, G. (2000). *Adaptive Thinking*. Oxford: Oxford University Press; Gigerenzer, G. (2007). *Gut Feelings*. New York: Viking Press; Gigerenzer, G. (2003). Why does framing influence judgment? *Journal of General Internal Medicine*, 18: 906–961.
61. Haidt, J. (2001). The emotional dog and its rational tail. *Psychological Review*, 108: 814–824; Haidt, J. (2007). The new synthesis in moral psychology. *Science*, 316: 998–1002.
62. Piaget, J. (1971, 1975). *Biology and Knowledge*. Chicago: University of Chicago Press.
63. Nussbaum, F.L. (1953). *The Triumph of Science and Reason*. New York: Harper and Row; Nussbaum, M.C. (1997). *Cultivating Humanity*. Cambridge, MA: Harvard University Press; Nussbaum, M.C. (2004). *Hiding from Humanity*. Princeton: Princeton University Press; Haidt, J. (2001). The emotional dog and its rational tail. *Psychological Review*, 108: 814–824; Haidt, J. (2007). The new synthesis in moral psychology. *Science*, 316: 998–1002.
64. Kagan, J. (1984). *The Nature of the Child*. New York: Basic Books; Kagan, J. (2002). *Surprise, Uncertainty and Mental Structure*. Cambridge, MA: Harvard University Press.
65. Darwin, C. (1871, 1874). *Descent of Man*. New York: Rand McNally and Company.
66. Moll, J., and Schulkin, J. (2009). Social attachment and aversion in human moral cognition. *Neuroscience and Biobehavioral Reviews*, 33: 456–465.
67. Moll. J., Eslinger, P.J., and De Oliveira-Souza, R. (2001). Frontopolar and anterior temporal cortex activation in a moral judgment: Preliminary functional MRI results in normal subjects. *Arquivos de Neuro-Psiquiatria*, 59: 657–664.
68. Greene, J.D., Sommerville, R.B., Nystrom, L.E., Darley, J.M., and Cohen, J.D. (2001). An fMRI investigation of emotional engagement in moral judgment. *Science*, 293: 2105–2108; Greene, J.D., Morelli, S.A., Lowenberg, K., Nystrom, L.E., and Cohen, J.D. (2008). Cognitive load selectively interferes with utilitarian moral judgement. *Cognition*, 107: 1144–1154.
69. Young, L., and Koenigs, M. (2007). Investigating emotion in moral cognition: A review of evidence from functional neuroimaging and neuropsychology. *British Medical Bulletin*, 84: 69–79.
70. Greene, J.D., Sommerville, R.B., Nystrom, L.E., Darley, J.M., and Cohen, J.D. (2001). An fMRI investigation of emotional engagement in moral judgment. *Science*, 293: 2105–2108.

71. Berthoz, A. (2000). *The Brain's Sense of Movement*. Cambridge, MA: Harvard University Press; Moll, J., de Oliveira-Souza, R., Moll, F.T., Ignácio, F.A., Bramati, I.E., Caparelli-Dáquer, E.M., and Eslinger, P.J. (2005). The moral affiliations of disgust: a functional MRI study. *Cognitive and Behavioral Neurology*, 18: 68–78; Greene, J.D., Nystrom, L.E., Engell, A.D., Darley, J.M., and Cohen, J.D. (2004). The neural bases of cognitive conflict and control in moral judgment. *Neuron*, 44: 389–400; Young, L., and Koenigs, M. (2007). Investigating emotion in moral cognition: A review of evidence from functional neuroimaging and neuropsychology. *British Medical Bulletin*, 84: 69–79.
72. Moll, J., De Oliveira-Souza, R., Eslinger, P.J., Bramati, I., Mourao-Miranda, J., Angelo Andreiuolo P., and Pessoa, L. (2002). The neural correlates of moral sensitivity: a functional magnetic resonance imaging investigation of basic and moral emotions. *Journal of Neuroscience*, 22: 2730–2736.
73. Passingham, R.E. (2008). *What Is Special about the Human Brain?* Oxford: Oxford University Press.
74. Morgan, M.A., Schulkin, J., and LeDoux, J.E. (2003). Ventral medial prefrontal cortex and emotional perseveration: The memory for prior extinction training. *Behavioral Brain Research*, 146: 121–130; Bechara, A., Damasio, A., and Damasio, H. (2000). Emotion, decision-making and the orbitofrontal cortex. *Cerebral Cortex*, 10: 295–307; Quirk, G.J., Likhtik, E., Pellerier, J.G., and Pare, D. (2003). Stimulation of medial prefrontal cortex decreases the responsiveness of central amygdala output neuron. *Journal of Neuroscience*, 23: 8800–8807.
75. MacLean, P.D. (1990). *The Triune Brain in Evolution: Role in Paleocerebral Functions*. New York: Plenum Press.
76. Broca, P. (1863). Localization des functions cerebrales. Siege du langage articule. *Bulletins de la Societe d'Anthropologie (Paris)*, 4: 200–203; MacLean, P.D. (1990). *The Triune Brain in Evolution: Role in Paleocerebral Functions*. New York: Plenum Press.
77. Herbert, J., and Schulkin J. (2002). Neurochemical coding of adaptive responses in the limbic system. In: *Hormones, Brain and Behavior* (edited by D. Pfaff). New York: Elsevier Press; Carter, C.S., Lederhendler, I.L., and Kirkpatrick, B. (1997, 1999). *The Integrative Neurobiology of Affiliation*. Cambridge, MA: MIT Press.
78. Zak, P.J., Kurzban, R., and Matzner, W.T. (2005). Oxytocin is associated with human trustworthiness. *Hormones and Behavior*, 48: 522–527.
79. Kirsch, P., Esslinger, C., Chen, Q., Mier, D., Lis, S., Siddhanti, S., Gruppe, H., Mattay, V.S., Gallhofer, B., and Meyer-Lindenberg, A. (2005). Oxytocin modulates neural circuitry for social cognition and fear in humans. *Journal of Neuroscience*, 25: 11489–11493.
80. Kosfeld, M., Heinrich, M, Zak, P.J., Fischbacher, U., and Fehr, E. (2005). Oxytocin increases trust in humans. *Science*, 435: 673–676.
81. Darwin, C. (1872, 1965). *The Expression of Emotions in Man and Animals*. Chicago: University of Chicago Press; Rozin, P. (1998). Evolution and development of brains and cultures. In: *Brain and Mind: Evolutionary Perspectives* (edited by M.S. Gazzaniga and J.S. Altman). Strasbourg, France: Human Frontiers Sciences Program.
82. Ibid.; Haidt, J. (2001). The emotional dog and its rational tail. *Psychological Review*, 108: 814–824.

83. Adolphs, R. (1999). Social cognition and the human brain. *Trends in Cognitive Sciences*, 3: 469–479.
84. Baron-Cohen, S. (1995, 2000). *Mindblindness*. Cambridge, MA: MIT Press.
85. Adolphs, R. (1999). Social cognition and the human brain. *Trends in Cognitive Sciences*, 3: 469–479; Young, L., and Koenigs, M. (2007). Investigating emotion in moral cognition: A review of evidence from functional neuroimaging and neuropsychology. *British Medical Bulletin*, 84: 69–79.
86. Nussbaum, F.L. (1953). *The Triumph of Science and Reason*. New York: Harper and Row; Nussbaum, M.C. (1997). *Cultivating Humanity*. Cambridge, MA: Harvard University Press; Nussbaum, M.C. (2004). *Hiding from Humanity*. Princeton: Princeton University Press; Moll, J., and Schulkin J. (2009). Social attachment and aversion in human moral cognition. *Neuroscience and Biobehavioral Reviews*, 33: 456–465.
87. Smith, A. (1759, 1882). *The Theory of Moral Sentiments*. Indianapolis: Liberty Classics; Hume, D. (1748, 1962). *An Inquiry Concerning Human Understanding*. New York: Washington Square Press; Locke, J. (1690, 1959). *An Essay Concerning Human Understanding*. New York: Dover Press; Locke, J. (1692–1704, 1955). *A Letter Concerning Toleration*. Indianapolis: Bobbs-Merrill Co.
88. Smith, A. (1759, 1882). *The Theory of Moral Sentiments*. Indianapolis: Liberty Classics.
89. Johnson, M. (1987, 1990). *The Body in the Mind*. Chicago: University of Chicago Press; Johnson, M. (1993). *Moral Imagination*. Chicago: University of Chicago Press; Johnson, M. (2007). *The Meaning of the Body*. Chicago: University of Chicago Press.
90. Rousseau, J.J. (1762, 1956). *Emile*. New York: Columbia University Press.
91. Kant, I. (1788, 1956). *Critique of Practical Reason* (translated by L.W. Beck). New York: Bobbs-Merrill; Rawls, J. (1971). *A Theory of Justice*. Cambridge, MA: Harvard University Press.
92. Adolphs, R. (1999). Social cognition and the human brain. *Trends in Cognitive Sciences*, 3: 469–479; Greene, J. D., and Haidt, J. (2002). How (and where) does moral judgment work? *Trends in Cognitive Science*, 6: 517–523; Moll, H., and Tomasello, M. (2007). Cooperation and human cognition: the Vygotskian intelligence hypothesis. *Philosophical Transactions of the Royal Society, B: Biological Sciences*, 362: 639–648.
93. Bechara, A. (2005). Decision-making, impulse control and loss of willpower to resist drugs: A neurocognitive perspective. *Nature Neuroscience*, 8: 1458–1463.
94. Damasio, A.R. (1996). The somatic marker hypothesis and the possible functions of the prefrontal cortex. *Philosophical Transactions of the Royal Society*, 354: 1413–1420; Cohen, J.D., McClure, S.M., and Yu, A.J. (2007). Should I stay or should I go? How the brain manages the trade-off between exploitation and exploration. *Philosophical Transactions of the Royal Society, B*, 362: 933–942.
95. Jackson, J.H. (1884, 1958). Evolution and dissolution of the nervous system. In: *Selected Writings of John Hughlings Jackson*, vol. II (edited by J. Taylor). London: Staples Press; James, W. (1890, 1952). *The Principles of Psychology*. New York: Henry Holt; Bechara, A. (2005). Decision-making, impulse control and loss of willpower to resist drugs: A neurocognitive perspective. *Nature Neuroscience*, 8: 1458–1463.

96. Jackson, J.H. (1884, 1958). Evolution and dissolution of the nervous system. In: *Selected Writings of John Hughlings Jackson*, vol. II (edited by J. Taylor). London: Staples Press; Greene, J.D., Sommerville, R.B., Nystrom, L.E., Darley, J.M., and Cohen, J.D. (2001). An fMRI investigation of emotional engagement in moral judgment. *Science*, 293: 2105–2108; Damasio, A.R. (1996). The somatic marker hypothesis and the possible functions of the prefrontal cortex. *Philosophical Transactions of the Royal Society*, 354: 1413–1420.
97. Egner, T., and Hirsch, J. (2005). Cognitive control mechanisms resolve conflict through cortical amplification of task-relevant information. *Nature Neuroscience*, 8: 1784–1790; Botvinick, M. M., Braver, T.S., Barch, D. M., Carter, C. S., and Cohen, J. D. (2001). Conflict monitoring and cognitive control. *Psychological Review*, 108: 624–652.
98. Diamond, A., and Goldman-Rakic, P.S. (1989). Comparison of human infants and rhesus monkeys on Piaget's AB task: evidence for dependence on dorsolateral prefrontal cortex. *Experimental Brain Research*, 74: 24–40.
99. Diamond, A. (2001). A model system for studying the role of dopamine in the prefrontal cortex during early development of humans: early and continuously treated phenylketonuria. In: *Handbook of Developmental Cognitive Neuroscience* (edited by C.A. Nelson and M. Luciana). Cambridge, MA: MIT Press: 433–472.
100. Ibid.
101. Jackson, J.H. (1884, 1958). Evolution and dissolution of the nervous system. In: *Selected Writings of John Hughlings Jackson*, vol. II (edited by J. Taylor). London: Staples Press.
102. Schutz, A., and Luckman, A. (1973). *The Structures of the Life-World*. Evanston, IL: Northwestern University Press; Wittgenstein, L. (1953, 1968). *Philosophical Investigations*. New York: Macmillan Publishing.
103. Peirce, C.S. (1899, 1992). *Reasoning and the Logic of Things* (edited by K.L. Ketner and H. Putnam). Cambridge, MA: Harvard University Press.
104. Aristotle. (1968). *De Anima*. Oxford: Oxford University Press.
105. Goldman, A.I. (1999). *Knowledge in a Social World*. Oxford: Clarendon Press.
106. Ekman, P. (1972). Universals and cultural differences in facial expressions of emotion. In: *Nebraska Symposium on Motivation, 1971* (edited by J. Cole). Lincoln: University of Nebraska Press.
107. Nussbaum, M.C. (1997). *Cultivating Humanity*. Cambridge, MA: Harvard University Press.
108. Levinson, S. (2006). Cognition at the heart of human interaction. *Discourse Studies*, 8: 85–93; Levinson, S. (2003). *Space in Language and Cognition*. Cambridge: Cambridge University Press.
109. Parrott, G.W., and Schulkin, J. (1993). Neuropsychology and the cognitive nature of emotions. *Cognition and Emotion*, 7: 43–59.
110. Clark, A. (1997). *Being There*. Cambridge, MA: MIT Press; Wheeler, M. and Clark, A. (2008). Culture, embodiment, and genes: Unravelling the triple helix. *Philosophical Transactions of the Royal Society, B*, 373: 3563–3575; Wheeler, M. and Clark, A. (1999). Genic representation: reconciling content and causal complexity. *British Journal for the Philosophy of Science*, 50: 103–135; Noe, A. (2004). *Action in Perception*. Cambridge, MA: MIT Press.

111. See Franks, D. (2010). *Neurosociology: The Nexus Between Neuroscience and Social Psychology.* New York: Springer; Sabini, J., and Schulkin, J. (1994). Biological realism and social constructivism. *Journal for the Theory of Social Behavior,* 224: 207–217.
112. See also Wittgenstein, L. (1953, 1968). *Philosophical Investigations.* New York: Macmillan Publishing; Goldman A.I. (1999). *Knowledge in a Social World.* Oxford: Clarendon Press.
113. Clark, A. (1997). *Being There.* Cambridge, MA: MIT Press; Wheeler, M., and Clark, A. (2008). Culture, embodiment, and genes: unravelling the triple helix. *Philosophical Transactions of the Royal Society, B,* 373: 3563–3575; Wheeler, M. and Clark, A. (1999). Genic representation: reconciling content and causal complexity. *British Journal for the Philosophy of Science,* 50: 103–135; Schull, J. (1988). Intelligence and mind in evolution. *Journal of General Evolution,* 23: 263–273; Schull, J. (1992). Selection: James's principal principle. In: *Reinterpreting the Legacy of William James* (edited by M.E. Donnelly). Washington, DC: American Psychological Assocation: 139–152.
114. Dewey, J. (1909, 1975). *Moral Principles in Education.* Carbondale, IL: Southern Illinois University Press: 31.
115. Dewey, J. (1934, 1970). *A Common Faith.* New Haven: Yale University Press.
116. Dewey, J. (1908, 1960). *Theory of the Moral Life.* New York: Holt, Rinehart and Winston: 80.
117. Dewey, J. (1929, 1966). *A Theory of Valuation.* Chicago: University of Chicago Press.
118. Neville, R.C. (1974). *The Cosmology of Freedom.* New Haven: Yale University Press.
119. Chomsky, N. (1965). *Aspects of the Theory of Syntax.* Cambridge, MA: MIT Press; Chomsky, N. (1972). *Language and Mind.* New York: Harcourt Brace Jovanovich.
120. Ibid.
121. Humbolt W. Von (1836, 1971). *Linguistic and Intellectual Development.* Philadelphia: University of Pennsylvania Press.
122. Lifton, R.J. (1986). *The Nazi Doctors.* New York: Basic Books.
123. Deichmann, U. (1996). *Biologists under Hitler* (transl. by T. Dunlap). Cambridge, MA: Harvard University Press.
124. Jones, J.H. (1981). *Bad Blood.* New York: Free Press.
125. Rosenberg, C.E. (1976, 1997). *No Other Gods: On Science and American Social Thought.* Baltimore: Johns Hopkins University Press.
126. Davenport, C.D. (1921). Research in eugenics. *Science,* 54: 341–347; Witkowski, J.A. (2008). *Davenport's Dream.* Woodbury, NY: Cold Spring Harbor Laboratory Press.
127. Schulkin, J. (1996). *The Delicate Balance.* Lanham, MD: University Press of America.
128. Collingwood, R.G. (1945, 1976). *The Idea of Nature.* Oxford: Oxford University Press; Toulmin, S. (1977). *Human Understanding.* Princeton: Princeton University Press. For a very important depiction of our ethical evolution and the broadening inclusion of others from a distinctly pragmatist/naturalist perspective, see: Kitcher, P. (2011). *The Ethical Project.* Cambridge, MA: Harvard University Press.

129. Smith, A. (1759, 1882). *The Theory of Moral Sentiments*. Indianapolis: Liberty Classics; Wilson, J.Q. (1993). *The Moral Sense*. New York: Free Press; Schulkin, J. (2004). *Bodily Sensibility: Intelligent Action*. Oxford: Oxford University Press.
130. Rawls, J. (1971). *A Theory of Justice*. Cambridge, MA: Harvard University Press; Kant, I. (1788, 1956). *Critique of Practical Reason* (transl. L.W. Beck). New York: Bobbs-Merrill.
131. Greene, J.D., Sommerville, R.B., Nystrom, L.E., Darley, J.M., and Cohen, J.D. (2001). An fMRI investigation of emotional engagement in moral judgment. *Science*, 293: 2105–2108; Greene, J.D., Morelli, S.A., Lowenberg, K., Nystrom, L.E., and Cohen, J.D. (2008). Cognitive load selectively interferes with utilitarian moral judgement. *Cognition*, 107: 1144–1154.
132. Ibid.; Haidt, J. (2001). The emotional dog and its rational tail. *Psychological Review*, 108: 814–824; Haidt J. (2007). The new synthesis in moral psychology. *Science*, 316: 998–1002.
133. Sabini, J., and Schulkin, J. (1994). Biological realism and social constructivism. *Journal for the Theory of Social Behavior*, 224: 207–217; Haidt J. (2001). The emotional dog and its rational tail. *Psychological Review*, 108: 814–824; Haidt J. (2007). The new synthesis in moral psychology. *Science*, 316: 998–1002; Moll, J., and Schulkin, J. (2009). Social attachment and aversion in human moral cognition. *Neuroscience and Biobehavioral Reviews*, 33: 456–465; Decety, J., and Jackson, P.W. (2006). A social neuroscience perspective on empathy. *Current Directions in Psychological Science*, 15: 54–58.
134. Dewey, J., and Tufts, J.H. (1910). *Ethics*. New York: Henry Holt.
135. Moreno, J.D. (1995). *Deciding Together*. Oxford: Oxford University Press; Moreno, J.D. (1999, 2003). Bioethics is a naturalism. In: *Pragmatic Bioethics* (edited by G. McGee). Cambridge, MA: MIT Press; Moreno, J.D. (2005). *Is There an Ethicist in the House: On the Cutting Edge of Bioethics*. Bloomington: Indiana University Press; Edel, A. (1955, 1964). *Ethical Judgment: The Use of Science in Ethics*. New York: Free Press; Flanagan, O. (2007). *The Really Hard Problem*. Cambridge, MA: MIT Press.

7 Evolution, Devolution, and Human Progress

1. Niebuhr, R. (1952). *The Irony of American History*. New York: Charles Scribner and Sons; Diggins, J.P. (1991). *The Promise of Pragmatism*. Chicago: University of Chicago Press. See also Diggins, J.P. (2011). *Why Niebuhr Now?* Chicago: University of Chicago Press.
2. Freud, S. (1919, 1961). *Beyond the Pleasure Principle*. New York: Norton; Fromm, E. (1973). *The Anatomy of Human Destructiveness*. New York: Holt, Rinehart and Winston.
3. Fodor, J. (1983). *The Modularity of Mind*. Cambridge, MA: MIT Press.
4. Dennett, D. (1987). *The Intentional Stance*. Cambridge, MA: MIT Press.
5. Dewey, J. (1925, 1989). *Experience and Nature*. La Salle, IL: Open Court.
6. Ryan, A. (1995). *John Dewey and the High Tide of Amerian Liberalism*. New York: Norton Press; Whitman, W. (1871, 1973). Democratic vistas. In: *The Portable Walt Whitman* (ed. M. Van Doren). New York: Penguin.

7. See, for instance, Hickman, L.A. (1980, 1992). *John Dewey's Pragmatic Technology*. Bloomington: Indiana University Press; Hickman, L.A. (2002). *Philosophical Tools for Technological Culture*. Bloomington: Indiana University Press; Hickman, L.A. (2007). *Pragmatism as Post-Postmodernism*. New York: Fordham University Press; Hickman, L.A. (1998). Dewey's theory of inquiry. In: *Reading Dewey* (edited by L. Hickman). Bloomington: Indiana University Press; Garrison, J. (1997). *Dewey and Eros: Wisdom and Desire in the Art of Teaching*. New York: Teachers College Press; Garrison, J. (2008, editor). *Reconstructing Democracy, Recontextualizing Dewey*. Albany: SUNY Press; Stuhr, J. (1997). *Geneological Pragmatism: Philosophy, Experience, and Community*. Albany: SUNY Press; Stuhr, J. (2002). *Pragmatism, Postmodernism, and the Future of Philosophy*. New York: Routledge; Stuhr, J. (2003). *Experience and Criticism: John Dewey's Reconstruction in Philosophy*. Nashville: Vanderbilt University Press; Stuhr J. (1987). *Classical American Philosophy*. Oxford: Oxford University Press; Stuhr, J. (1980) Santayana's unnatural naturalism. In: *Two Centuries of Philosophy in America* (edited by P. Caw). Totowa, NJ: Rowman and Littlefield.
8. Dewey's insight is exemplified (albeit without these authors' awareness) in: Noe, A. (2004). *Action in Perception*. Cambridge, MA: MIT Press; Wheeler, M., and Clark, A. (2008). Culture, embodiment, and genes: unravelling the triple helix. *Philosophical Transactions of the Royal Society, B*, 373: 3563–3575; Wheeler, M., and Clark, A. (1999). Genic representation: reconciling content and causal complexity. *British Journal of Philosophical Science*, 50: 103–135; Clark, A. (1997). *Being There*. Cambridge, MA: MIT Press.
9. Gallagher, S. (2005). *How the Body Shapes the Mind*. Oxford: Oxford University Press; Heelan, P.A., and Schulkin, J. (1998). Hermeneutical philosophy and pragmatism: a philosophy of the science. *Synthese*, 115: 269–302; Johnson, M. (1987, 1990). *The Body in the Mind*. Chicago: University of Chicago Press; Johnson, M. (1993). *Moral Imagination*. Chicago: University of Chicago Press; Johnson, M. (2007) *The Meaning of the Body*. Chicago: University of Chicago Press; Damasio, A.R. (1994). *Descartes' Error: Emotion, Reason, and the Human Brain*. New York: Grosset/Putnam.
10. Shelley, M. (1817, 1976). *Frankenstein*. New York: Pyram Pub.
11. Hickman, L.A. (1980, 1992). *John Dewey's Pragmatic Technology*. Bloomington: Indiana University Press; Hickman, L.A. (2002). *Philosophical Tools for Technological Culture*. Bloomington: Indiana University Press; Hickman, L.A. (2007). *Pragmatism as Post-Postmodernism*. New York: Fordham University Press; Hickman, L.A. (1998). Dewey's Theory of Inquiry. In: *Reading Dewey* (edited by L. Hickman). Bloomington: Indiana University Press.
12. Rosenberg, C.E. (1976, 1997). *No Other Gods: On Science and American Social Thought*. Baltimore: Johns Hopkins University Press; Weidman, N.M. (1999). *Constructing Scientific Psychology*. Cambridge: Cambridge University Press.
13. Sumner, W.G. (1963). *Social Darwinism: Selected Essays*. Englewood, NJ: Prentice Hall.
14. Degler, C.N. (1991). *In Search of Human Nature*. Oxford: Oxford University Press.
15. Menand, L. (2001). *The Metaphysical Club*. New York: Farrar, Straus and Giroux; Rosenberg, C.E. (1976, 1997). *No Other Gods: On Science and American Social Thought*. Baltimore: Johns Hopkins University Press.

16. Rosenberg, C.E. (1976, 1997). *No Other Gods: On Science and American Social Thought*. Baltimore: Johns Hopkins University Press.
17. Morgenbesser, S. (1977, editor). *Dewey and His Critics: Essays from the Journal of Philosophy*. New York: Hackett.
18. Davenport, C.D. (1921). Research in eugenics. *Science*, 54: 341–347.
19. Witkowski, J.A. (2008). *Davenport's Dream*. Woodbury, NY: Cold Spring Harbor Laboratory Press.
20. Watson, J.D. (2001). *A Passion for DNA*. Oxford: Oxford University Press.
21. Richards, R.J. (2008). *The Tragic Sense of Life: Ernst Haeckel and the Struggle over Evolutionary Thought*. Chicago: University of Chicago Press.
22. Haeckel, E. (1900, 1992). *The Riddle of the Universe*. Buffalo, NY: Prometheus Books.
23. Gould, S.J. (1977). *Ontogeny and Phylogeny*. Cambridge, MA: Harvard University Press.
24. Gillham, N.W. (2001). *Sir Francis Galton: From African Exploration to the Birth of Eugenics*. Oxford: Oxford University Press.
25. Richards, R.J. (2008). *The Tragic Sense of Life: Ernst Haeckel and the Struggle over Evolutionary Thought*. Chicago: University of Chicago Press.
26. See Gillham, N.W. (2001). *Sir Francis Galton: From African Exploration to the Birth of Eugenics*. Oxford: Oxford University Press.
27. See Porter, T.M. (2004). *Karl Pearson: The Scientific Life in a Statistcal Age*. Princeton: Princeton University Press.
28. Gould, S.J. (1977). *Ontogeny and Phylogeny*. Cambridge, MA: Harvard University Press; Simpson, G.G. (1949). *The Meaning of Evolution*. New Haven: Yale University Press.
29. Schulkin, J. (2005). *Curt Richter: A Life in the Laboratory*. Baltimore: Johns Hopkins University Press.
30. Darwin, C. (1859, 1958). *The Origin of Species*. New York: Mentor Books.
31. Gould, S.J. (2002). *The Structure of Evolutionary Theory*. Cambridge, MA: Harvard University Press.
32. Darwin, C. (1859, 1958). *The Origin of Species*. New York: Mentor Books: 38.
33. Donaldson, H.H. (1915). The rat data and reference tables for the albino and Norway rat. In: *Memoirs of the Wistar Institute of Anatomy and Biology*, no. 6. Philadelphia: Wistar Institute.
34. Ibid. iii.
35. Darwin, C. (1868). *The Variation of Animals and Plants under Domestication*, 2 vols. London: John Murray: vol. 1, 2.
36. Darwin, C. (1868). *The Variation of Animals and Plants under Domestication*, 2 vols. London: John Murray: vol. 2, 89.
37. Richter, C.P. (1952). Domestication of the Norway rat and the implications for the study of genetics in man. *American Journal of Human Genetics*, 4: 273.
38. Ibid.
39. Ibid. 275.
40. Darwin, C. (1859, 1958). *The Origin of Species*. New York: Mentor Books; James, W. (1890, 1952). *The Principles of Psychology*. New York: Henry Holt.
41. Wheeler, M., and Clark, A. (2008). Culture, embodiment, and genes: unravelling the triple helix. *Philosophical Transactions of the Royal Society, B*, 373: 3563–3575; Wheeler, M., and Clark, A. (1999) Genic representation:

reconciling content and causal complexity. *British Journal for the Philosophy of Science*, 50: 103–135.

42. Putnam, H. (1990). *Realism with a Human Face*. Cambridge, MA: Harvard University Press; Putnam, H. (2000). *The Collapse of the Fact/Value Distinction*. Cambridge, MA: Harvard University Press; Putnam, H. (1995). *Pragmatism: An Open Question*. Malden, MA: Blackwell.
43. Brown, M. (2011). Science as socially disturbed cognition: bridging philosophy and sociology of science. In: *Foundations of the Formal Sciences VII* (edited by K. François, B. Löwe, T. Müller, and B. van Kerkhove). London: College Publications.
44. Schull, J. (1988). Intelligence and mind in evolution. *Journal of General Evolution*, 23: 263–273; Schull, J. (1992). Selection: James's principal principle. In: *Reinterpreting the Legacy of William James* (edited by M.E. Donnelly). Washington, DC: American Psychological Assocation: 139–152.
45. Keverne, E.B., and Curley, J.P. (2008). Epigenetics, brain evolution, and behavior. *Neuroendocrinology*, 29: 398–412.
46. Meaney, M.J. (2001). Maternal care, gene expression, and the transmission of individual differences in stress reactivity across generations. *Annual Review of Neuroscience*, 24: 1161–1192.
47. Schachar-Dadon, A., Schulkin, J., and Leshem, M. (2009). Adversity before conception will affect adult progeny in rats. *Developmental Psychology*, 45(4): 9–16.
48. Curley, J.P., Davidson, S., Bateson, P., and Champagne, F.A. (2009). Social enrichment during postnatal development induces transgenerational effects on emotional and reproductive behavior in mice. *Frontiers in Behavioral Neuroscience*, 3: 25.
49. Holliday, R. (2002). Epigenetics comes of age in the twenty-first century. *Journal of Genetics*, 81: 1–4; Holliday, R. (2006). Epigenetics: a historical overview. *Epigenetics*, 1: 76–80; Crews, D. (2008). Epigenetics and its implications for behavioral endocrinology. *Frontiers in Neuroendocrinology*, 29: 344–357; Dobzhansky, T.C. (1962). *Mankind Evolving*. New Haven: Yale University Press.
50. Keller, E.F. (1983). *A Feeling for the Organism*. New York: Freeman and Company.
51. Holliday, R. (2002). Epigenetics comes of age in the twenty-first century. *Journal of Genetics*, 81: 1–4.
52. Keverne, E.B., and Curley, J.P. (2008). Epigenetics, brain evolution, and behavior. *Neuroendocrinology*, 29: 398–412.
53. Carter, C.S., Lederhendler, I.L., and Kirkpatrick, B. (1997, 1999). *The Integrative Neurobiology of Affiliation*. Cambridge, MA: MIT Press; Schulkin, J. (2011). *Adaptation and Well-Being: Social Allostasis*. Cambridge: Cambridge University Press.
54. Weaver, I.C., Cervoni, N., Champagne, F.A., D'Alessio, A.C., Sharma, S., Seckl, J.R., Dymov, S., Szyf, M., and Meaney, M.J. (2004). Epigenetic programming by maternal behavior. *Nature Neuroscience*, 7: 847–854; Weaver, I.C., Champagne, F.A., Brown, S.E., Dymov, S., Sharma, S., Meaney, M.J., and Szyf, M. (2005). Reversal of maternal programming of stress responses in adult offspring through methyl supplementation: altering epigenetic marking later in life. *Journal of Neuroscience*, 25: 11045–11054.

55. Dewey, J. (1931). *Philosophy and Civilization*. New York: Minton, Balch and Co.: 44.
56. Dunbar, R.I.M. (1996). *Grooming, Gossip and the Evolution of Language*. Cambridge, MA: Harvard University Press; Dunbar, R.I.M. (1992). Neocortex size as a constraint on group size in primates. *Journal of Human Evolution*, 22: 469–493; Dunbar, R.I.M., and Shultz, S. (2007). Understanding primate evolution. *Philosophical Transactions of the Royal Society*, 362: 649–658; Barton, R.A. (2004). Binocularity and brain evolution in primates. *Proceedings of the National Academy of Sciences*, 101: 10113–10115; Barton, R.A. (2006). Primate brain evolution: integrating comparative neurophysiological and ethological data. *Evolutionary Anthropology*, 15: 224–236.
57. Richards, R.J. (1992, 1995). *The Meaning of Evolution*. Chicago: University of Chicago Press.
58. Ruse, M. (1996). *Monad to Man*. Cambridge, MA: Harvard University Press; Ruse, M. (1995). *Evolutionary Naturalism*. London: Routledge; Ruse, M. (2009, editor). *Philosophy After Darwin*. Princeton: Princeton University Press; Bowler, P.J. (1983). *The Eclipse of Darwinism*. Baltimore: Johns Hopkins University Press; Bowler, P.J. (1988). *The Non-Darwinian Revolution*. Baltimore: Johns Hopkins University Press; Bowler, P.J. (2009). *Science for All*. Chicago: Chicago University Press.
59. Richards, R.J. (1992, 1995). *The Meaning of Evolution*. Chicago: University of Chicago Press.
60. Gould, S.J. and Eldridge, N. (1977) Punctuated equilibria: the tempo and mode of evolution reconsidered. *Paleobiology*, 3: 115–151.
61. Gould, S.J. (2002). *The Structure of Evolutionary Theory*. Cambridge, MA: Harvard University Press: 767.
62. Gould, S.J. (2002). *The Structure of Evolutionary Theory*. Cambridge, MA: Harvard University Press; McHenry, H.M. (1994). Tempo and mode in human evolution. *Proceedings of the National Academy of Sciences*, 91: 6780–6786; McHenry, H.M. (2009). Human evolution. In: *Evolution: The First Four Billion Years* (edited by M. Ruse and J. Travis). Cambridge, MA: Harvard University Press: 256–280.
63. Mayr, E. (1942, 1982). *Systemics and the Origin of Species*. New York: Columbia Univ. Press; Mayr, E. (1963). *Animal Species and Evolution*. Cambridge, MA: Harvard University Press.
64. Gould, S.J. (2002). *The Structure of Evolutionary Theory*. Cambridge, MA: Harvard University Press; Gould, S.J., and Eldridge, N. (1977). Punctuated equilibria: the tempo and mode of evolution reconsidered. *Paleobiology*, 3: 115–151.
65. Gould, S.J., and Lewontin, R.C. (1979). The spandrels of San Marco and the Panglossian paradigm: A critique of the adaptationist programme. *Proceedings of the Royal Society, B: Biological Sciences*, 205: 581–598.
66. Corballis, M.C. (2002). *From Hand to Mouth*. Princeton: Princeton University Press; Pinker, S. (1994). *The Language Instinct*. New York: William Morrow and Co.
67. Bury, J.B. (1933, 1960). *The Idea of Progress*. New York: Dover Press.
68. Locke, J. (1692–1704, 1955). *A Letter Concerning Toleration*. Indianapolis: Bobbs-Merrill Co; Bury, J.B. (1933, 1960). *The Idea of Progress*. New York: Dover Press.

69. Comte, A.C. (1842, 1975). *A General View of Positivism* (translated by J. Bridges). New York: Robert Speller and Sons.
70. Langer, S.K. (1962). *Philosophical Sketches*. New York: Mentor Books: 29.
71. Schulkin, J. (2009). *Cognitive Adaptation: A Pragmatist Perspective*. Cambridge: Cambridge University Press.
72. Skidelsky, E. (2008). *Ernst Cassirer: The Last Philosopher of Culture*. Princeton: Princeton University Press.
73. See Cassirer, E. (1957). *The Philosophy of Symbolic Forms*, vol. III. New Haven: Yale University Press.
74. Langer, S.K. (1937). *Philosophy in a New Key*. Cambridge, MA: Harvard University Press.
75. Whitehead, A.N. (1938, 1967). *Modes of Thought*. New York: Free Press.
76. Innis, R.E. (2009). *Susanne Langer in Focus: The Symbolic Mind*. Bloomington: Indiana University Press.
77. Lovejoy, A.O. (1936, 1978). *The Great Chain of Being*. Cambridge, MA: Harvard Univ. Press; Lovejoy, A.O. (1955). *Essays in the History of Ideas*. New York: George Braziller Inc.; Gundling, T. (2005). *First In Line*. New Haven: Yale University Press.
78. Whitehead, A.N. (1929, 1958). *The Function of Reason*. Boston: Beacon Press.
79. Malthus, T.R. (1798, 1970). *An Essay on the Principle of Population*. Baltimore: Penguin Books.
80. Sarokin, D. and Schulkin, J. (1994). Co-evolution of rights and environmental justice. *The Environmentalist*, 14: 121–129.
81. Whitehead, A.N. (1927, 1953). *Symbolism*. New York: Macmillan.
82. Peirce, C.S. (2000). *The Writings of C.S. Peirce*, vol. 6. Bloomington: Indiana University Press: 181.
83. Dewey, J. (1925, 1989). *Experience and Nature*. La Salle, IL: Open Court; Whitehead, A.N. (1938, 1967). *Modes of Thought*. New York: Free Press.
84. Rosenthal, S.B. (1986, 1990). *Speculative Pragmatism*. La Salle, IL: Open Court; Rosenthal, S.B. (1995). Paul Weiss and Pragmatism: A Dialogue. In: *The Philosophy of Paul Weiss* (edited by L.E. Hahn). LaSalle, IL: Open Court; Weissman, D. (1993). *Truth's Debt to Value*. New Haven: Yale University Press.
85. Weiss, P. (1947, 1967). *Nature and Man*. Carbondale: South Illinois Univ. Press; Weiss, P. (1965). Charles S. Peirce: philosopher. In: *Perspectives on Peirce* (edited by R.J. Bernstein). New Haven: Yale University Press; Weissman, D. (1989). *Hypothesis and the Spiral of Reflection*. Albany, NY: SUNY Press; Weissman, D. (1993). *Truth's Debt to Value*. New Haven: Yale University Press; Weissman, D. (2000). *A Social Ontology*. New Haven: Yale University Press; Grange, J. (1977). *Nature: An Enviromental Cosmology*. Albany, NY: SUNY.
86. Weissman, D. (1989). *Hypothesis and the Spiral of Reflection*. Albany, NY: SUNY Press; Weissman, D. (1993). *Truth's Debt to Value*. New Haven: Yale University Press; Weissman, D. (2000). *A Social Ontology*. New Haven: Yale University Press; Buchler, J. (1939, 1966). *Charles Peirce's Empiricism*. New York: Octagon Books; Buchler, J. (1955, 1966). *Nature and Judgment*. New York: Grossett and Dunlap; Sellars, W. (1956, 1997). *Empiricism and the Philosophy of Mind*. Cambridge, MA: Harvard University Press; Sellars, W. (1962). *Science, Perception, and Reality*. New York: Routledge; Sellars, W. (1968). *Science and Metaphysics*. New York: Humanities Press.

87. Langer, S.K. (1937). *Philosophy in a New Key*. Cambridge, MA: Harvard University Press; Langer, S.K. (1962). *Philosophical Sketches*. New York: Mentor Books; Langer, S.K. (1972). *Mind: an Essay on Human Feeling*. Baltimore, MD: Johns Hopkins University Press; Whitehead, A.N. (1927, 1953). *Symbolism*. New York: Macmillan Company.
88. See some of the work of the philosopher of science, Godfrey-Smith, P. (2002). Dewey on naturalism, realism and science. *Philosophy of Science*, 69: S1–S11; Godfrey-Smith, P. (1998). *Complexity and Function of Mind in Nature*. Cambridge: Cambridge University Press; Godfrey-Smith, P. (2009). *Darwinian Populations and Natural Selection*. Oxford: Oxford University Press; Godfrey-Smith, P. (2008). Induction, samples, and kinds. In: *Carving Nature at its Joints: Topics in Contemporary Philosophy*, vol. 8. Cambridge, MA: MIT Press.
89. Neville, R.C. (1974). *The Cosmology of Freedom*. New Haven: Yale University Press; Weiss, P. (1947, 1967) *Nature and Man*. Carbondale: South Illinois University Press; Weiss, P. (1965). *Charles S. Peirce: Philosopher*. In: *Perspectives on Peirce* (edited by R.J. Bernstein). New Haven: Yale University Press; Weissman, D. (2000). *A Social Ontology*. New Haven: Yale University Press.
90. Schulkin, J. (1992) *The Pursuit of Inquiry*. Albany, NY: SUNY Press.
91. Diggins, J.P. (1991). *The Promise of Pragmatism*. Chicago: University of Chicago Press; Neville, R.C. (1992). *The Highroad around Modernism*. New York: SUNY Press; Livingston, J. (2010). Pragmatism, nihilism, and democracy: what is called thinking at the end of modernity? In: *100 Years of Pragmatism: William James's Revolutionary Philosophy* (edited by John Stuhr). Bloomington: Indiana University Press.
92. Galileo, G. (1610, 1957). The starry messenger. In: *Discoveries and Opinions of Galileo* (edited by S. Drake). New York: Doubleday.
93. Butterfield, H. (1957, 1965). *The Origins of Modern Science*. New York: Free Press.
94. Lasch, C. (1991). *The True and Only Heaven: Progress and its Critics*. New York: Norton.
95. Croce, B. (1933, 1963). *History of Europe in the Nineteenth Century*. New York: Harcourt, Brace and World; Dilthey, W. (1883). *Introduction to the Human Sciences*. Princeton, NJ: Princeton University Press; Vico, G. (1744, 1970). *The New Sciences*. Ithaca: Cornell University Press; Berlin, I. (1976). *Vico and Herder*. New York: Vintage Press; Berlin, I. (1999). *The Roots of Romanticism*. Princeton: Princeton University Press.
96. Butterfield, H. (1981). *The Origins of History*. New York: Basic Books.
97. Comte, A.C. (1842, 1975). *A General View of Positivism*. (translated by J. Bridges). New York: Robert Speller and Sons.
98. Whitehead, A.N. (1929, 1958). *The Function of Reason*. Boston: Beacon Press.
99. Ibid. 4.
100. Ibid.
101. Hall, D.L. (1973). *The Civilization of Experience*. New York: Fordham University Press.
102. Whitehead, A.N. (1929, 1958). *The Function of Reason*. Boston: Beacon Press: 4.
103. Ibid.
104. Neville, R.C. (1974). *The Cosmology of Freedom*. New Haven: Yale University Press.

105. Whitehead, A.N. (1925, 1953). *Science and the Modern World*. New York: Free Press: 49.
106. Varela, F., Thompson, E., and Rosch, E. (1991). *The Embodied Mind*. Cambridge: Cambridge University Press; Lakoff, G., and Johnson, M. (1999). *Philosophy in the Flesh*. New York: Basic Books.
107. Whitehead, A.N. (1925, 1953). *Science and the Modern World*. New York: Free Press: 91.
108. Kahneman, D., Slovic, P., and Tversky, A. (1982, editors). *Judgment under Uncertainty: Heuristics and Biases*. New York: Cambridge University Press; Loewenstein, G. (2006). The pleasures and pains of information. *Science*, 312: 704–706.
109. Whitehead, A.N. (1933, 1961). *Adventures of Ideas*. New York: Free Press: 274; Hall, D.L. (1973). *The Civilization of Experience*. New York: Fordham University Press.
110. Hall, D.L. (1973). *The Civilization of Experience*. New York: Fordham University Press.
111. Whitehead A.N. (1939, 1989). John Dewey and his influence. In: *The Philosophy of John Dewey*. La Salle, IL: Open Court: 477.
112. Dewey, J. (1939, 1963). *Freedom and Culture*. New York: Capricorn Books.
113. Niebuhr, R. (1952). *The Irony of American History*. New York: Charles Scribner and Sons; Rorty, R. (1999). *Philosophy and Social Hope*. New York: Penguin.
114. Dewey, J. (1935, 1963). *Liberalism and Social Action*. New York: Capricorn Books.
115. Ibid. 59.
116. Ibid. 66.
117. Moreno, J.M. (1995). *Deciding Together*. Oxford: Oxford University Press; Kitcher, P. (2001). *Science, Truth, and Democracy*. New York: Oxford University Press.
118. Dewey, J. (1929, 1960). *The Quest for Certainty*. New York: Capricorn Books.
119. Jaspers, K. (1913, 1997). *General Psychopathology*, 2 vols (translated by J. Hoenig and M.W. Hamilton). Baltimore: Johns Hopkins University Press; Jaspers, K. (1951, 1954). *Way to Wisdom*. New Haven: Yale University Press.
120. Dewey, J. (1920, 1948). *Reconstruction in Philosophy*. Boston: Beacon Press.
121. Jonas, H. (1966). *The Phenomenon of Life*. Chicago: University of Chicago Press.
122. Clark, A. (1997). *Being There*. Cambridge, MA: MIT Press; Schulkin, J. (2009). *Cognitive Adaptation: A Pragmatist Perspective*. Cambridge: Cambridge University Press.
123. Fromm, E. (1947). *Man for Himself*. New York: Rinehart and Company; Fromm, E. (1956, 1975). *The Art of Loving*. New York: Harper and Row; Humphrey, N. (2007). The society of selves. *Philosophical Transactions of the Royal Society, B*, 362: 745–754.

8 Conclusion: Adaptation, Well-being, and Social Hope

1. Conference on Medical Decision Making, Frankfurt, Germany: October 2009.

2. Sarokin, D., and Schulkin, J. (1992). The role of pollution in large-scale population disturbances. Part 1: aquatic populations. *Environmental Science and Technology*, 26: 1476–1483.
3. Dewey, J. (1939, 1963). *Freedom and Culture*. New York: Capricorn Books; Wolin, S.S. (2008). *Managed Democracy and the Specter of Inverted Totalitarianism*. Princeton: Princeton University Press.
4. Sarokin, D., and Schulkin, J. (1994). Co-evolution of rights and environmental justice. *The Environmentalist*, 14: 121–129; Pratt, S.L. (2002). *Native Pragmatism: Rethinking the Roots of American Pragmatism*. Bloomington: Indiana University Press; Kitcher, P. (2011). *The Ethical Project*. Cambridge, MA: Harvard University Press.
5. Kant, I. (1792, 1951). *Critique of Judgment*. New York: Haffner Press; Dewey, J. (1934, 1958). *Art as Experience*. New York: Capricorn Press.
6. Butterfield, H. (1957, 1965). *The Origins of Modern Science*. New York: Free Press; Butterfield, H. (1981). *The Origins of History*. New York: Free Press.
7. Kitcher, P. (1996). *The Lives To Come*. New York: Simon and Schuster.
8. Bellah, R.N., Madsen, R., Sullivan, W.M., Swidler, A., and Tipton, S.M. (1985). *Habits of the Heart: Individualism and Commitment in American Life*. New York: Harper and Row.
9. Appiah, K.A. (2006). *Cosmopolitanism*. New York: Norton; Moreno, J.D. (1995). *Deciding Together*. Oxford: Oxford University Press.
10. Pinker, S. (2011). *The Better Angels of Our Nature: Why Violence Has Declined*. New York: Viking.
11. Moll, J., and Schulkin, J. (2009). Social attachment and aversion in human moral cognition. *Neuroscience and Biobehavioral Reviews*, 33: 456–465.
12. For example, Jaspers, K. (1913, 1997). *General Psychopathology*, 2 vols (translated by J. Hoenig and M.W. Hamilton). Baltimore: Johns Hopkins University Press; Jaspers, K. (1951, 1954). *Way to Wisdom*. New Haven: Yale University Press; Fromm, E. (1947). *Man for Himself*. New York: Rinehart and Company; Nussbaum, M.C. (2001). *Upheavals of Thought: The Intelligence of Emotion*. Cambridge: Cambridge University Press; Nussbaum, M.C. (1997). *Cultivating Humanity*. Cambridge, MA: Harvard University Press.
13. Darwin, C. (1859, 1958). *The Origin of Species*. New York: Mentor Books; Dewey, J. (1925, 1989). *Experience and Nature*. LaSalle, IL: Open Court.
14. Hofer, M.A. (1973). The role of nutrition in the physiological and behavioral effects of early maternal separation on infant rats. *Psychosomatic Medicine*, 35: 350–359; Carter, C.S., Lederhendler, I.L., and Kirkpatrick, B. (1997, 1999). *The Integrative Neurobiology of Affiliation*. Cambridge, MA: MIT Press; Keverne, E.B. (2004). Understanding well-being in the evolutionary context of brain development. *Philosophical Transactions of the Royal Society*, 359: 1349–1358; Kagan, J. (1998). *Three Seductive Ideas*. Cambridge, MA: Harvard University Press.
15. Galison, P. (1988). History, philosophy and the central metaphor. *Science in Context*, 3: 197–212; Lakoff, G., and Johnson, M. (1999). *Philosophy in the Flesh: The Embodied Mind and its Challenge to Western Thought*. New York: Basic Books; Heelan, P.A., and Schulkin, J. (1998). Hermeneutical philosophy and pragmatism: a philosophy of the science. *Synthese*, 115: 269–302.
16. Gibson, J.J. (1966). *The Senses Considered as Perceptual Systems*. New York: Houghton-Mifflin; Garcia, J., Hankins, W.G., and Rusiniak, K.W. (1974).

Behavioral regulation of the milieu interne in man and rat. *Science*, 185: 824–831; Rozin, P. (1976). The evolution of intelligence and access to the cognitive unconscious. In: *Progress in Psychobiology and Physiological Psychology* (edited by J. Sprague and A.N. Epstein). New York: Academic Press; Rosati, A.G., Stevens, J.R., Hare, B., and Hauser, M.D. (2007). The evolutionary origins of human patience: temporal preferences in chimpanzees, bonobos and human adults. *Current Biology*, 17: 1663–1668.
17. Greene, J.D., and Haidt, J. (2002). How (and where) does moral judgment work? *Trends in Cognitive Science*, 6: 517–523.
18. Moll, J., and Schulkin, J. (2009). Social attachment and aversion in human moral cognition. *Neuroscience and Biobehavioral Reviews*, 33: 456–465.
19. Perrett, D., Harries, M., Bevan, R., Thomas, S., Benson, P., Mistlin, A., Chitty, A., Hietanen, J., and Ortega, J. (1989). Frameworks of analysis for the neural representation of animate objects and actions. *Journal of Experimental Biology*, 146: 87–113.
20. Greene, J. D., and Haidt, J. (2002). How (and where) does moral judgment work? *Trends in Cognitive Science*, 6: 517–523; Moll, J., and Schulkin, J. (2009). Social attachment and aversion in human moral cognition. *Neuroscience and Biobehavioral Reviews*, 33: 456–465.
21. Moreno, J.D. (1995). *Deciding Together*. Oxford: Oxford University Press.
22. Baron-Cohen, S. (1995, 2000). *Mindblindness*. Cambridge, MA: MIT Press.
23. For example, Adolphs, R. (1999). Social cognition and the human brain. *Trends in Cognitive Sciences*, 3: 469–479; Greene, J.D., and Haidt, J. (2002). How (and where) does moral judgment work? *Trends in Cognitive Science*, 6: 517–523.
24. Jaspers, K. (1913, 1997). *General Psychopathology*, 2 vols (transl. J. Hoenig and M.W. Hamilton). Baltimore: Johns Hopkins University Press; Singer P. (1981, 2011). *The Expanding Circle*. Princeton: Princeton University Press.
25. Rorty, R. (1999). *Philosophy and Social Hope*. New York: Penguin Books; Bernstein, R.J. (2010). *The Pragmatic Turn*. Malden, MA: Polity Press.
26. Jaspers, K. (1913, 1997). *General Psychopathology*, 2 vols (transl. J. Hoenig and M.W. Hamilton). Baltimore: Johns Hopkins University Press.
27. Kahneman, D., Slovic, P., and Tversky, A. (1982, editors). *Judgment under Uncertainty: Heuristics and Biases*. New York: Cambridge University Press; Loewenstein, G. (1994). The psychology of curiosity. *Psychological Bulletin*, 116: 75–98.
28. Jaspers, K. (1913, 1997). *General Psychopathology*, 2 vols (transl. J. Hoenig and M.W. Hamilton). Baltimore: Johns Hopkins University Press.
29. See Spinoza, B. (1668, 1955). *On the Improvement of the Understanding* (transl. by R.H.M. Elwes). New York: Dover Press.
30. Mead, M. (1964). *Continuities in Cultural Evolution*. New Haven: Yale University Press.
31. Durkheim, E. (1974). *Sociology and Philosophy* (transl. D.F. Pocock). New York: Free Press.
32. Cacioppo, J.T., Hawkley, L., Ernst, J.M., Burleson, M., Berntson, G.G., Nouriani, B., and Spiegel, D. (2006). Loneliness within a nomological net: an evolutionary perspective. *Journal of Research in Personality*, 40: 1054–1085.
33. Aristotle. (1962). *The Politics* (translated by T. Sinclair). New York: Penguin Books; Nussbaum, F.L. (1953). *The Triumph of Science and Reason*. New York:

Harper and Row; Nussbaum, M.C. (1997). *Cultivating Humanity*. Cambridge, MA: Harvard University Press; Nussbaum, M.C. (2004). *Hiding from Humanity*. Princeton: Princeton University Press.

34. Schulkin, J. (1992). *The Pursuit of Inquiry*. Albany: SUNY Press; Schulkin, J. (2009). *Cognitive Adaptation: A Pragmatist Perspective*. Cambridge: Cambridge University Press. See also Kitcher, P. (2011). *The Ethical Project*. Cambridge, MA: Harvard University Press with Brandom, R.B. (2011). *Perspectives on Pragmatism*. Cambridge, MA: Harvard University Press for the two strands within pragmatism; one emphasizes an experimental sensibility, the other an orientation to concepts and the inferences and the consequences and responsibility for our conceptual orientations. These strains need not inherently be at odds with one each other.
35. Rorty, R. (1999). *Philosophy and Social Hope*. New York: Penguin Books; Tillich, P. (1966). *On the Boundary*. New York: Scribner and Sons.
36. Moreno, J.D. (1995). *Deciding Together*. Oxford: Oxford University Press; Moreno, J.D. (1999, 2003). Bioethics is a naturalism. In: *Pragmatic Bioethics* (edited by G. McGee). Cambridge, MA: MIT Press; Moreno, J.D. (2005). *Is There an Ethicist in the House? On the Cutting Edge of Bioethics*. Bloomington: Indiana University Press; McDermott, J.J. (2007). *The Drama of Possibility: Experience as Philosophy of Culture* (edited by D.R. Anderson). New York: Fordham University Press; Habermas, J. (1967, 1988). *On the Logic of the Social Sciences* (translated by S.W. Nicholson and J.A. Stark). Cambridge, MA: MIT Press; Habermas, J. (1996). *The Liberating Power of Symbols*. Cambridge, MA: MIT Press; Putnam, H. (1990). *Realism with a Human Face*. Cambridge, MA: Harvard University Press; Putnam, H. (2000). *The Collapse of the Fact/Value Distinction*. Cambridge, MA: Harvard University Press; Koopman, C. (2009). *Pragmatism as Transition*. New York: Columbia University Press.
37. Dewey, J. (1925, 1989). *Experience and Nature*. New York: Dover Press.
38. Cooper, D.E. (2002, 2007). *The Measure of Things*. Oxford: Clarendon Press.
39. Niebuhr, R. (1952). *The Irony of American History*. New York: Charles Scribner and Sons.
40. Dewey, J. (1920, 1948). *Reconstruction in Philosophy*. Boston: Beacon Press; Margolis, J. (2002). *Reinventing Pragmatism: American Philosophy at the End of the Twentieth Century*. Ithaca, NY: Cornell University Press; Margolis, J. (2003). *The Unraveling of Scientism: American Philosophy at the End of the Twentieth Century*. Ithaca, NY: Cornell University Press; Putnam, H. (1990). *Realism with a Human Face*. Cambridge, MA: Harvard Univ. Press; Putnam, H. (2000). *The Collapse of the Fact/Value Distinction*. Cambridge, MA: Harvard University Press; Putnam, H. (1995). *Pragmatism: An Open Question*. Malden, MA: Blackwell; Rockmore, T., and Singer, B.J. (1992). *Antifoundationalism Old and New*. Philadelphia: Temple University Press.
41. See also, Bernstein, R.J. (2010). *The Pragmatic Turn*. Malden, MA: Polity Press.
42. Hook, S. (1934, 1961). *The Quest for Being*. New York: St. Martins; Hook, S. (1974). *Pragmatism and the Tragic Sense of Life*. New York: Basic Books; Hook, S. (1987). *Out of Step*. New York: Harper and Row.
43. Ibid.
44. Bellah, R.N., Madsen, R., Sullivan, W.M., Swidler, A., and Tipton, S.M. (1985). *Habits of the Heart: Individualism and Commitment in American Life*. New York: Harper and Row.
45. Appiah, K.A. (2006). *Cosmopolitanism*. New York: Norton.

46. Sullivan, W.M. (1986). *Reconstructing Public Philosophy*. Berkeley: University of California Press; Bellah, R.N., Madsen, R., Sullivan, W.M., Swidler, A., and Tipton, S.M. (1985). *Habits of the Heart: Individualism and Commitment in American Life*. New York: Harper and Row.
47. Dewey, J. (1939, 1963). *Freedom and Culture*. New York: Capricorn Books.
48. Kloppenberg, J.T. (1986). *Uncertain Victory: Social Democracy and Progressivism in European and American Thought, 1870–1920*. New York: Oxford University Press.
49. Whitman, W. (1855, 1965). *Leaves of Grass*. New York: Norton.
50. Diggins, J.P. (2000). *On Hallowed Ground: The Foundation of American History*. New Haven: Yale University Press.
51. Habermas, J. (1967, 1988). *On the Logic of The Social Sciences* (transl. S.W. Nicholson and J.A. Stark). Cambridge, MA: MIT Press; Habermas, J. (1996). *The Liberating Power of Symbols*. Cambridge, MA: MIT Press; Bhaskar, R. (1998). *The Possibility of Naturalism*. London: Routledge; Hubner, K. (1983). *Critique of Scientific Reason*. Chicago: University of Chicago Press; Rouse, J. (2002). *How Scientific Practices Matter: Reclaiming Philosophical Naturalism*. Chicago: University of Chicago Press.
52. Lovejoy, A.O. (1936, 1978). *The Great Chain of Being*. Cambridge, MA: Harvard Univ. Press; Lovejoy, A.O. (1955). *Essays in the History of Ideas*. New York: George Braziller Inc.; French, R. (2008). *Medicine Before Science*. Cambridge: Cambridge University Press.
53. Margolis, J. (2002). *Reinventing Pragmatism: American Philosophy at the End of the Twentieth Century*. Ithaca: Cornell University Press; Margolis, J. (2003). *The Unraveling of Scientism: American Philosophy at the End of the Twentieth Century*. Ithaca: Cornell University Press.
54. Polanyi, M. (1946, 1964). *Science, Faith and Society*. Chicago: University of Chicago Press.
55. Clark, A. (1997). *Being There*. Cambridge, MA: MIT Press.
56. Wilson, R.A. (2004). *Boundaries of the Mind*. Cambridge: Cambridge University Press.
57. Pratt, S.L. (2002). *Native Pragmatism: Rethinking the Roots of American Pragmatism*. Bloomington: Indiana University Press; Prado, C.G. (1987). *The Limits of Pragmatism*. Atlantic Highlands: Humanities Press International; Gunn, G. (1992). *Thinking across the American Grain*. Chicago: University of Chicago Press.
58. Du Bois, W.E.B. (1903, 1982). *The Souls of Black Folk*. New York: Penguin; Du Bois, W.E.B. (1940, 1968). *Dusk of Dawn*. London: Transactions Publishers.
59. Hobbes, T. (1651, 1949). *De Cive*. New York: Appleton-Century Crafts.
60. Du Bois, W.E.B. (1968, 1991). *The Autobiography of W.E.B. Du Bois*. New York: International Publishers.
61. Ibid. 143.
62. Royce, J. (1900, 1959). *The World and the Individual*. New York: Dover Press; Royce, J. (1908, 1995). *The Philosophy of Loyalty*. Nashville: Vanderbilt University Press.
63. Royce (1959); Murphey, M. (2005). *C.I. Lewis: The Last Great Pragmatist*. Albany: SUNY Press.
64. Hammington, M. (2009). *The Social Philosophy of Jane Addams*. Champaign, IL: Illinois University Press; Margolis, J. (2002). *Reinventing Pragmatism:*

American Philosophy at the End of the Twentieth Century. Ithaca: Cornell University Press; Margolis, J. (2003). *The Unraveling of Scientism: American Philosophy at the End of the Twentieth Century*. Ithaca: Cornell University Press; Schulkin, J. (2009). *Cognitive Adaptation: A Pragmatist Perspective*. Cambridge: Cambridge University Press.

65. Lawson, B.E. and Koch, D. (2004, editors). *Pragmatism and the Problem of Race*. Bloomington: Indiana University Press.
66. Hickman, L.A. (2002). *Philosophical Tools for Technological Culture*. Bloomington: Indiana University Press; Stuhr, J. (2002). *Experience and Criticism: John Dewey's Reconstruction in Philosophy*. Nashville: Vanderbilt University Press.
67. Dewey, J. (1939, 1961). *Freedom and Culture*. New York: Capricorn Books; Addams, J. (1911). *Democracy and Social Ethics*. London: Macmillan; Addams, J. (1915). *Newer Ideals of Peace*. London: Macmillan; Addams, J. (1916, 2002). *The Long Road of Woman's Memory* (edited by C.H. Seigfried). Urbana: Univ. of Illinois Press; Habermas, J. (1967, 1988). *On the Logic of The Social Sciences* (transl. S.W. Nicholson and J.A. Stark). Cambridge, MA: MIT Press; Habermas, J. (1996). *The Liberating Power of Symbols*. Cambridge, MA: MIT Press; Kloppenberg, J.T. (1998). *The Virtues of Liberalism*. New York: Oxford University Press.
68. Donald, D.H. (1947, 1971). *Lincoln Reconsidered: Essays on the Civil War Era*. New York: Vintage Press.
69. Tocqueville, A. de (1830, 1945). *Democracy in America*. New York: Vintage Press: 203. See also a beautiful little book: Elster, J. (2009). *Alexis de Tocqueville: The First Social Scientist*. Cambridge: Cambridge University Press.
70. Hayek, F.A. (1944, 1972). *The Road to Serfdom*. Chicago: University of Chicago Press.
71. Smith, A. (1759, 1882). *The Theory of Moral Sentiments*. Indianapolis: Liberty Classics; Wilson, J.Q. (1993). *The Moral Sense*. New York: Free Press.
72. Milgram, S. (1974, 2009). *Obedience to Authority: An Experimental View*. New York: Harper Perennial Modern Classics; Sabini, J., and Silver, M. (1982). *Moralities of Everyday Life*. Oxford: Oxford University Press.
73. Fromm. E. (1973). *The Anatomy of Human Destructiveness*. New York: Holt, Rinehart and Winston.
74. Mellars, P. (2006). Why did modern human populations disperse from Africa ca. 60,000 years ago? *Proceedings of the National Academy of Sciences*, 103: 9381–9386.
75. Lovelock. J. (1988). *The Ages of Gaia*. New York: Norton.
76. Peirce, C.S. (1893, 1992). Evolutionary love. In: *The Essential Peirce*, vol. I (ed. N. Houser and C. Kloesel). Bloomington: Indiana University Press.
77. Barrow, M.V. (2009). *Nature's Ghosts*. Chicago: University of Chicago Press.
78. See also Locke, J. (1692–1704, 1955). *A Letter Concerning Toleration*. Indianapolis: Bobbs-Merrill Co.
79. Fromm. E. (1973). *The Anatomy of Human Destructiveness*. New York: Holt, Rinehart and Winston; Derrida, J. (1973, 1980). *The Archeology of the Frivolous*. Lincoln, NE: University of Nebraska Press; Derrida, J. (2008). *Islam and the West* (edited by M. Cherif). Chicago: University of Chicago Press; Jaspers, K. (1913, 1997). *General Psychopathology*, 2 vols (translated by J. Hoenig and M.W. Hamilton). Baltimore: Johns Hopkins University Press; Appiah, K.A. (2006). *Cosmopolitanism*. New York: Norton; Dewey, J. (1908, 1960). *Theory of the Moral Life*. New York: Holt, Rinehart and Winston.

80. West, C. (1989). *The American Evasion of Philosophy*. Madison: University of Wisconsin Press; Hamington, M. (2009). Feminist prophetic pragmatism. *Journal of Speculative Philosophy*, 25: 83–92.
81. Pratt, S.L. (2002). *Native Pragmatism: Rethinking the Roots of American Philosophy*. Bloomington: Indiana University Press.
82. Addams, J. (1911). *Democracy and Social Ethics*. London: Macmillan; Addams, J. (1915). *Newer Ideals of Peace*. London: Macmillan; Addams, J. (1916, 2002). *The Long Road of Woman's Memory* (ed. C.H. Seigfried). Urbana: University of Illinois Press.
83. Shields, P.M. (2006). Democracy and the social feminist ethics of Jane Addams: A vision for public administration. *Administrative Theory and Praxis*, 28: 418–443.
84. Seigfried. C.H. (1996). *Pragmatism and Feminism: Reweaving the Social Fabric*. Chicago: University of Chicago Press; Mahowald, M.B. (1997). What classical American philosophers mind: Jane Addams, critical pragmatism and cultural feminism. *The Journal of Value Inquiry*, 31: 39–54.
85. Spinoza, B. (1668, 1955). *On the Improvement of the Understanding* (transl. by R.H.M. Elwes). New York: Dover Press; Israel, J.I. (2001). *Radical Enlightenment*. Oxford: Oxford University Press; Israel, J.I. (2010). *A Revolution of the Mind: Radical Enlightenment and the Intellectual Origins of Modern Democracy*. Princeton: Princeton University Press.
86. Dewey, J. (1939, 1963). *Freedom and Culture*. New York: Capricorn Books; Rorty, R. (1999). *Philosophy and Social Hope*. New York: Penguin Books; Neville, R.C. (1974). *The Cosmology of Freedom*. New Haven: Yale University Press; Bernstein, R.J. (2010). *The Pragmatic Turn*. Malden, MA: Polity Press.
87. Dewey, J. (1908, 1960). *Theory of the Moral Life*. New York: Holt, Rinehart and Winston; Fromm, E. (1973). *The Anatomy of Human Destructiveness*. New York: Holt, Rinehart and Winston; Merlau-Ponty, M. (1968, 1970). *Themes*. Evanston, IL: Northwestern University Press.
88. Moreno, J. (1995). *Deciding Together*. Oxford: Oxford University Press; McAfee, N. (2008). *Democracy and the Political Unconscious*. New York: Columbia University Press; McAfee, N. (2005). Two feminisms. *Journal of Speculative Philosophy*, 19: 140–149; Ryan, A. (1995). *John Dewey and the High Tide of American Liberalism*. New York: Norton Press.
89. Whitehead, A.N. (1929, 1958). *The Function of Reason*. Boston: Beacon Press; Whitehead, A.N. (1929, 1978). *Process and Reality*. New York: Free Press.
90. Dewey, J. (1934, 1970). *A Common Faith*. New Haven:Yale University Press.
91. Dewey, J. (1939, 1963). *Freedom and Culture*. New York: Capricorn Books; Sullivan, M. (2007). *Legal Pragmatism: Community, Rights and Democracy*. Bloomington: Indiana University Press; Sullivan, S. (2001). *Living Across and Through Skins: Transactional Bodies, Pragmatism and Feminism*. Bloomington: Indiana University Press; Sullivan, W.M. (1986). *Reconstructing Public Philosophy*. Berkeley: University of California Press; Green, C.D. (2009). Darwinian theory, functionalism, and the first American revolution. *American Psychologists*, 64: 75–84.
92. Whitehead, A.N. (1938, 1967). *Modes of Thought*. New York: Free Press.
93. Pratt, S.L. (2002). *Native Pragmatism: Rethinking the Roots of American Philosophy*. Bloomington: Indiana University Press.

Bibliography

Abela, P. (2002). *Kant's Empirical Realism*. Oxford: Oxford University Press.
Adams, H. (1906, 1990). *The Education of Henry Adams*. New York: Vintage Books.
Addams, J. (1911). *Democracy and Social Ethics*. London: Macmillan.
Addams, J. (1915). *Newer Ideals of Peace*. London: Macmillan.
Addams, J. (1916, 2002). *The Long Road of Woman's Memory* (edited by C.H. Seigfried). Urbana: University of Illinois Press.
Adolphs, R. (1999). Social cognition and the human brain. *Trends in Cognitive Sciences*, 3: 469–479.
Adolphs, R., Denburg, N.L., and Tranel, D. (2001). The amygdala's role in long-term declarative memory for gist and detail. *Behavioral Neuroscience*, 112: 983–992.
Aggleton, J. (1992, 2000). *The Amygdala*. Oxford: Oxford University Press.
Aiello, L.C., and Dunbar, R.I.B. (1993). Neocortex size, group size and the evolution of language. *Current Anthropology*, 34: 184–192.
Alembert J. (1751, 1963). *Preliminary Discourse to the Encyclopedia of Diderot*. New York: Library of Liberal Arts.
Alexander, T.M. (1987). *John Dewey's Theory of Art, Experience, and Nature: The Horizons of Feeling*. Albany, NY: SUNY Press.
Alexander, T.M. (1996). The fourth world of American philosophy: the philosophical significance of Native American culture. *Transactions of the Charles S. Peirce Society*, 32.
Altman, J. (1966). Autoradiographic and histological studies of postnatal neurogenesis. *Journal of Comparative Neurology*, 124: 431–474.
Anderson, D.R. (1987). *Creativity and the Philosophy of C.S. Peirce*. New York: Kleuwer.
Anderson, D.R. (1995). *Science of Discovery*. West Lafayette: Purdue University Press.
Anderson, D.R. (1995). *Strands of System: The Philosophy of C.S. Peirce*. West Lafayette: Purdue University Press.
Anderson, D.R. (1997). John E. Smith and the heart of experience. In: *The Recovery of Philosophy in America: Essays in Honor of John Edwin Smith* (edited by T.P. Kasulis and R.C. Neville). Albany: SUNY Press: 115–130.
Apel, K.O. (1981). *Charles S. Peirce: From Pragmatism to Pragmaticism* (trans. J.M. Krois). Amherst: University of Massachusetts Press.
Appiah, K.A. (2006). *Cosmopolitanism*. New York: Norton.
Aristotle. (1962). *The Politics* (transl. T. Sinclair). New York: Penguin Books.
Aristotle. (1968). *De Anima* (transl. D. Hamylin). Oxford: Oxford University Press.
Atran, S. (1990, 1996). *Cognitive Foundations of Natural History*. New York: Cambridge University Press.
Atran, A., Medin, D.L., and Ross, N.O. (2005). The cultural mind. *Psychological Review*, 112: 744–776.

Atran, S., Medin, D., Ross, N., Lynch, E., Coley, J., Ucan Ek', E., and Vapnarsky, V. (1999). Folkecology and commons management in the Maya Lowlands. *Proceedings of the National Academy of Sciences*, 96: 7598–7603.
Bacon, F. (1620, 1887). *Novum Organum*. New York: P.F. Collier and Sons.
Barger, N., Stefanacci, L., and Semendeferi, K. (2007). A comparative volumetric analysis of the amygdaloid complex and basolateral division in the human and ape brain. *American Journal of Physical Anthropology*, 134: 392–403.
Baron, J. (1988, 2008). *Thinking and Deciding*. Cambridge: Cambridge University Press.
Baron-Cohen, S. (1995, 2000). *Mindblindness*. Cambridge, MA: MIT Press.
Barrow, M.V. (2009). *Nature's Ghosts*. Chicago: University of Chicago Press.
Barsalou, L.W. (2003). Abstraction in perceptual symbol systems. *Philosophical Transactions of the Royal Society B*, 358: 1177–1187.
Barton, R.A. (2004). Binocularity and brain evolution in primates. *Proceedings of the National Academy of Sciences*, 101: 10113–10115.
Barton, R.A. (2006). Primate brain evolution: Integrating comparative neurophysiological and ethological data. *Evolutionary Anthropology*, 15: 224–236.
Barton, R.A., Aggleton, J.P., and Grenyer, R. (2003). Evolutionary coherence of the mammalian amygdala. *Proceedings of the Royal Society B: Biological Sciences*, 270: 539–543.
Barzun, J. (1983). *A Stroll with William James*. New York: Harper and Row.
Beard, C.A. (1913, 1935). *An Economic Interpretation of the Constitution of the United States*. New York: Free Press.
Bechara, A. (2005). Decision-making, impulse control and loss of willpower to resist drugs: A neurocognitive perspective. *Nature Neuroscience*, 8: 1458–1463.
Bechara, A., Damasio, A., and Damasio, H. (2000). Emotion, decision-making and the orbitofrontal cortex. *Cerebral Cortex*, 10: 295–307.
Becker, C.L. (1915, 1967). *Beginning of the American People*. Ithaca: Cornell University Press.
Becker, C.L. (1932). *The Heavenly City of the Eighteenth Century Philosophers*. New Haven: Yale University Press.
Beebe, W. (1944, 1988). *The Book of Naturalists*. Princeton: Princeton University Press.
Beiser, F.C. (2002). *German Idealism*. Cambridge, MA: Harvard Univesity Press.
Bellah, R.N., Madsen, R., Sullivan, W.M., Swidler, A., and Tipton, S.M. (1985). *Habits of the Heart: Individualism and Commitment in American Life*. New York: Harper and Row.
Bergson, H. (1911, 1998). *Creative Evolution*. New York: Dover Press.
Bergson, H. (1919, 1946). *The Creative Mind* (translated by M. Andison). New York: Citadel Press.
Berlin, I. (1976). *Vico and Herder*. New York: Vintage Press.
Berlin, I. (1999). *The Roots of Romanticism*. Princeton: Princeton University Press.
Bernard, C. (1865, 1957). *An Introduction to the Study of Experimental Medicine*. New York: Dover Publications.
Bernstein, R.J. (1961). Charles Sanders Peirce and the nation. *The Antioch Review*, 21: 15–25.
Bernstein, R.J. (1961, editor). *Perspectives on Peirce*. New Haven: Yale University.
Bernstein, R.J. (2010). *The Pragmatic Turn*. Malden, MA: Polity Press.

Berthoz, A. (2000). *The Brain's Sense of Movement*. Cambridge, MA: Harvard University Press.
Berthoz, A., and Petite, J.L. (2006, 2008). *The Physiology and Phenomenology of Action*. Oxford: Oxford University Press.
Bhaskar, R. (1998). *The Possibility of Naturalism*. London: Routledge.
Bjork, D.W. (1983). *The Compromised Scientist: William James in the Development of American Psychology*. New York: Columbia University Press.
Bliss, M. (1999). *William Osler: A Life in Medicine*. Oxford: Oxford University Press.
Bliss, M. (2003). *Harvey Cushing: A Life in Surgery*. Oxford: Oxford University Press.
Bliss, M. (2011). *The Making of Modern Medicine: Turning Points in the Treatment of Disease*. Chicago: The University of Chicago Press.
Boas, F. (1911, 1968). *The Mind of Primitive Man*. New York: Free Press.
Bordogna, R. (2008). *William James at the Boundaries*. Chicago: University of Chicago Press.
Botvinick, M.M., Braver, T.S., Barch, D.M., Carter, C.S., and Cohen, J.D. (2001). Conflict monitoring and cognitive control. *Psychological Review*, 108: 624–652.
Bowler, P.J. (1983). *The Eclipse of Darwinism*. Baltimore: Johns Hopkins University Press.
Bowler, P.J. (1988). *The Non-Darwinian Revolution*. Baltimore: Johns Hopkins University Press.
Bowler, P.J. (2009). *Science for All*. Chicago: Chicago University Press.
Boyd, R. (1999). Homeostasis, species and higher taxa. In: *Species: New Interdisciplinary Essays* (edited by R.A. Wilson). Cambridge, MA: MIT Press.
Boyd, R., and Richerson, P. (2005). *Not by Genes Alone: How Culture Transformed Human Evolution*. Chicago: University of Chicago Press.
Bradford, P.H., and Blume, H. (1992). *The Pygmy in the Zoo*. New York: St Martins Press.
Brandom, R.B. (2009). *Reason in Philosophy*. Cambridge, MA: Harvard University Press.
Brandom, R.B. (2011). *Perspectives on Pragmatism*. Cambridge, MA: Harvard University Press.
Breisach, E.A. (1993). *American Progressive History*. Chicago: University of Chicago Press.
Brent, J. (1993). *Charles Sanders Peirce*. Bloomington: Indiana University Press.
Brinkley, D. (2009). *The Wilderness Warrior: Theodore Roosevelt and the Crusade for America*. New York: HarperCollins.
Broca, P. (1863). Localization des functions cerebrales. Siege du langage articule. *Bulletins de la Societe d'Anthropologie (Paris)*, 4: 200–203.
Brodel, M. (1946). *Three Unpublished Drawings of the Anatomy of the Human Ear*. Philadelphia: W.B. Sanders Co.
Brooks, V.W. (1957). *The Flowering of New England*. New York: E.P. Dutton and Co.
Brown, M. (2011). Science as Socially Disturbed Cognition: Bridging Philosophy and Sociology of Science. In: *Foundations of the Formal Sciences VII* (edited by K. François, B. Löwe, T. Müller, and B. van Kerkhove). London: College Publishing.

Buchler, J. (1939, 1966). *Charles Peirce's Empiricism*. New York: Octagon Books.
Buchler, J. (1955, 1966). *Nature and Judgment*. New York: Grossett and Dunlap.
Buffon, G.L. (1749–1767). *Histoire naturelle generale et particuliere*, 15 vols. Paris: Imprimerie Royale.
Burckhardt, J. (1929, 1958). *The Civilization of the Renaissance in Italy*. New York: Colophon Books.
Burckhardt, R. (1977, 1995). *The Spirit of System: Lamarck and Evolutionary Biology*. Cambridge, MA: Harvard University Press.
Burke, T. (1994). *Dewey's New Logic: A Reply to Russell*. Chicago: University of Chicago Press.
Bury, J.B. (1933, 1960). *The Idea of Progress*. New York: Dover Press.
Butterfield, H. (1957, 1965). *The Origins of Modern Science*. New York: Free Press.
Butterfield, H. (1981). *The Origins of History*. New York: Basic Books.
Buzsaki, G. (2006). *Rhythms of the Brain*. Oxford: Oxford University Press.
Bynum, W. (2008). *The History of Medicine*. Oxford: Oxford University Press.
Byrne, R.W., and Corp, N. (2004). Neocortex size predicts deception rate in primates. *Proceedings of the Royal Society*, 271: 1693–1699.
Cacioppo, J.T., Hawkley, L., Ernst, J.M., et al. (2006). Loneliness within a nomological net: an evolutionary perspective. *Journal of Research in Personality*, 40: 1054–1085.
Cadwallader, T.C. (1975). Peirce as an experimental psychologist. *Transactions of the Charles S. Peirce Society*, 11: 167–186.
Cadwallader, T.C., and Cadwallader, J.V. (1972). American's first modern psychologist: William James or Charles S. Peirce. *Proceedings of the American Psychological Association*, 773–774.
Campbell, J. (1995). *Understanding John Dewey: Nature and Cooperative Intelligence*. Chicago: Open Court.
Cannon, W.B. (1915, 1929, 1963). *Bodily Changes in Pain, Hunger, Fear and Rage*. New York: Harper Torchbooks.
Cannon, W.B. (1932, 1963). *The Wisdom of the Body*. New York: W.W. Norton.
Canquilhem, G. (1991). *The Normal and the Pathological*. New York: Zone Books.
Carey, S. (1985, 1987). *Conceptual Change in Childhood*. Cambridge, MA: MIT Press.
Carey, S. (2004). Bootstrapping and the origins of concepts. *Daedalus*, Winter 133: 59–68.
Carey, S. (2009). *On the Origins of Concepts*. Oxford: Oxford University Press.
Carnap, R. (1928, 1969). *The Logical Structure of the World*. Berkeley: University of California Press.
Carson, R. (1962, 1987). *Silent Spring*. Boston: Houghton Mifflin Company.
Carter, C.S., Lederhendler, I.L., and Kirkpatrick, B. (1997, 1999). *The Integrative Neurobiology of Affiliation*. Cambridge, MA: MIT Press.
Carver, C.F. (2007) *Explaining the Brain*. Cambridge MIT Press.
Cassirer, E. (1918, 1981). *Kant's Life and Thought*. New Haven: Yale University Press.
Cassirer, E. (1926, 1963). *The Individual and the Cosmos in Renaissance Philosophy* (translated by M. Domandi). New York: Harper and Row.
Cassirer, E. (1944, 1978). *An Essay on Man*. New Haven: Yale University Press.
Cassirer, E. (1951). *The Philosophy of the Enlightenment*. Princeton: Princeton University Press.

Cassirer, E. (1957). *The Philosophy of Symbolic Forms*, vol. III. New Haven: Yale University Press.
Chan, W.J. (1963). *A Source Book in Chinese Philosophy.* Princeton: Princeton University Press.
Chomsky, N. (1965). *Aspects of the Theory of Syntax*. Cambridge, MA: MIT Press.
Chomsky, N. (1972). *Language and Mind*. New York: Harcourt Brace Jovanovich.
Cicero (1960). *Selected Works* (translated by M. Grant). New York: Penguin Books.
Clark, A. (1997). *Being There*. Cambridge, MA: MIT Press.
Clark, R.W. (1983). *Benjamin Franklin*. New York: DaCapo Press.
Cohen, I.B. (1980, editor). *Benjamin Peirce: Father of Pure Mathematics in America*. New York: Arno Press.
Cohen, M.R. (1923). *Chance, Logic and Love*. New York: Harcourt.
Cohen, M.R. (1931, 1959). *Reason and Nature*. New York: Dover Press.
Cohen, J.D., McClure, S.M., and Yu, A.J. (2007). Should I stay or should I go? How the brain manages the trade-off between exploitation and exploration. *Philosophical Transactions of the Royal Society, B*, 362: 933–942.
Colapterio, V. (1996, editor). *Peirce's Philosophical Perpectives*. NY: Fordham University Press.
Coleridge, S.T. (1840, 1956). *Confessions of an Inquiring Spirit*. Stanford: Stanford University Press.
Coleridge, S.T. (1956). *Selected Poems*. New York: Appleton-Century Crofts.
Collingwood, R.G. (1945, 1976). *The Idea of Nature*. Oxford: Oxford University Press.
Comfort, N.C. (1999). The real point is control: the reception of Barbara McClintock's controlling elements. *Journal of the History of Biology*, 32: 133–162.
Comte, A.C. (1842, 1975). *A General View of Positivism* (translated by J. Bridges). New York: Robert Speller and Sons.
Cook, D.J., Murrow, C.D., and Haynes, R.B. (1997). Systematic reviews: synthesis of the best evidence in clinical medicine. *Annals of Internal Medicine*, 126: 377–391.
Cooper, D.E. (2002, 2007). *The Measure of Things*. Oxford: Clarendon Press.
Corballis, M.C. (2002). *From Hand to Mouth*. Princeton: Princeton University Press.
Cosmides, L., and Tooby, J. (1992). Cognitive adaptations for social exchange. In: *The Adapted Mind* (edited by J. Barkow, L. Cosmides, and J. Tooby). New York: Oxford University Press.
Craig, W. (1918). Appetites and aversions as constituents of instinct. *Biological Bulletin*, 34: 91–107.
Craver, C.F. (2009). Mechanisms and natural kinds. *Philosophical Psychology*, 22: 575–594.
Crews, D. (2008). Epigenetics and its implications for behavioral endocrinology. *Frontiers in Neuroendocrinology*, 29: 344–357.
Critchley, H.D. (2005). Neural mechanisms of autonomic, affective, and cognitive integration. *Journal of Comparative Neurology*, 493: 154–166.
Croce, B. (1933, 1963). *History of Europe in the Nintheenth Century.* New York: Harcourt, Brace and World.
Cronon, W. (1983, 1991). *Changes in the Land*. New York: Hill and Wang.

Crosby, R.W., and Cody, J. (1991). *Max Brodel: The Man Who Put Art into Medicine.* New York: Springer Verlag.
Crowe, S.J. (1957). *Halsted of Johns Hopkins.* Springfield, IL: Charles C. Thomas.
Curley, J.P., Davidson, S., Bateson, P., and Champagne, F.A. (2009). Social enrichment during postnatal development induces transgenerational effects on emotional and reproductive behavior in mice. *Frontiers in Behavioral Neuroscience,* 3: 25.
Da Vinci, L. (1980). *The Notebooks of Leonardo da Vinci.* Oxford: Oxford University Press.
Dalton, T.C. (2002). *Becoming John Dewey: Dilemmas of a Philosopher and Naturalist.* Bloomington, IN: Indiana University Press.
Damasio, A.R. (1994). *Descartes' Error: Emotion, Reason, and the Human Brain.* New York: Grosset/Putnam.
Damasio, A.R. (1996). The somatic marker hypothesis and the possible functions of the prefrontal cortex. *Philosophical Transactions of the Royal Society,* 351: 1413–1420.
Damasio, A.R. (1999). *The Feeling of What Happens.* New York: Harcourt Press.
Daniels, N. (1974, 1989). *Thomas Reid's Inquiry.* Palo Alto: Stanford University Press.
Darwin, C. (1859, 1958). *The Origin of Species.* New York: Mentor Books.
Darwin, C. (1868). *The Variation of Animals and Plants under Domestication,* 2 vols. London: John Murray.
Darwin, C. (1871, 1874). *Descent of Man.* New York: Rand McNally and Company.
Darwin, C. (1872, 1965). *The Expression of Emotions in Man and Animals.* Chicago: University of Chicago Press.
Darwin, E. (1801, 1994). *Zoonomia, or the Law of Organic Life* (3rd edition). London: J. Johnson.
Davenport, C.D. (1921). Research in eugenics. *Science,* 54: 341–347.
Dear, P. (1995). *Discipline and Experience.* Chicago: University of Chicago Press.
Dear, P. (2006). *The Intelligibility of Nature.* Chicago: University of Chicago Press.
Decety, J., and Jackson, P.W. (2006). A social neuroscience perspective on empathy. *Current Directions in Psychological Science,* 15: 54–58.
Degler, C.N. (1991). *In Search of Human Nature.* Oxford: Oxford University Press.
Dehaene, S. (1997). *The Number Sense.* Oxford: Oxford University Press.
Deheane, S., Izard, V., Pica, P., and Spelke, E. (2006). Core knowledge of geometry in an Amazonian Indigene group. *Science,* 311: 381–384.
Deichmann, U. (1996). *Biologists under Hitler* (translated by T. Dunlap). Cambridge, MA: Harvard University Press.
Dennett, D. (1987). *The Intentional Stance.* Cambridge, MA: MIT Press.
Derrida, J. (1973, 1980). *The Archeology of the Frivolous.* Lincoln, NE; University of Nebraska Press.
Derrida, J. (2008). *Islam and the West* (edited by M. Cherif). Chicago: University of Chicago Press.
Descartes, R. (1637, 1993). *Discourse on Method.* Indianapolis: Hackett Publishing Co.
Descartes, R. (1649, 1989). *The Passions of the Soul* (translated by S. Voss). Cambridge, MA: Hackett Publishing Co.

Desmurget, M., Reilly, K.T., Richard, N., Szathmari, A., Mottolese, C., and Sirigu, A. (2009). Movement intention after parietal cortex stimulation in humans. *Science*, 324: 811–813.
deWaal, F., and Lanting, F. (1997). *Bonobo: The Forgotten Ape*. Berkeley: University of California Press.
Dewey, J. (1896). The reflex arc concept in psychology. *Psychological Review*, 3: 357–370.
Dewey, J. (1902, 1974). *The Child and the Curriculum and the School and Society*. Chicago: University of Chicago Press.
Dewey, J. (1908, 1960). *Theory of the Moral Life*. New York: Holt, Rinehart and Winston.
Dewey, J. (1909, 1975). *Moral Principles in Education*. Carbondale, IL: Southern Illinois University Press.
Dewey, J. (1910). *How We Think*. New York: DC Heath and Publ.
Dewey, J. (1910, 1965). *The Influence of Darwin on Philosophy*. Bloomington, IN: Indiana University Press.
Dewey, J. (1916). *Essays in Experimental Logic*. Chicago: University of Chicago Press.
Dewey, J. (1920, 1948). *Reconstruction in Philosophy*. Boston: Beacon Press.
Dewey, J. (1922, 1957). *Human Nature and Conduct*. New York: Random House.
Dewey, J. (1925, 1989). *Experience and Nature*. La Salle, IL: Open Court.
Dewey, J. (1929, 1960). *The Quest for Certainty*. New York: Capricorn Books.
Dewey, J. (1929, 1962). *Individualism, Old and New*. New York: Capricorn Books.
Dewey, J. (1929, 1966). *A Theory of Valuation*. Chicago: University of Chicago Press.
Dewey, J. (1931). *Philosophy and Civilization*. New York: Minton, Balch and Co.
Dewey, J. (1934, 1958). *Art as Experience*. New York: Capricorn Books.
Dewey, J. (1934, 1970). *A Common Faith*. New Haven: Yale University Press.
Dewey, J. (1935, 1963). *Liberalism and Social Action*. New York: Capricorn Books.
Dewey, J. (1938). *Logic: The Theory of Inquiry*. New York: Holt, Rinehart.
Dewey, J. (1938, 1973). *Experience and Education*. New York: Collier.
Dewey, J. (1939, 1963). *Freedom and Culture*. New York: Capricorn Books.
Dewey, J. (1967). John Dewey, Psychology. In: *John Dewey: The Early Works 1882–1898*. Carbondale, IL: Southern Illinois University Press.
Dewey, J., and Tufts, J.H. (1910). *Ethics*. New York: Henry Holt.
DeWulf, M. (1907, 1956). *Scholastic Philosophy*. New York: Dover Press.
Diamond, A. (2001). A model system for studying the role of dopamine in the prefrontal cortex during early development of humans: early and continuously treated phenylketonuria. In: *Handbook of Developmental Cognitive Neuroscience* (edited by C.A. Nelson and M. Luciana). Cambridge, MA: MIT Press: 433–472.
Diamond, A., and Goldman-Rakic, P.S. (1989). Comparison of human infants and rhesus monkeys on Piaget's AB task: evidence for dependence on dorsolateral prefrontal cortex. *Experimental Brain Research*, 74: 24–40.
Diamond, J. (1998). *Guns, Germs and Steel*. New York: Norton.
Diderot, D. (1755, 1964). The Encyclopedia. In: *Rameau's Nephew and Other Works*. New York: The Library of Liberal Arts.
Diggins, J.P. (1991). *The Promise of Pragmatism*. Chicago: University of Chicago Press.

Diggins, J.P. (2000). *On Hallowed Ground: The Foundation of American History.* New Haven: Yale University Press.
Diggins, J.P. (2011). *Why Niebuhr Now?* Chicago: University of Chicago Press.
Dilthey, W. (1883). *Introduction to the Human Sciences.* Princeton, NJ: Princeton University Press.
Dobzhansky, T.C. (1962). *Mankind Evolving.* New Haven: Yale University Press.
Domes, G., Heinrichs, M., Glascher, J., Büchel, C., Braus, D.F., and Herpertz, S.C. (2007). Oxytocin attenuates amygdala responses to emotional faces regardless of valence. *Biological Psychiatry* 62: 1187–1190.
Donald, D.H. (1947, 2001). *Lincoln, Reconsidered: Essays on the Civil War Era.* New York: Vintage Books.
Donald, M. (1991). *Origins of Modern Man.* Cambridge, MA: Harvard University Press.
Donald, M. (2004). Hominid enculturation and cognitive evolution. In: *The Development of the Mediated Mind* (edited by J.M. Luraciello, J.A. Hudson, R. Fivush, and P.J. Bauer). Mawash, NJ: Erlbaum Press.
Donaldson, H.H. (1915). The rat data and reference tables for the albino and Norway rat. In: *Memoirs of the Wistar Institute of Anatomy and Biology,* no. 6. Philadelphia: Wistar Institute.
Dretkse, F. (1995). *Naturalizing the Mind.* Cambridge, MA: MIT Press.
Du Bois, W.E.B. (1903, 1982). *The Souls of Black Folk.* New York: Penguin.
Du Bois, W.E.B. (1940, 1968). *Dusk of Dawn.* London: Transactions Publishers.
Du Bois, W.E.B. (1968, 1991). *The Autobiography of W.E.B. Du Bois.* New York: International Publishers.
Duhem, P. (1915, 1991). *German Science.* La Salle, IL: Open Court Press.
Dunbar, R.I.M. (1992). Neocortex size as a constraint on group size in primates. *Journal of Human Evolution,* 22: 469–493.
Dunbar, R.I.M. (1996). *Grooming, Gossip and the Evolution of Language.* Cambridge, MA: Harvard University Press.
Dunbar, R.I.M. (2003). The social brain. *Annual Review of Anthropology,* 32: 163–181.
Dunbar, R.I.M., and Shultz, S. (2007). Understanding primate evolution. *Philosophical Transactions of the Royal Society,* 362: 649–658.
Dupre, J. (1981). Natural kinds and biological taxa. *The Philosophical Review,* 90: 66–90.
Durkheim, E. (1974). *Sociology and Philosophy* (translated by D.F. Pocock). New York: Free Press.
Durst, A. (2010). *Women Educators in the Progressive Era: The Women behind Dewey's Laboratory School.* Basingstoke: Palgrave Macmillan.
Eames, S.M. (1977). *Pragmatic Naturalism.* Carbondale, IL: Southern Illinois University Press.
Edel, A. (1944, 1949). Naturalism and ethical theory. In: *Naturalism and the Human Spirit* (edited by Y.H. Krikorian). New York: Columbia University Press.
Edel, A. (1955, 1965). *Ethical Judgment: The Use of Science in Ethics.* New York: Free Press.
Edel, A. (2001). *Ethical Theory and Social Change: The Evolution of John Dewey's Ethics, 1908–1932.* New Brunswick, NJ: Transaction Publishers.
Edie, J.M. (1987). *William James and Phenomenology.* Bloomington, IN: Indiana University Press.

Egner, T., and Hirsch, J. (2005). Cognitive control mechanisms resolve conflict through cortical amplification of task-relevant information. *Nature Neuroscience*, 8: 1784–1790.
Eichenbaum, H., and Cohen, N.J. (2001). *From Conditioning to Conscious Recollection*. Oxford: Oxford University Press.
Eisele, C. (1957). The scientist-philosopher C.S. Peirce at the Smithsonian. *Journal of the History of Ideas*, 18: 537–547.
Eisele, C. (1979). *Studies in the Scientific Philosophy of C.S. Peirce*. Paris: Mouton Publ.
Ekman, P. (1972). Universals and cultural differences in facial expressions of emotion. In: *Nebraska Symposium on Motivation, 1971* (edited by J. Cole). Lincoln: University of Nebraska Press.
Elder, C.L. (2009). Biological species are natural kinds. *Southern Journal of Philosophy*, 46: 339–362.
Eldridge, M. (1998). *Transforming Experience: John Dewey's Cultural Instrumentalism*. Nashville: Vanderbilt University Press.
Elster, J. (1979, 1988). *Ulysses and the Sirens*. Cambridge: Cambridge University Press.
Elster, J. (1983). *Explaining Technical Change*. Cambridge: Cambridge University Press.
Elster, J. (2009). *Alexis de Tocqueville: The First Social Scientist*. Cambridge: Cambridge University Press.
Elstein, A.S. (2004). On the origins and development of evidence-based medicine and medical decision-making. *Inflammation Research*, 53: 8184–8189.
Elstein, A.S., Christensen, C., Cottrell, J.J., Polson, A., and Ng, M. (1999). Effects of prognosis, perceived benefit and decision style upon decision making in critical care. *Critical Care Medicine*, 27: 58–65.
Emerson, R.W. (1855, 1883). *Nature, Addresses and Lectures*. Cambridge, MA: The Riverside Press.
Esposito, J.L. (1980). *Evolutionary Metaphysics: The Development of Peirce's Theory of Categories*. Athens, OH: Ohio State University.
Fakhry, M. (1970, 1983). *A History of Islamic Philosophy*. New York: Columbia University Press.
Farber, P.L. (2000). *Finding Order in Nature*. Baltimore: Johns Hopkins University Press.
Farquhar, C.M., Kofa, E., Power, M.L., Zinberg, S., and Schulkin, J. (2002). Clinical practice guidelines as educational tools for obstetrician-gynecologists. *Journal of Reproductive Medicine*, 47: 897–902.
Feinstein, H.M. (1984). *Becoming William James*. Ithaca: Cornell University Press.
Fiala, A. (2002). Toleration and pragmatism. *Journal of Speculative Philosophy*, 16: 103–116.
Fisch, M.A. (1954). Alexander Bain and the genealogy of pragmatism. *Journal of the History of Ideas*, 15: 413-444.
Fisch, M.H. (1986). Evolution in American philosophy. In: *Peirce, Semiotic and Pragmatism* (edited by K.L. Ketner and J.W. Kloesel). Bloomington: Indiana University Press.
Fisch, M.H. (1986). Peirce at the Johns Hopkins University. In: *Peirce, Semiotic and Pragmatism* (edited by K.L. Ketner and J.W. Kloesel). Bloomington: Indiana University Press.

Flanagan, O. (2007). *The Really Hard Problem*. Cambridge, MA: MIT Press.
Flexner, A. (1910). *Medical Education in the United States and Canada: A Report to the Carnegie Foundation for the Advancement of Teaching. Bulletin Number Four.* Boston: Updyke.
Flexner, S. (1941). *William Welsh and the Heroic Age of American Medicine*. New York: Viking Press.
Flower, E., and Murphy, M.G. (1977). *A History of Philosophy in America*, vol. 1 and 2. New York: Capricorn Books.
Fodor, J. (1983). *The Modularity of Mind*. Cambridge, MA: MIT Press.
Foley, R.A. (2001). The evolutionary consequences of increased carnivory in hominids. In: *Meat-Eating and Human Evolution* (edited by C.B. Stanford and H.T. Bunn). New York: Oxford University Press.
Foley, R.A. (2006). The Emergence of culture in the context of hominid evolutionary patterns. In: *Evolution and Culture* (edited by S.C. Levinson and P. Jaisson). Cambridge, MA: MIT Press.
Foley, R.A., and Lee, P.D. (1991). Ecology and energetics of encephalization in hominid evolution. *Philosophical Transactions of the Royal Society, Series B*, 334: 223–232.
Franks, D. (2010). *Neurosociology: The Nexus between Neuroscience and Social Psychology*. New York: Springer.
Franklin, B. (1987). *Essays, Articles, Bagatelles, and Letters, Poor Richards Almanac, Autobiography*. New York: Library Classics.
French, R. (2008). *Medicine before Science*. Cambridge: Cambridge University Press.
Freud, S. (1919, 1961). *Beyond the Pleasure Principle*. New York: Norton.
Freud, S. (1924, 1960). *A General Introduction to Psychoanalysis*. New York: Washington Square Press.
Friedman, M. (1992). *Kant and the Exact Sciences*. Cambridge, MA: Harvard University Press.
Friedman, M. (1998). *The Consequences of Logical Positivism*. Cambridge: Cambridge University Press.
Friedman, M. (2000). *A Parting of the Ways; Carnap, Cassirer and Heidegger*. Chicago: Open Court.
Friedman, M. (2001). *Dynamics of Reason*. Stanford: Stanford University Press.
Fromm, E. (1947). *Man for Himself*. New York: Rinehart and Company.
Fromm, E. (1956, 1975). *The Art of Loving*. New York: Harper and Row.
Fromm, E. (1973). *The Anatomy of Human Destructiveness*. New York: Holt, Rinehart and Winston.
Fuller, M. (1845). *Woman in the Nineteenth Century*. New York: Greeley and McElrath.
Fulton, J.F. (1946). *Harvey Cushing: A Biography*. Springfield, IL: Charles Thomas.
Furumoto, L. (1992). Joining separate spheres – Christine Ladd-Franklin. *American Psychologist*, 47: 175–181.
Gabbe, S.G. (1996). The alphabet of academic medicine. *Obstetrics & Gynecology*, 88: 479–481.
Gadamer, H.G. (1981, 1986). *Reason in the Age of Science* (translated by F.G. Lawrence). Cambridge, MA: MIT Press.
Gale, R.M. (1999). *The Philosophy of William James: An Introduction*. Cambridge: Cambridge University Press.
Galen (1968). *On the Usefulness of the Parts of the Body*, vol. 1. Ithaca: Cornell University Press.

Galileo, G. (1610, 1957). The Starry Messenger. In: *Discoveries and Opinions of Galileo* (edited by S. Drake). New York: Doubleday.
Galileo, G. (1957). *Discoveries and Opinions*. New York: Anchor Books.
Galison, P. (1988). History, philosophy and the central metaphor. *Science in Context*, 3: 197–212.
Gallagher, S. (2005). *How the Body Shapes the Mind*. Oxford: Oxford University Press.
Gallistel, C.R. (1980). *The Organization of Action: A New Synthesis*. Hillsdale, NJ: Lawrence Erlbaum.
Gallistel, C.R. (1993). *The Organization of Learning*. Cambridge, MA: MIT Press.
Gallistel, C.R., Gelman, R.M., and Cordes, I.S. (2006). The cultural and evolutionary history of the real numbers. In: *Culture and Evolution* (edited by S. Levinson and P. Jaisson). Cambridge, MA: MIT Press.
Gandhi, M. (1948, 1983). *Autobiography: The Story of my Experiments with Truth*. New York: Dover Press.
Garcia, J., Hankins, W.G., and Rusiniak, K.W. (1974). Behavioral regulation of the milieu interne in man and rat. *Science*, 185: 824–831.
Garrison, J. (1997). *Dewey and Eros: Wisdom and Desire in the Art of Teaching*. New York: Teachers College Press.
Garrison, J. (2008, editor). *Reconstructing Democracy, Recontextualizing Dewey*. Albany: State University of New York Press.
Gibson, J.J. (1966). *The Senses Considered as Perceptual Systems*. New York: Houghton-Mifflin.
Gibson, K.R., and Ingold, T. (1993, editors). *Tools, Language and Cognition in Human Evolution*. Cambridge: Cambridge University Press.
Gigerenzer, G. (2000). *Adaptive Thinking*. Oxford: Oxford University Press.
Gigerenzer, G. (2003). Why does framing influence judgment? *Journal of General Internal Medicine*, 18: 906–961.
Gigerenzer, G. (2007). *Gut Feelings*. New York: Viking Press.
Gigerenzer, G., Gaismaiser W., Milcke, E.K., Schwartz, L.M., and Woloshin, S. (2008). Helping doctors and patients make sense of health statistics. *Psychological Science in the Public Interest*, 8: 53–96.
Gigerenzer, G., and Selten, R. (2001). *Bounded Rationality*. Cambridge, MA: MIT Press.
Gigerenzer, G., Todd, R.M., and ABC Research Group (1999). *Simple Heuristics That Make Us Smart*. New York: Oxford University Press.
Gillham, N.W. (2001). *Sir Francis Galton: From African Exploration to the Birth of Eugenics*. Oxford: Oxford University Press.
Gilligan, C. (1982, 1993). *In a Different Voice: Psychological Theory and Women's Development*. Cambridge, MA: Harvard University Press.
Gillispie, C.C. (1960). *The Edge of Objectivity*. Princeton: Princeton University Press.
Gliboff, S. (2008). *H.G. Bronn, Ernst Haeckel, and the Origins of German Darwinism*. Cambridge, MA: MIT Press.
Godfrey-Smith, P. (1998). *Complexity and Function of Mind in Nature*. Cambridge: Cambridge University Press.
Godfrey-Smith, P. (2002). Dewey on naturalism, realism and science. *Philosophy of Science*, 69: S1–S11.

Godfrey-Smith, P. (2008). Induction, samples, and kinds. In: *Carving Nature at its Joints: Topics in Contemporary Philosophy, Volume 8*. Cambridge, MA: MIT Press.

Godfrey-Smith, P. (2009). *Darwinian Populations and Natural Selection*. Oxford: Oxford University Press.

Goethe, J.W. von (1786–1788, 1999). *The Flight to Italy*. Oxford: Oxford University Press.

Goethe, J.W. von (1790, 2009). *The Metamorphosis of Plants*. Cambridge, MA: MIT Press.

Goethe, J.W. von (1988). *Scientific Studies*. New York. Shurkamp.

Goldman, A.I. (1999). *Knowledge in a Social World*. Oxford: Clarendon Press.

Goldsmith, R.B. (1940, 1982). *The Material Basis of Evolution*. New Haven, CT: Yale University Press.

Goldstein, D.G., and Gigerenzer, G. (2002). Models of ecological rationality: the recognition heuristic. *Psychological Review*, 109: 75–90.

Goodman, N. (1955, 1978). *Fact, Fiction and Forecast*. New York: Bobbs-Merrill Co.

Goodman, R.B. (1990). *American Philosophy and the Romantic Tradition*. Cambridge: Cambridge University Press.

Goodman, R.B. (1998). Wittgenstein and pragmatism. *Paralas*, 4: 91–105.

Goodman, R.B. (2002). *Wittgenstein and William James*. Cambridge: Cambridge University Press.

Gloor, P. (1978). *Inputs and Outputs of the Amygdala: What the Amygdala is Trying to Tell the Rest of the Brain*. New York: Wiley Press.

Gloor, P. (1997). *The Temporal Lobe and Limbic System*. New York: Oxford University Press.

Gould, E., Beylin, A., Tanapat, P., Reeves, A., and Shors, T.J. (1999). Learning enhances adult neurogenesis in the hippocampal formation. *Nature Neuroscience*, 2: 260–265.

Gould, S.J. (1977). *Ontogeny and Phylogeny*. Cambridge, MA: Harvard University Press.

Gould, S.J. (1989). *Church, Humboldt and Darwin: The Tension and Harmony of Art and Science* (edited by Franklin Kelly). Washington DC: National Gallery of Art.

Gould, S.J. (2002). *The Structure of Evolutionary Theory*. Cambridge, MA: Harvard University Press.

Gould, S.J., and Eldridge, N. (1977). Punctuated equilibria: the tempo and mode of evolution reconsidered. *Paleobiology*, 3: 115–151.

Gould, S.J., and Lewontin, R.C. (1979). The spandrels of San Marco and the Panglossian paradigm: A critique of the adaptationist programme. *Proceedings of the Royal Society, B: Biological Sciences*, 205: 581–598.

Grange, J. (1977). *Nature: An Enviromental Cosmology*. Albany, NY: SUNY Press.

Grant, E. (2007). *A History of Natural Philosophy*. Cambridge: Cambridge University Press.

Green, C.D. (2009). Darwinian theory, functionalism, and the first American revolution. *American Psychologists*, 64: 75–84.

Greene, J.D., and Haidt, J. (2002). How (and where) does moral judgment work? *Trends in Cognitive Science*, 6: 517–523.

Greene, J.D., Morelli, S.A., Lowenberg, K., Nystrom, L.E., and Cohen, J.D. (2008). Cognitive load selectively interferes with utilitarian moral judgement. *Cognition*, 107: 1144–1154.

Greene, J.D., Nystrom, L.E., Engell, A.D., Darley, J.M., and Cohen, J.D. (2004). The neural bases of cognitive conflict and control in moral judgment. *Neuron*, 44: 389–400.

Greene, J.D., Sommerville, R.B., Nystrom, L.E., Darley, J.M., and Cohen, J.D. (2001). An fMRI investigation of emotional engagement in moral judgment. *Science*, 293: 2105–2108.

Green, J.M. (2004). Building a cosmopolitan world future through pragmatist mutual hospitality. In: *Pragmatism and the Problem of Race* (edited by D. Koch and B Lawson). Bloomington: Indiana University Press.

Green, J.M. (2009). *Pragmatism and Social Hope*. New York: Columbia University Press.

Grene, M. (1995). *A Philosophical Testament*. LaSalle, IL: Open Court Press.

Gundling, T. (2005). *First in Line*. New Haven: Yale University Press.

Gunn, G. (1992). *Thinking across the American Grain*. Chicago: University of Chicago Press.

Haack, S. (1998). *Manifesto of a Passionate Moderate*. Chicago: University of Chicago Press.

Habermas, J. (1967, 1988). *On the Logic of the Social Sciences* (translated by S.W. Nicholson and J.A. Stark). Cambridge, MA: MIT Press.

Habermas, J. (1996). *The Liberating Power of Symbols*. Cambridge, MA: MIT Press.

Hacking, I. (1964). *Logic of Statistical Inference*. Cambridge: Cambridge University Press.

Hacking, I. (1975). *The Emergence of Probability*. Cambridge: Cambridge University Press.

Hacking, I. (1999). *The Taming of Chance*. Cambridge: Cambridge University Press.

Haeckel, E. (1866). *General Morphology of Organisms*. Berlin: G. Reimer.

Haeckel, E. (1900, 1992). *The Riddle of the Universe*. Buffalo, NY: Prometheus Press.

Haidt, J. (2001). The emotional dog and its rational tail. *Psychological Review*, 108: 814–824.

Haidt, J. (2007). The new synthesis in moral psychology. *Science*, 316: 998–1002.

Hall, D.L. (1973). *The Civilization of Experience*. New York: Fordham University Press.

Hall, M.B. (1965). *Robert Boyle on Natural Philosophy: An Essay with Selections from His Writings*. Bloomington: Indiana University Press.

Hamington, M. (2009). Feminist prophetic pragmatism. *Journal of Speculative Philosophy*, 25: 83–92.

Hamington, M. (2009). *The Social Philosophy of Jane Addams*. Champaign, IL: University of Illinois Press.

Hanna, R. (2001). *Kant and the Foundations of Analytic Philosophy*. Oxford: Oxford University Press.

Hanson, N.R. (1958, 1972). *Patterns of Discovery*. Cambridge: Cambridge University Press.

Hanson, N.R. (1971). *Observation and Explanation*. New York: Harper Press.
Harrington, A. (1987). *Medicine, Mind and the Double Brain*. Princeton: Princeton University Press.
Harvey, W. (1651, 1965). Anatomical exercises on the generation of animals. In: *Works of William Harvey*. New York: Johnson Reprint Corp.
Haskins, C.H. (1923, 1957). *The Rise of the Universities*. Ithaca: Cornell University Press.
Haskins, C.H. (1957, 1966). *The Renaissance of the 12th Century*. New York: Meridian Press.
Hauser, M.D. (1997). *The Evolution of Communication*. Cambridge, MA: MIT Press.
Hauser, M.D., Chomsky, N., and Tecumseh Fitch, W. (2002). The faculty of language: What is it, who has it, and how did it evolve? *Science*, 298: 1569–1576.
Hausman, C. (1993). *The Evolutionary Philosophy of Charles S. Peirce*. Cambridge: Cambridge University Press.
Hayek, F.A. (1944, 1972). *The Road to Serfdom*. Chicago: University of Chicago Press.
Heelan, P.A. (1983). *Space Perception and the Philosophy of Science*. Berkeley, CA: University of California Press.
Heelan, P.A. (2001). The Lifeworld and Scientific Interpretation. In: *Handbook of Phenomenology and Medicine*. Waco, TX: Baylor University.
Heelan, P.A., and Schulkin, J. (1998). Hermeneutical philosophy and pragmatism: A philosophy of the science. *Synthese*, 115: 269–302.
Heidegger, M. (1927, 1962). *Being and Time*. New York: Harper and Row.
Heidegger, M. (1934–1935, 1942, 1996). *Holderlin's Hymn "The Ister."* Bloomington, IN: Indiana University Press.
Heidegger, M. (1962, 1977). *The Question Concerning Technology*. New York: Harper and Row.
Heinrichs, M., and Domes, G. (2008). Neuropeptides and social behaviour: effects of oxytocin and vasopressin in humans. In: *Progress in Brain Research*, vol. 170 (edited by I. D. Neumann and R. Landgraf). Amsterdam: Elsevier Science: 337–350.
Helmholtz, H. von (1867, 1963). *Handbook of Physiological Optics*. New York: Dover Press.
Helmhotlz, H. von (1995). *Science and Culture* (edited by D. Cahan). Chicago: University of Chicago Press.
Herbert, J. (1993). Peptides in the limbic system: Neurochemical codes for coordinated adaptive responses to behavioral and physiological demand. *Progress in Neurobiology*, 41: 723–791.
Hickman, L.A. (1980, 1992). *John Dewey's Pragmatic Technology*. Bloomington, IN: Indiana University Press.
Hickman, L.A. (1998). Dewey's Theory of Inquiry. In: *Reading Dewey* (edited by L. Hickman). Bloomington, IN: Indiana University Press.
Hickman, L.A. (2002). *Philosophical Tools for Technological Culture*. Bloomington, IN: Indiana University Press.
Hickman, L.A. (2007). *Pragmatism as Post-Postmodernism*. New York: Fordham University Press.
Hempel, C.G. (1965). *Aspects of Scientific Explanation*. New York: Free Press.

Hendricks, M. (2000). Applying art to medicine. *Johns Hopkins Magazine*, 12: 20–26.
Herbert, J., and Schulkin, J. (2002). Neurochemical Coding of adaptive responses in the limbic system. In: *Hormones, Brain and Behavior* (edited by D. Pfaff). New York: Elsevier Press.
Herrmann, E., Call, J., Hérnandez-Lloreda, M.V., Hare, B., and Tomasello, M. (2007). Humans have evolved specialized skills of social cognition. *Science*, 317: 1360–1366.
Hildebrand, D.L. (1996). Genuine doubt and the community in Peirce's Theory of Inquiry. *Southwest Philosophy Review*, 12: 33–43.
Hildebrand, D.L. (2000). Putnam, Pragmatism and Dewey. *Transactions of C.S. Peirce Society*, 36: 109–123.
Hinde, R.A. (1968). Critique of energy models of motivation. In: *Motivation* (edited by D. Bindra and J. Stewart). Baltimore, MD: Penguin Books.
Hinde, R.A. (1970). *Animal Behavior* (2nd edition). New York: McGraw-Hill.
Hippocrates. (1983). *Hippocratic Writings*. Harmondsworth: Penguin Books.
Hobart, M.E., and Schiffman, Z.S. (1998). *Information Ages*. Baltimore: Johns Hopkins University Press.
Hobbes, T. (1651, 1949). *De Cive*. New York: Appleton-Century Crafts.
Hofer, M.A. (1973). The role of nutrition in the physiological and behavioral effects of early maternal separation on infant rats. *Psychosomatic Medicine*, 35: 350–359.
Hofstadter, R. (1963). *Anti-Intellectualism in American Life*. New York: Knopf.
Hofstadter, R. (1944, 1958). *Social Darwinism in American Thought*. Boston: Beacon Press.
Holmes, O.W. (1881, 1991). *The Common Law*. New York: Dover Press.
Holliday, R. (2002). Epigenetics comes of age in the twenty-first century. *Journal of Genetics*, 81: 1–4.
Holliday, R. (2006). Epigenetics: a historical overview. *Epigenetics*, 1: 76–80.
Hollinger, D.A. (1975). *Morris Cohen and the Scientific Ideal*. Cambridge, MA: MIT Press.
Hollinger, D.A. (2005). The one drop and the one hate rule. *Daedalus*, Winter 134: 18–28.
Hollinger, R., and Depew, D. (1999, editors). *Pragmatism: From Progressivism to Post Modernism*. Santa Barbara, CA: Praeger.
Hook, S. (1934, 1961). *The Quest for Being*. New York: St. Martins.
Hook, S. (1974). *Pragmatism and the Tragic Sense of Life*. New York: Basic Books.
Hook, S. (1987). *Out of Step*. New York: Harper and Row.
Hookway, C. (1984). Naturalism, fallibilism, and evolutionary epistemology. In: *Minds, Machines, and Evolution* (edited by C. Hookway). Cambridge: Cambridge University Press.
Hookway, C. (2000). *Truth, Rationality, and Pragmatism: Themes from Peirce*. Oxford: Clarendon Press.
Houser, N., Eller, J.R., Lewis, A.C., De Tienne, A., Clark, C.L., and Bront Davis, D. (1998). *The Essential Peirce, Volume 2 (1893–1913)*. Bloomington, IN: Indiana University Press.
Houser, N., and Kloesel, C. (1992, 1998). *The Essential Peirce, Volume 1 (1867–1893)*. Bloomington, IN: Indiana University Press.
Houser, N., Roberts, D.D., and Van Evra, J. (1997, editors). *Studies in the Logic of Charles Sanders Peirce*. Indianapolis: University of Indiana Press.

Hubner, K. (1983). *Critique of Scientific Reason*. Chicago: University of Chicago Press.

Hull, D.L. (1988). *Science as a Process*. Chicago: University of Chicago Press.

Humboldt, A. Von (1859, 1993). *Cosmos*. New York: George Bell and Sons.

Humbolt, W. Von (1836, 1971). *Linguistic and Intellectual Development*. Philadelphia: University of Pennsylvania Press.

Hume, D. (1748, 1962). *An Inquiry Concerning Human Understanding*. New York: Washington Square Press.

Humphrey, N. (1976). The social function of intellect. In: *Growing Points in Ethology* (edited by P.P.G. Bateson and R.A. Hinde). Cambridge: Cambridge University Press: 307–317.

Humphrey, N. (2007). The society of selves. *Philosophical Transactions of the Royal Society, B*, 362: 745–754.

Huxley, T.H. (1863). *Man's Place in Nature*. London: Macmillan.

Innis, R.E. (2009). *Susanne Langer in Focus: The Symbolic Mind*. Bloomington, IN: Indiana University Press.

Israel, J.I. (2001). *Radical Enlightenment*. Oxford: Oxford University Press.

Israel, J.I. (2010). *A Revolution of the Mind: Radical Enlightenment and the Intellectual Origins of Modern Democracy*. Princeton: Princeton University Press.

Izard, V., Sann, C., Spelke, E.S., and Steri, A. (2009). Newborn infants perceive abstract numbers. *Proceedings of the National Academy of Sciences*, 106: 10382–10385.

Jackson, J.H. (1884, 1958). Evolution and dissolution of the nervous system. In: *Selected Writings of John Hughlings Jackson*, vol. II (edited by J. Taylor). London: Staples Press.

Jackson, P.L., and Decety, J. (2004). Motor cognition: a new paradigm to self and other interactions. *Current Opinion in Neurobiology*, 14: 259–263.

James, W. (1887). Some human instincts. *Popular Science Monthly*, 31: 160–176.

James, W. (1890, 1952). *The Principles of Psychology*, vols 1 and 2. New York: Henry Holt.

James, W. (1896, 1956). *The Will to Believe, Human Immortality*. New York: Dover Press.

James, W. (1902, 1974). *The Varieties of Religious Experience: A Study in Human Nature*. New York: Collier Books, Macmillan Publishing Co.

James, W. (1907, 1969). *Pragmatism*. Cleveland: Meridian Books, The World Publishing Co.

James, W. (1910, 1968). The Moral Equivalent of War. In: *The Writings of William James* (edited by John J. McDermott). New York: Modern Library: 660–670.

James, W. (1912, 1967). *Radical Empiricism* (edited by R.B. Perry). New York: Peter Smith.

Jaspers, K. (1913, 1997). *General Psychopathology*, 2 vols (transl. J. Hoenig and M.W. Hamilton). Baltimore: Johns Hopkins University Press.

Jaspers, K. (1951, 1954). *Way to Wisdom*. New Haven: Yale University Press.

Jeannerod, M. (1997). *The Cognitive Neuroscience of Action*. Oxford: Blackwell Publishers.

Jeannerod, M. (1999). To act or not to act: perspectives on the representation of action. *Quarterly Journal of Experimental Psychology*, 52: 1–29.

Jefferson, T. (1787, 1982). *Notes on the State of Virginia*. New York: Norton.

Joas, H. (1985). *G.H. Mead: A Contemporary Reexamination of his Thought*. Cambridge, MA: MIT Press.

Joas, H. (1993). *Pragmatism and Social Theory*. Chicago: University of Chicago Press.
Johnson, M. (1987, 1990). *The Body in the Mind*. Chicago: University of Chicago Press.
Johnson, M. (1993). *Moral Imagination*. Chicago: University of Chicago Press.
Johnson, M. (2007). *The Meaning of the Body*. Chicago: University of Chicago Press.
Jonas, H. (1966). *The Phenomenon of Life*. Chicago: University of Chicago Press.
Jones, J.A. (1998). *Intensity: An Essay in Whiteheadian Ontology*. Nashville: Vanderbilt University Press.
Jones, J.H. (1981). *Bad Blood*. New York: Free Press.
Kagan, J. (1984). *The Nature of the Child*. New York: Basic Books.
Kagan, J. (1994). *Galen's Prophecy*. New York: Basic Books.
Kagan, J. (1998). *Three Seductive Ideas*. Cambridge, MA: Harvard University Press.
Kagan, J. (2002). *Surprise, Uncertainty and Mental Structure*. Cambridge, MA: Harvard University Press.
Kagan, J. (2009). *The Three Cultures: Natural Sciences, Social Sciences and the Humanities in the 21st Century*. Cambridge: Cambridge University Press.
Kahneman, D. (2011). *Thinking Fast and Slow*. New York: Farrar, Straus and Giroux.
Kahneman, D., Slovic, P., and Tversky, A. (1982, editors). *Judgment under Uncertainty: Heuristics and Biases*. New York: Cambridge University Press.
Kant, I. (1787, 1965). *Critique of Pure Reason* (translated by L.W. Beck). New York: St. Martin's Press.
Kant, I. (1788, 1956). *Critique of Practical Reason* (translated by L.W. Beck). New York: Bobbs-Merrill.
Kant, I. (1792, 1951). *Critique of Judgment*. New York: Haffner Press.
Kasulis, T.P., and Neville, R.C. (1997). *The Recovery of Philosophy in America: Essays in Honor of John Edwin Smith*. Albany: SUNY Press.
Kaufman, M. (1976). *American Medical Education: The Formative Years, 1765–1910*. New York: Greenwood Press.
Keil, F.C. (1979). *Semantic and Conceptual Development: An Ontological Perspective*. Cambridge, MA: Harvard University Press.
Keil, F.C. (1989). *Concepts, Kinds and Cognitive Development*. Cambridge, MA: MIT Press.
Keil, F.C., and Wilson, R.A. (2000). *Explanation and Cognition*. Cambridge, MA: MIT Press.
Keller, E.F. (1983). *A Feeling for the Organism*. New York: Freeman and Company.
Kelly, S.D. (2002). Merleau-Ponty on the body. *Ratio*, 4: 0034–0006.
Ketner, K.L. (1998). *His Glassy Essence: An Autobiography of C.S. Peirce*. Nashville: Vanderbilt University Press.
Keverne, E.B. (2004). Understanding well-being in the evolutionary context of brain development. *Philosophical Transactions of the Royal Society*, 359: 1349–1358.
Keverne, E.B., and Curley, J.P. (2008). Epigenetics, brain evolution, and behavior, *Neuroendocrinology*, 29, 398–412.
Keynes, J.M. (1921, 1957). *A Treatise on Probability*. New York: Harper and Row.

Kim, J. (2003). The American origins of philosophical naturalism. *Journal of Philosophical Research, the APA, Centennial Volume*, 83–98.

Kimura, D. (1993). *Neuromotor Mechanism in Human Communication*. Oxford: Oxford University Press.

Kirsch, P., Esslinger, C., Chen, Q., Mier, D., Lis, S., Siddhanti, S., Gruppe, H., Mattay, V.S., Gallhofer, B., and Meyer-Lindenberg, A. (2005). Oxytocin modulates neural circuitry for social cognition and fear in humans. *Journal of Neuroscience*, 25: 11489–11493.

Kitcher, P. (1990). *Kant's Transcendental Psychology*. Oxford: Oxford University Press.

Kitcher, P. (1992). The naturalists return. *Philosophical Review*, 101: 3–51.

Kitcher, P. (1993). *The Advancement of Science*. Oxford: Oxford University Press.

Kitcher, P. (1996). *The Lives to Come*. New York: Simon and Schuster.

Kitcher, P. (2001). *Science, Truth, and Democracy*. New York: Oxford University Press.

Kitcher, P. (2011). *The Ethical Project*. Cambridge, MA: Harvard University Press.

Kosfeld, M., Heinrichs, M., Zak, P.J., Fischbacher, U., and Fehr, E. (2005). Oxytocin increases trust in humans. *Science*, 435: 673–676.

Klerman, G. (1979). The psychobiology of affective states: the legacy of Adolf Meyer. In: *Research in the Psychobiology of Human Behavior* (edited by E. Meyer III and J.V. Brady). Baltimore: Johns Hopkins University Press.

Kline, M. (1959). *Mathematics and the Physical World*. New York: Dover Press.

Kloppenberg, J.T. (1986). *Uncertain Victory: Social Democracy and Progressivism in European and American Thought, 1870–1920*. New York: Oxford University Press.

Kloppenberg, J.T. (1998). *The Virtues of Liberalism*. New York: Oxford University Press.

Kloppenberg, J.T. (2010). James's pragmatism and American culture, 1907–2007. In: *100 Years of Pragmatism: William James's Revolutionary Philosophy* (edited by John Stuhr). Bloomington: Indiana University Press.

Kloppenberg, J.T., and Wightman Fox, R. (1995, editors). *A Companion to American Thought*. Cambridge: Blackwell Publishers.

Knops, A., Thirion, B., Hubbard, E.M., Michel, V., and Dehaene, S. (2009). Recruitment of an area involved in eye movements during mental arithmetic. *Science*, 324: 1583–1585.

Kohler, R.E. (1991). *Partners in Science*. Chicago: University of Chicago Press.

Koopman, C. (2009). *Pragmatism as Transition*. New York: Columbia University Press.

Kornblith, H. (1987, editor). *Naturalizing Epistemology*. Cambridge, MA: MIT Press.

Kornblith, H. (1993). *Inductive Inference and its Natural Ground*. Cambridge, MA: MIT Press.

Kornblith, H. (1994). Naturalism: both metaphysical and epistemelogical. *Midwest Studies in Philosophy*, 19: 39–52.

Koyre, A. (1961). *Renaissance Thought: The Classic, Scholastic, and Humanist Strains*. New York: Harper Torchbooks.

Koyre, A. (1968). *Metaphysics and Measurement: Essays in Scientific Revolution*. Cambridge, MA: Harvard University Press.

Krikorian, Y.H. (1944, 1949). *Naturalism and the Human Spirit*. New York: Columbia University Press.
Kripke, S. (1980). *Naming and Necessity*. Cambridge, MA: Harvard University Press.
Kristeller, P.O. (1955, 1961). *Renaissance Thought*. New York: Harper and Row.
Kuhn, T.S. (1962). *The Structure of Scientific Revolution*. Chicago: University of Chicago Press.
Kuhn, T.S. (2000). *The Road Since Structure*. Chicago: University of Chicago Press.
Kukla, A. (1998). *Studies in Scientific Realism*. Oxford: Oxford University Press.
Kuklick, B. (2001). *A History of Philosophy in America*. Oxford: Oxford University Press.
Kurtz, P. (1990). *Philosophical Essays in Pragmatic Naturalism*. New York: Prometheus Books.
Ladd-Franklin, C. (1887). A method for the experimental determination of the horopter. *The American Journal of Psychology*, 1: 99–111.
Ladd-Franklin, C. (1889). On some characteristics of symbolic logic. *The American Journal of Psychology*, 2: 543–567.
Ladd-Franklin, C. (1916). Charles S. Peirce at Johns Hopkins. *The Journal of Philosophy*, 13: 715–722.
Ladd-Franklin, C. (1927). Visible radiation from excited nerve fiber: the reddish blue arcs and the reddish blue glow of the retina. *Science*, 66: 239–241.
Lahr, M.M., and Foley, R. (1998). Towards a theory of modern human origins: geography, demography, and diversity in recent human evolution. *Yearbook of Physical Anthropology*, 41: 137–176.
Lahr, M.M., and Foley, R. (2004). Human evolution writ small. *Nature*, 431: 1043–44.
Lakoff, G., and Johnson, M. (1999). *Philosophy in the Flesh: The Embodied Mind and its Challenge to Western Thought. New York: Basic Books.*
Lakoff, G., and Núñez, R.E. (2000). *Where Mathematics Comes From*. New York: Basic Books.
Lamarck, J.B. (1809, 1984). *Zoological Philosophy* (translated by H. Elliot). Chicago: University of Chicago Press.
Landau, B., and Gleitman, L. (1985). *Languge and Experience*. Cambridge, MA: Harvard University Press.
Langer, S.K. (1937). *Philosophy in a New Key*. Cambridge, MA: Harvard University Press.
Langer, S.K. (1962). *Philosophical Sketches*. New York: Mentor Books.
Langer, S.K. (1972). *Mind: An Essay on Human Feeling*. Baltimore: Johns Hopkins University Press.
Lasch, C. (1991). *The True and Only Heaven: Progress and its Critics*. New York: Norton.
Laudan, L. (1977). *Progress and Its Problems*. Berkeley: University of California Press.
Laudan, L. (1997, 2003). Progress or rationality? the prospects for normative naturalism. In: *The Philosophy of Science* (edited by D. Papineau). Oxford: Oxford University Press.
Lawrence, C. (1998). Medical minds: surgical bodies. In: *Science Incarnate* (edited by C. Lawrence and S. Shapin). Chicago: University of Chicago Press.

Lawson, B.E., and Koch, D.F. (2004). *Pragmatism and the Problem of Race*. Bloomington, Indiana: Indiana University press.
LeDoux, J.E. (1996). *The Emotional Brain*. New York: Simon and Schuster.
Leibniz, G. (1764, 1996). *New Essays on Human Understanding*. Cambridge: Cambridge University Press.
Leslie, A., Gelman, R., and Gallistel, C.R. (2008). The generative basis of natural number concepts. *Trends in Cognitive Science*, 12: 213–218.
Levi, I. (1967). *Gambling with Truth*. Cambridge, MA: MIT Press.
Levi, I. (2004). Beware of syllogism: reasoning and conjecture according to Peirce. In: *The Cambridge Companion to Peirce*. Cambridge: Cambridge University Press.
Levinson, S. (2003). *Space in Language and Cognition*. Cambridge: Cambridge University Press.
Levinson, S. (2006). Cognition at the heart of human interaction. *Discourse Studies*, 8: 85–93.
Lewis, C.L. (1929, 1956). *Mind and the World Order*. New York: Dover Press.
Lewis, C.L. (1946, 1971). *An Analysis of Knowledge and Valuation*. LaSalle, IL: Open Court.
Leys, R. (1984). Meyer, Watson, and the dangers of behaviorism. *Journal of the History of the Behavioral Sciences*, 20: 128–149.
Lieberman, P. (2000). *Human Language and our Reptilian Brain*. Cambridge, MA: Harvard University Press.
Lifton. R.J. (1986). *The Nazi Doctors*. New York: Basic Books.
Linas, R.R. (2001). *I of the Vortex*. Cambridge: MIT Press.
Linnaeus, C. (1735). *Systema Naturae*. Leiden: Haak.
Livingston, J. (2010). Pragmatism, nihilism, and democracy: what is called thinking at the end of modernity? In: *100 Years of Pragmatism: William James's Revolutionary Philosophy* (ed. John Stuhr). Bloomington: Indiana University Press.
Locke, J. (1690, 1959). *An Essay Concerning Human Understanding*. New York: Dover Press.
Locke, J. (1692–1704, 1955). *A Letter Concerning Toleration*. Indianapolis: Bobbs-Merrill Co.
Loewenstein, G. (1994). The psychology of curiosity. *Psychological Bulletin*, 116: 75–98.
Loewenstein, G. (1996). Out of control: visceral influences on behavior. *Organizational Behavior and Human Decision Processes*, 65: 272–292.
Loewenstein, G. (2006). The pleasures and pains of information. *Science*, 312: 704–6.
Longo, L.D. (1980). Obstetrics and gynecology. In: *The Education of American Physicians* (edited by R.L. Numbers). Berkeley: University of California Press.
Lovejoy, A.O. (1936, 1978). *The Great Chain of Being*. Cambridge, MA: Harvard University Press.
Lovejoy, A.O. (1955). *Essays in the History of Ideas*. New York: George Braziller Inc.
Lovelock, J. (1988). *The Ages of Gaia*. New York: Norton.
Ludmerer, K.L. (1985). *Learning to Heal*. New York: Basic Books.
Ludmerer, K.L. (1999). *Time to Heal*. Oxford: Oxford University Press.
Lurie, E. (1988). *Louis Agassiz: A Life in Science*. Baltimore, MD: Johns Hopkins University Press.

Lyell, C. (1830, 2010). *Principles of Geology.* New York: Nabu Press.
Lysaken, J.T. (2008). *Emerson and Self-Culture.* Bloomington, IN: Indiana University Press.
Machiavelli, M. (1525, 1988). *Florentine Histories* (translated by L.F. Banfield and H.C. Mansfield, Jr). Princeton: Princeton University Press.
Maclean, I. (2002). *Logic, Signs and Nature in the Renaissance.* Cambridge: Cambridge University Press.
MacLean, P.D. (1990). *The Triune Brain in Evolution: Role in Paleocerebral Functions.* New York: Plenum Press.
Madden, E.H. (1963, 1990). *Chauncey Wright and the Foundations of Pragmatism.* Seattle, WA: University of Washington Press.
Madden, E.H. (1964). *Chauncey Wright.* New York: Washington Square Press.
Madden, E.H. (1968). Civil *Disobedience and Moral Law in 19th Century American Philosophy.* Seattle, WA: University of Washington Press.
Mahowald, M.B. (1997). What classical American philosophers mind: Jane Addams, critical pragmatism and cultural feminism. *The Journal of Value Inquiry,* 31: 39–54.
Malthus, T.R. (1798, 1970). *An Essay on the Principle of Population.* Baltimore: Penguin Books.
Margolis, J. (2002). *Reinventing Pragmatism: American Philosophy at the End of the Twentieth Century.* Ithaca: Cornell University Press.
Margolis, J. (2003). *The Unraveling of Scientism: American Philosophy at the End of the Twentieth Century.* Ithaca: Cornell University Press.
Marler, P.R., and Hamilton, W.J. (1966). *Mechanisms of Animal Behavior.* New York: Wiley and Sons.
Martin, A. (2007). The representation of object concepts in the brain. *Annual Review of Psychology,* 58: 25–45.
Martin, A., and Weisber, J. (2003). Neural foundations for understanding social and mechanical concepts. *Cognitive Neuropsychology,* 20: 575–87.
Martin, A., Wiggs, C.L., Ungerleider, L.G., and Haxby, J.V. (1996). Neural correlates of category specific knowledge. *Nature,* 379: 649–652.
Matthiessen, F.O. (1947, 1980). *The James Family.* New York: Vintage Press.
Mayr, E. (1942, 1982). *Systemics and the Origin of Species.* New York: Columbia University Press.
Mayr, E. (1963). *Animal Species and Evolution.* Cambridge, MA: Harvard University Press.
Mayr, E. (1991). *One Long Argument.* Cambridge, MA: Harvard University Press.
McAfee, N. (2005). Two feminisms. *Journal of Speculative Philosophy,* 19: 140–9.
McAfee, N. (2008). *Democracy and the Political Unconscious.* New York: Columbia University Press.
McDermott, J.J. (2007). *The Drama of Possibility: Experience as Philosophy of Culture* (edited by D.R. Anderson). New York: Fordham University Press.
McEwen, B.S. (1998). Stress, adaptation, and disease: allostasis and allostatic load. *Ann NY Acad Sci,* 840: 33–44.
McHenry, H.M. (1994). Tempo and mode in human evolution. *Proceedings of the National Academy of Sciences,* 91: 6780–6786.
McHenry, H.M. (2009). Human evolution. In: *Evolution: The First Four Billion Years* (edited by M. Ruse and J. Travis). Cambridge, MA: Harvard University Press: 256–280.

McMullen, E.T. (1995). Anatomy of a physiological discovery: William Harvey and the circulation of the blood. *Journal of the Royal Society of Medicine*, 88: 491–98.
Mead, G.H. (1928, 1972). *The Philosophy of the Act*, vol. 3. Chicago: University of Chicago Press.
Mead, G.H. (1932, 1980). *The Philosophy of the Present*. Chicago: University of Chicago Press.
Mead, G.H. (1934, 1964). Evolution becomes a general idea. In: *Selected Papers*. Chicago: University of Chicago Press.
Mead, G.H. (1934, 1972). *Mind, Self, and Society*, vol. 1. Chicago: University of Chicago Press.
Mead, G.H. (1936). *Movements of Thought in the Nineteenth Century* (edited by C.W. Morris). Chicago: University of Chicago Press.
Mead, M. (1964). *Continuities in Cultural Evolution*. New Haven: Yale University Press.
Meaney, M.J. (2001). Maternal care, gene expression, and the transmission of individual differences in stress reactivity across generations. *Annual Review of Neuroscience*, 24: 1161–1192.
Meinecke, F. (1957, 1977). *The Age of German Liberation*. Berkeley, CA: University of California Press.
Mellars, P. (2006). Why did modern human populations disperse from Africa ca. 60,000 years ago? *Proceedings of the National Academy of Sciences*, 103: 9381–9386.
Menand, L. (1997). *Pragmatism: A Reader*. New York: Random House.
Menand, L. (2001). *The Metaphysical Club*. New York: Farrar, Straus and Giroux.
Menand, L. (2002). *American Studies*. New York: Farrar, Straus and Giroux.
Merchant, C. (1989). *Ecological Revolutions*. Chapel Hill: University of North Carolina Press.
Merck, F. (1963). *Manifest Destiny and Mission in American History*. New York: Vintage Press.
Merleau-Ponty, M. (1942, 1967). *The Structure of Behavior*. Boston: Beacon Press.
Merleau-Ponty, M. (1968, 1970). *Themes*. Evanston, IL: Northwestern University Press.
Merton, P.K. (1973). *The Sociology of Science*. Chicago: University of Chicago Press.
Mettrie, J.O. de La (1748, 1912). *Man a Machine*. La Salle, IL: Open Court Press.
Meyer, A. (1915). Objective psychology or psychobiology with subordination of the medically useless contrast of mental and physical. *Journal of the American Medical Association*, 65: 860–863.
Meyer, A. (1931, 1957). *Psychobiology*. Springfield, IL: Charles C. Thomas.
Meyer, A. (1935). Scope and teaching of psychobiology. *Journal of the American Medical Association*, 10: 93–98.
Meyer, A. (1951). *The Collected Papers of Adolf Meyer*, 4 vols (edited by E.E. Winters). Baltimore: Johns Hopkins University Press.
Midgley, M. (1979, 1995). *Beast and Man*. London: Routledge.
Milgram, S. (1974, 2009). *Obedience to Authority: An Experimental View*. New York: Harper Perennial Modern Classics.
Mill, J.S. (1843, 1873). *A System of Logic*. London: Longmans, Green, Reader and Dyer.

Miller, C.A. (1988). *Jefferson and Nature*. Baltimore: Johns Hopkins University Press.

Miller, P. (1965). *The Life of the Mind in America from the Revolution to the Civil War*. New York: Harcourt Brace and Jovanovich.

Misak, C.J. (1991). *Truth and the End of Inquiry: A Peirceian Account of Truth*. Oxford: Clarendon Press.

Misak, C.J. (2004). *The Cambridge Companion to Peirce*. Cambridge: Cambridge University Press.

Mithen, S. (1996). *The Prehistory of the Mind. The Cognitive Origins of Art and Science*. London: Thames and Hudson, Ltd.

Mithen, S. (2006). *The Singing Neanderthal*. Cambridge, MA: Harvard University Press.

Moll, J., De Oliveira-Souza, R., Eslinger, P.J., Bramati, I., Mourao-Miranda, J., Angelo Andreiuolo P., and Pessoa, L. (2002). The neural correlates of moral sensitivity: A functional magnetic resonance imaging investigation of basic and moral emotions. *Journal of Neuroscience*, 22: 2730–2736.

Moll, J., De Oliveira-Souza, R., Moll, F.T., Ignácio, F.A., Bramati, I.E., Caparelli-Dáquer, E.M., and Eslinger, P.J. (2005). The moral affiliations of disgust: A functional MRI study. *Cognitive and Behavioral Neurology*, 18: 68–78.

Moll, J., De Oliveira-Souza, R., and Zahn, R. (2008). The neural basis of moral cognition: Sentiments, concepts, and values. *Annals of the New York Academy of Sciences*, 1124: 161–180.

Moll, J., Eslinger, P.J., and De Oliveira-Souza, R. (2001). Frontopolar and anterior temporal cortex activation in a moral judgment: Preliminary functional MRI results in normal subjects. *Arquivos de Neuro-Psiquiatria*, 59: 657–664.

Moll, J., and Schulkin, J. (2009). Social attachment and aversion in human moral cognition. *Neuroscience and Biobehavioral Reviews*, 33: 456–465.

Moll, H., and Tomasello, M. (2007). Cooperation and human cognition: The Vygotskian intelligence hypothesis. *Philosophical Transactions of the Royal Society, B: Biological Sciences*, 362: 639–648.

Moore, D.A., Cain, D.M., Loewenstein, G., and Bazerman, M.H. (2005). *Conflicts of Interest*. Cambridge: Cambridge University Press.

Moreno, J.D. (1995). *Deciding Together*. Oxford: Oxford University Press.

Moreno, J.D. (1999, 2003). Bioethics is a naturalism. In: *Pragmatic Bioethics* (edited by G. McGee). Cambridge, MA: MIT Press.

Moreno, J.D. (2005). *Is There an Ethicist in the House? On the Cutting Edge of Bioethics*. Bloomington: Indiana University Press.

Moreno, J.D., and Bergerm, S. (2010). *Progress in Bioethics: Science, Policy and Politics*. Cambridge, MA: MIT Press.

Morgan, M.A., Schulkin, J., and LeDoux, J.E. (2003). Ventral medial prefrontal cortex and emotional perseveration: The memory for prior extinction training. *Behavioral Brain Research*, 146: 121–130.

Morgenbesser, S. (1977, editor). *Dewey and His Critics: Essays from the Journal of Philosophy*. New York: Hackett.

Morris, C. (1970). *The Pragmatic Movement in American Philosophy*. New York: George Braziller.

Mounce, H.O. (1997). *The Two Pragmatisms*. London: Routledge.

Moyal, A. (2004). *Platypus*. Baltimore: Johns Hopkins University Press.

Muir, J. (1912, 1962). *The Yosemite*. New York: Doubleday.

Murphey, M. (2005). *C.I. Lewis: The Last Great Pragmatist*. Albany: SUNY Press.
Murphy, G.L. (2002). *The Big Book of Concepts*. Cambridge, MA: MIT Press.
Nagel, E. (1954). *Sovereign Reason*. New York: The Free Press.
Nash, R. (1967). *Wilderness and the American Mind*. New Haven: Yale University Press.
Neville, R.C. (1974). *The Cosmology of Freedom*. New Haven: Yale University Press.
Neville, R.C. (1992). *The Highroad around Modernism*. New York: SUNY Press.
Neville, R.C. (2004) Whitehead and pragmatism. In: *Whitehead's Philosophy: Points of Connection* (edited by J. A. Piolamowski and D. W. Sherburne). Albany, NY.
Neville, R.C. (undated). *Pragmatism, Metaphysics, Comparison and Realism*. Unpublished manuscript.
Nicholls, S. (2009). *Paradise Found*. Chicago: University of Chicago Press.
Niebuhr, R. (1952). *The Irony of American History*. New York: Charles Scribner and Sons.
Nietzsche, F. (1878, 1984). *Human, All Too Human* (transl. M. Faber and S. Lehmann). Lincoln: University of Nebraska Press.
Noe, A. (2004). *Action in Perception*. Cambridge, MA: MIT Press.
Noordzij, M.L., Newman-Norlund, S.E., de Ruiter, J. P., Hagoort, P., Levinson, S.C., and Toni, I. (2009). Brain mechanisms underlying human communication. *Frontiers in Human Neuroscience*, 3: 1–13.
Nordmann, A. (2010). Critical realism, critical idealism, critical commonsensism. In: *The Kantian Legacy in Nineteenth-Century Science* (edited by M. Friedman and A. Nordmann). Cambridge, MA: MIT Press.
Norrell, R.J. (2009). *Up from History: The Life of Booker T. Washington*. Cambridge, MA: Harvard University Press.
Northrop, F.S.C. (1962). *Man, Nature and God*. New York: Simon and Schuster.
Novak, B. (1980). *Nature and Culture*. Oxford: Oxford University Press.
Novak, B. (2007). *Voyages of the Self*. Oxford: Oxford University Press.
Nussbaum, F.L. (1953). *The Triumph of Science and Reason*. New York: Harper and Row.
Nussbaum, M.C. (1997). *Cultivating Humanity*. Cambridge, MA: Harvard University Press.
Nussbaum, M.C. (2001). *Upheavals of Thought: The Intelligence of Emotion*. Cambridge: Cambridge University Press.
Nussbaum, M.C. (2004). *Hiding from Humanity*. Princeton: Princeton University Press.
Oelschlaeger, M. (1991). *The Idea of the Wilderness*. New Haven: Yale University Press.
Oshinksy, D.M. (2005). *Polio: An American Story*. Oxford: Oxford University Press.
Osler, W. (1892). *The Principles and Practices of Medicine*. New York: D. Appleton and Company.
Osler, W. (1923). *The Evolution of Modern Medicine*. New Haven: Yale University Press.
Otis, L. (2007). *Muller's Lab*. Oxford: Oxford University Press.
Panksepp, J. (1998). *Affective Neuroscience: The Foundations of Human and Animal Emotions*. New York: Oxford University Press.

Papineau, D. (1993). *Philosophical Naturalism*. Oxford: Blackwell.
Pappas, G.F. (2008). John Dewey's ethics. In: *Reading Dewey* (edited by Larry Hickman). Bloomington: Indiana University Press.
Parrott, G.W., and Schulkin, J. (1993). Neuropsychology and the cognitive nature of emotions. *Cognition and Emotion*, 7: 43–59.
Passingham, R.E. (2008). *What is Special About the Human Brain?* Oxford: Oxford University Press.
Pavlov, I.P. (1897, 1902). *The Work of the Digestive Glands* (trans. W.H. Thompson). London: Charles Griffin.
Pavlov, I.P. (1928, 1963). *Lectures on Conditioned Reflexes*. New York: International Publishing Co.
Pearl, R. (1925). Vital statistics of the National Academy of Sciences II: elections of young men. *Proceedings of the National Academy, vol. II*: 757–760.
Peirce, C.S. (1866, 1992). On a method for search for categories. In: *The Essential Peirce*, vol. I., (edited by N. Houser and C. Kloesel). Bloomington: Indiana University Press.
Peirce, C.S. (1868, 1992). Questions concerning certain faculties claimed for man. In: *The Essential Peirce*, vol I. Bloomington: Indiana University Press.
Peirce, C.S. (1877). The fixation of belief. *Popular Science Monthly*, 12: 1–15.
Peirce, C.S. (1878). Deduction, induction and hypothesis. *Popular Science Monthly*, 13: 470–482.
Peirce, C.S. (1883). *Studies in Logic, by Members of the Johns Hopkins University*. Boston: Little Brown.
Peirce, C.S. (1892). The architecture of theories. *The Monist*, 1: 61–76.
Peirce, C.S. (1893, 1992). Evolutionary love. In: *The Essential Peirce*, vol. I (edited by N. Houser and C. Kloesel). Bloomington: Indiana University Press.
Peirce, C.S. (1899, 1992). *Reasoning and the Logic of Things* (edited by K.L. Ketner and H. Putnam). Cambridge, MA: Harvard University Press.
Peirce, C.S. (1932, 1933) *Collected Papers* vols 1–4 (edited by C. Hartshorne and P. Weiss) and vols 7–8 (edited A.W. Burks). Cambridge, MA: Harvard University Press.
Peirce, C.S. (1975, 1978) *Contribution to the Nation* (edited by K.L. Ketner and J.E. Cook). Lubbock: Texas Tech University.
Peirce, C.S. (2000). *Writings of C.S. Peirce*, vols 1–6. Bloomington: Indiana University Press.
Peirce, C.S. (2009). *Writings of C.S. Peirce*, vol. 8. Bloomington: Indiana University Press.
Peirce, C.S., and Jastrow, J. (1885). On small differences of sensation. *Memoirs of the National Academy of Sciences*, 3: 75–83.
Peirce, C.S., and Victoria, Lady Welby (1903–1912, 1977). *Semiotic and Significs: The Correspondence between Charles S. Peirce and Victoria, Lady Welby* (edited by C.S. Hardwick). Bloomington: Indiana University Press.
Pellegrino, E.D., and Thomasma, D.C. (1988). *For the Patient's Good*. Oxford: Oxford University Press.
Percy, W. (1954, 1975). *The Message in the Bottle*. New York: Noonday Press.
Perrett, D., Harries, M., Bevan, R., Thomas, S., Benson, P., Mistlin, A., Chitty, A., Hietanen, J., and Ortega, J. (1989). Frameworks of analysis for the neural representation of animate objects and actions. *Journal of Experimental Biology*, 146: 87–113.

Perry, R.B. (1912). *Present Philosophical Tendencies, a Critical Survey of Naturalism, Idealism, Pragmatism, and Realism Together with a Synopsis of the Philosophy of William James*. New York: Longmans, Green.
Perry, R.B. (1935). *The Thought and Character of William James*, vols 1 and 2. Boston: Little Brown and Co.
Perry, R.B. (1938, 1958). *In the Spirit of William James*. Bloomington: Indiana University Press.
Piaget, J. (1971, 1975). *Biology and Knowledge*. Chicago: University of Chicago Press.
Piazza, M., and Izard, V. (2009). How humans count. *The Neuroscientist*, 15: 261–273.
Pinault, J.R. (1992). *Hippocratic Lives and Legends*. Leiden: Brill Academic Publishers.
Pinker, S. (1994). *The Language Instinct*. New York: William Morrow and Co.
Pinker, S. (2011). *The Better Angels of Our Nature: Why Violence Has Declined*. New York: Viking.
Plato (1985). *Meno*. Oxford: Aris and Phillips.
Polanyi, M. (1946, 1964). *Science, Faith and Society*. Chicago: University of Chicago Press.
Popper, K.R. (1934, 1968). *The Logic of Scientific Discovery*. New York: Harper Torch Books.
Popper, K.R. (1962, 1971). *Conjectures and Refutations*. New York: Harper and Row.
Porter, R. (1998). *The Greatest Benefit to Mankind*. New York: Norton.
Porter, T.M. (2004). *Karl Pearson: The Scientific Life in a Statistical Age*. Princeton: Princeton University Press.
Potter, M. (2000). *Reason's Nearest Kin*. Oxford: Oxford University Press.
Power, M., and Schulkin, J. (2008). Anticipatory physiological regulation in feeding biology: Cephalic phase responses. *Appetite*, 50: 194–206.
Power, M., and Schulkin, J. (2009). *The Evolution of Obesity*. Baltimore: Johns Hopkins University Press.
Powley, T.L. (1977). The ventralmedial hypothalamic syndrome, satiety and cephalic phase. *Psychological Review*, 84: 89–126.
Prado, C.G. (1987). *The Limits of Pragmatism*. Atlantic Highlands: Humanities Press International.
Pratt, S.L. (2002). *Native Pragmatism: Rethinking the Roots of American Philosophy*. Bloomington: Indiana University Press.
Price, E.O. (2002). *Animal Domestication and Behavior*. New York: CABI Publishing.
Prinz, J.J. (2002). *Furnishing the Mind*. Cambridge, MA: MIT Press.
Putnam, H. (1990). *Realism With a Human Face*. Cambridge, MA: Harvard University Press.
Putnam, H. (1995). *Pragmatism: An Open Question*. Malden, MA: Blackwell.
Putnam, H. (2000). *The Collapse of the Fact/Value Distinction*. Cambridge, MA: Harvard University Press.
Pylshyn, Z.W. (2007). *Things and Places*. Cambridge, MA: MIT Press.
Quine, W.V.O. (1951). Two dogmas of empiricism. *The Philosophical Review*, 60: 20–43.
Quine, W.V.O. (1953, 1961). *From a Logical Point of View*. New York: Harper Torchbooks.

Quine, W.V.O. (1969). Epistemology naturalized. In: *Ontological Relativity and Other Essays*. New York: Columbia University Press.
Quine, W.V.O. (1969). Natural kinds. In: *Ontological Relativity and Other Essays*. New York: Columbia University Press.
Quine, W.V.O. (1974). *The Roots of Reference*. La Salle, IL: Open Court.
Quirk, G.J., Likhtik, E., Pellerier, J.G., and Pare, D. (2003). Stimulation of medial prefrontal cortex decreases the responsiveness of central amygdala output neuron. *Journal of Neuroscience*, 23: 8800–8807.
Randall, J.H. (1977). *Philosophy After Darwin*. New York: Columbia University Press.
Rao, M.S. (1966). The history of medicine in India and Burma. *The Symposium on The History of Medicine in the Commonwealth*. London.
Rawls, J. (1971). *A Theory of Justice*. Cambridge, MA: Harvard University Press.
Ray, J. (1660). *Catalogus plantarum circa Cantabrigiam Nascentium*. Cambridge: Field.
Reader, S.M., and Laland, K.N. (2002). Social intelligence, innovation, and enhanced brain size in primates. *Proceedings of the National Academy of Sciences*, 99: 4436–4441.
Reck, A.J. (1972). *Speculative Philosophy*. Albuquerque: University of New Mexico Press.
Reid, T. (1785, 1969). *Essays on the Intellectual Powers of Man*. Cambridge, MA: MIT Press.
Reschler, N. (1992). *A System of Pragmatic Idealism*. Princeton: Princeton University Press.
Richards, R.J. (1992, 1995). *The Meaning of Evolution*. Chicago: University of Chicago Press.
Richards, R.J. (2008). *The Tragic Sense of Life: Ernst Haeckel and the Struggle over Evolutionary Thought*. Chicago: University of Chicago Press.
Richardson, R.D. (1995). *Emerson*. Berkeley, CA: University of California Press.
Richter, C.P. (1941). The biology of drive. *Psychosomatic Medicine*, 3: 105–110.
Richter, C.P. (1943). *Total Self-Regulatory Functions in Animals and Man*. New York: Harvey Lecture Series.
Richter, C.P. (1952). Domestication of the Norway rat and the implications for the study of genetics in man. *American Journal of Human Genetics*, 4: 273–285.
Richter, C.P. (1965, 1979). *Biological Clocks in Medicine and Psychiatry*. Springfield, IL: Charles C. Thomas.
Rizzolatti, G., and Luppino, G. (2001). The cortical motor system. *Neuron*, 31: 889–901.
Rockmore, T., and Singer, B.J. (1992). *Antifoundationalism Old and New*. Philadelphia: Temple University Press.
Rorty, R. (1979). *Philosophy and the Mirror of Nature*. Chicago: University of Chicago Press.
Rorty, R. (1982). *Consequences of Pragmatism*. Minneapolis: University of Minnesota Press.
Rorty, R. (1999). *Philosophy and Social Hope*. New York: Penguin Books.
Rosati, A.G., Stevens, J.R., Hare, B., and Hauser, M.D. (2007). The evolutionary origins of human patience: temporal preferences in chimpanzees, bonobos and human adults. *Current Biology*, 17: 1663–1668.

Rosen, J., and Schulkin, J. (1998). From normal fear to pathological anxiety. *Psychological Review*, 105: 325–350.
Rosenberg, A. (1985). *The Structure of Biological Science*. Cambridge: Cambridge University Press.
Rosenberg, A. (1994). *Instrumental Biology or the Disunity of Science*. Chicago: University of Chicago.
Rosenberg, C.E. (1976, 1997). *No Other Gods: On Science and American Social Thought*. Baltimore: Johns Hopkins University Press.
Rosenthal, S.B. (1986, 1990). *Speculative Pragmatism*. La Salle, IL: Open Court.
Rosenthal, S.B. (1995). Paul Weiss and pragmatism: a dialogue. In: *The Philosophy of Paul Weiss* (ed. L.E. Hahn). La Salle, IL: Open Court.
Rosenthal, S.B. (2007). *C.I. Lewis in Focus: The Pulse of Pragmatism*. Bloomington: Indiana University Press.
Rouse, J. (2002). *How Scientific Practices Matter: Reclaiming Philosophical Naturalism*. Chicago: University of Chicago Press.
Rousseau, J.J. (1762, 1956). *Emile*. New York: Columbia University Press.
Rowland, L.P. (2009). *The Legacy of Tracy J. Putnam and H. Houson Merrit: Modern Neurology in the United States*. Oxford: Oxford University Press.
Royce, J. (1900, 1959). *The World and the Individual*. New York: Dover Press.
Royce, J. (1908, 1995). *The Philosophy of Loyalty*. Nashville: Vanderbilt University Press.
Rozin, P. (1976). The evolution of intelligence and access to the cognitive unconscious. In: *Progress in Psychobiology and Physiological Psychology* (edited by J. Sprague and A.N. Epstein). New York: Academic Press.
Rozin, P. (1998). Evolution and development of brains and cultures. In: *Brain and Mind: Evolutionary Perspectives* (edited by M.S. Gazzaniga and J.S. Altman). Strasbourg, France: Human Frontiers Sciences Program.
Rozin, P. (1999). The process of moralization. *Psychological Science*, 10: 218–221.
Rozin, P. (2005). The meaning of natural: process more important than content. *Psychological Science*, 16: 652–658.
Rozin, P., and Fallon, A.E. (1987). A perspective on disgust. *Psychological Review*, 94: 23–41.
Rozin, P., Haidt, J., and McCauley, C.R. (1993). Disgust. In: *Handbook of Emotions* (edited by M. Lewis and J.M. Haviland). New York: Guilford.
Rucker, D. (1969). *The Chicago Pragmatists*. Minneapolis: University of Minnesota Press.
Ruse, M. (1995). *Evolutionary Naturalism*. London: Routledge.
Ruse, M. (1996). *Monad to Man*. Cambridge, MA: Harvard University Press.
Ruse, M. (2009, editor). *Philosophy after Darwin*. Princeton: Princeton University Press.
Ryan, A. (1995). *John Dewey and the High Tide of American Liberalism*. New York: Norton Press.
Sabini, J., and Schulkin, J. (1994). Biological realism and social constructivism. *Journal for the Theory of Social Behavior*, 224: 207–217.
Sabini, J., and Silver, M. (1982). *Moralities of Everyday Life*. Oxford: Oxford University Press.
Sacks, K.S. (2003). *Understanding Emerson*. Princeton: Princeton University Press.

Sagoff, M. (1988). *The Economy of the Earth*. Cambridge: Cambridge University Press.
Sanger, M. (1926). The Function of sterilization. *Birth Control Review*, October.
Santayana, G. (1923, 1955). *Skepticism and Animal Faith*. New York: Dover Press.
Santayana, G. (1932, 1967). *Character and Opinion in the United States*. New York: Norton.
Santayana, G. (1967). *Animal Faith and Spiritual Life* (edited by J. Lachs). New York: Appleton, Century, Crofts.
Santayana, G. (1988). Some gleanings from The Life of Reason. *Journal of Philosophy*, 51: 46–49.
Sarokin, D., and Schulkin, J. (1992). The role of pollution in large-scale population disturbances. Part 1: Aquatic populations. *Environmental Science and Technology*, 26: 1476–1483.
Sarokin, D., and Schulkin, J. (1994). Co-evolution of rights and environmental justice. *The Environmentalist*, 14: 121–129.
Scarry, E. (1985). *The Body in Pain*. Oxford: Oxford University Press.
Scerri, E.R. (2002). *The Periodic Table*. Oxford: Oxford University Press.
Schachar-Dadon, A., Schulkin, J., and Leshem, M. (2009). Adversity before conception will affect adult progeny in rats. *Developmental Psychology*, 45(4): 9–16.
Scheler, M. (1928, 1976). *Man's Place in Nature*. New York: Noonday Press.
Schiller, F. (1795, 1980). *On the Aesthetic Education of Man*. New York: Frederick Ungar, Publ.
Schiller, F.C.S. (1966). *Humanistic Pragmatism: The Philosophy of F.C.S. Schiller* (edited by R. Abel). New York: Free Press.
Schlegel, F. (1800, 1991). *Philosophical Fragments*. Minneapolis: University of Minnesota Press.
Schmidt, L., and Schulkin, J. (1999). *Extreme Fear, Shyness and Social Behavior*. Oxford: Oxford University Press.
Schneider, H.W. (1946, 1963). *A History of American Philosophy*. NY: Columbia University Press.
Schneider, H.W. (1964). *Sources of Contemporary Philosophical Realism in America*. New York: Bobbs Merrill.
Scotus, D. (1962). *Philosophical Writings*. Indianapolis, IN: Bobbs-Merrill.
Schulkin, J. (1992). *The Pursuit of Inquiry*. Albany, NY: SUNY Press.
Schulkin, J. (1996). *The Delicate Balance*. Lanham, MD: University Press of America.
Schulkin, J. (2000). Decision sciences and evidence-based medicine. Two intellectual movements to support clinical decision making. *Academic Medicine*, 75: 816–818.
Schulkin, J. (2000). *Roots of Social Sensibility and Neural Function*. Cambridge, MA: MIT Press.
Schulkin, J. (2003). *Rethinking Homeostasis*. Cambridge, MA: MIT Press.
Schulkin, J. (2004). *Bodily Sensibility: Intelligent Action*. Oxford: Oxford University Press.
Schulkin, J. (2005). *Curt Richter: A Life in the Laboratory*. Baltimore: Johns Hopkins University Press.
Schulkin, J. (2007). *Effort: A Behavioral Neuroscience Perspective on the Will*. Mahwah, NJ: Erlbaum Press.

Schulkin, J. (2008). *Medical Decisions, Estrogen, and Aging*. New York: Springer.
Schulkin, J. (2009). *Cognitive Adaptation: A Pragmatist Perspective*. Cambridge: Cambridge University Press.
Schulkin, J. (2011). *Adaptation and Well-Being: Social Allostasis*. Cambridge: Cambridge University Press.
Schull, J. (1988). Intelligence and mind in evolution. *Journal of General Evolution*, 23: 263–273.
Schull, J. (1992). Selection: James's principal principle. In: *Reinterpreting the Legacy of William James* (edited by M.E. Donnelly). Washington, DC: American Psychological Assocation: 139–152.
Schultz, S. (2002). William Harvey and the circulation of the blood: the birth of a scientific revolution and modern physiology. *News in Psychological Sciences*, 17: 175–180.
Schumpeter, J.A. (1934). *Theory of Economic Development*. Cambridge, MA: Harvard University Press.
Schutz, A. (1932, 1967). *The Phenomenology of the Social World* (transl. G. Walsh and F. Lehnert). Evanston, IL: Northwestern University Press.
Schutz, A. (1967). *The Phenomenology of the Social World*. Evanston, IL: Northwestern University Press.
Schutz, A., and Luckman, A. (1973). *The Structures of the Life-World*. Evanston: Northwestern University.
Schwartz, J. (2000). *Fighting Poverty with Virtue*. Bloomington: Indiana University Press.
Sehon, S. (2005). *Teleological Realism: Mind, Agency and Explanation*. Cambridge, MA: MIT Press.
Seigfried, C.H. (1984). The positivist foundation in Williams James' *Principles*. *Review of Metaphysics*, 38, 579–593.
Seigfried, C.H. (1996). *Pragmatism and Feminism: Reweaving the Social Fabric*. Chicago: University of Chicago Press.
Seigfried, C.H. (1998). John Dewey's pragmatist feminism. In: *Reading Dewey* (edited by L. Hickman). Bloomington: Indiana Univ. Press.
Seigfried, C.H. (2002, editor). *Feminist Interpretations of John Dewey*. University Park, PA: Pennsylvania State University Press.
Sellars, R.W. (1922). *Evolutionary Naturalism*. Chicago: Open Court Press.
Sellars, W. (1956, 1997). *Empiricism and the Philosophy of Mind*. Cambridge, MA: Harvard University Press.
Sellars, W. (1962). *Science, Perception, and Reality*. New York: Routledge.
Sellars, W. (1968). *Science and Metaphysics*. New York: Humanities Press.
Shapin, S. (1995). *A Social History of Truth*. Chicago: University of Chicago Press.
Shapin, S. (1996). *The Scientific Revolution*. Chicago: University of Chicago Press.
Shapin, S. (2008). *The Scientific Life*. Chicago: University of Chicago Press.
Shapin, S. (2010). *Never Pure*. Baltimore: Johns Hopkins University Press.
Shapin, S., and Schaffer, S. (1985). *Leviathan and the Air-Pump*. Princeton: Princeton University Press.
Shelley, M. (1817, 1976). *Frankenstein*. New York: Pyram Pub.
Shields, P.M. (2006). Democracy and the social feminist ethics of Jane Addams: a vision for public administration. *Administrative Theory and Praxis*, 28: 418–443.

Shook, J.R. (1998). *Pragmatism: An Annotated Bibliography.* Amsterdam: Rodopi.
Shook, J.R. (2000). *Dewey's Empirical Theory of Knowledge and Reality.* Nashville: Vanderbilt University Press.
Shook, J.R. (2003, editor). *Pragmatic Naturalism and Realism.* Amherst, NY: Prometheus Press.
Shook, J.R. (2000–2006, editor). *Pragmatism in American Thought* series. Bristol, UK: Thoemmes.
Shook, J.R., and Margolis, J. (2006). *A Companion to Pragmatism.* Oxford: Blackwell Publishing.
Shusterman, R. (1997). *Practicing Philosophy: Pragmatism and the Philosophical Life.* New York: Routledge.
Simon, H.A. (1982). *Models of Bounded Rationality.* Cambridge, MA: MIT Press.
Simpson, G.G. (1949). *The Meaning of Evolution.* New Haven: Yale University Press.
Singer, B. (1999). *Pragmatism, Rights and Democracy.* New York: Fordham University Press.
Singer, C. (1957). *A Short History of Anatomy and Physiology from the Greeks to Harvey.* New York: Dover Press.
Singer, P. (1981, 2011). *The Expanding Circle.* Princeton: Princeton University Press.
Skidelsky, E. (2008). *Ernst Cassirer: The Last Philosopher of Culture.* Princeton: Princeton University Press.
Sleeper, R.W. (1986). *The Necessity of Pragmatism: John Dewey's Conception of Philosophy.* New Haven: Yale University Press.
Slotten, R.A. (2004). *The Life of Alfred Russel Wallace.* New York: Columbia University Press.
Smith, A. (1759, 1882). *The Theory of Moral Sentiments.* Indianapolis: Liberty Classics.
Smith, G.P. (2000). Pavlov and integrative physiology. *American Journal of Physiology: Regulatory, Integrative and Comparative Physiology,* 279: R743–R755.
Smith, J.E. (1970). *Themes in American Philosophy.* New York: Harper and Row.
Smith, J.E. (1978). *Purpose and Thought.* New Haven: Yale University Press.
Smith, J.E. (1985). Experience in Peirce, James and Dewey. *Monist,* 68: 538–554.
Smith, W.J. (1977). *The Behavior of Communicating: An Ethological Approach.* Cambridge, MA: Harvard University Press.
Snow, C.P. (1959, 1998). *The Two Cultures.* Cambridge: Cambridge University Press.
Snow, C.P. (1961). *Science and Government.* Cambridge, MA: Harvard University Press.
Spinoza, B. (1668, 1955). *On the Improvement of the Understanding* (transl. by R.H.M. Elwes). New York: Dover Press.
Starr, P. (1984). *The Social Transformation.* New York: Basic Books.
Stephan, H., Frahm, H.D., and Baron, G. (1981). New and revised data on volumes of brain structure in insectivores and primates. *Folia Primatologica,* 35: 1–29.
Stengers, I. (2011). *Thinking with Whitehead: A Free and Wild Creation of Concepts.* (trans. M. Chase). Cambridge, MA: Harvard University Press.
Sterelny, K. (2000). *The Evolution of Agency and Other Essays.* Cambridge: Cambridge University Press.

Sterelny, K. (2003). *Thought in a Hostile World*. New York: Blackwell.

Sterelny, K. (2004). Genes, memes and human history. *Mind and Language*, 19: 249–257.

Sterling, P. (2004). Principles of allostasis. In: *Allostasis, Homeostasis and the Costs of Physiological Adaptation* (edited by J. Schulkin). Cambridge: Cambridge University Press.

Strand, F. L. (1999). *Neuropeptides: Regulators of Physiological Processes*. Cambridge, MA: MIT Press.

Stroud, B. (1996). The charm of naturalism. *Proceedings and Addresses of the American Philosophical Association*, 70: 43–45.

Stuhr, J. (1980). Santayana's unnatural naturalism. In: *Two Centuries of Philosophy in America* (edited by P. Caw). Totowa, NJ: Rowman and Littlefield.

Stuhr J. (1987). *Classical American Philosophy*. Oxford: Oxford University Press.

Stuhr, J. (1997). *Geneological Pragmatism: Philosophy, Experience, and Community*. Albany: SUNY Press.

Stuhr, J. (2002). *Pragmatism, Postmodernism, and the Future of Philosophy*. New York: Routledge.

Stuhr, J. (2003). *Experience and Criticism: John Dewey's Reconstruction in Philosophy*. Nashville: Vanderbilt University Press.

Stuhr, J. (2010). *100 Years of Pragmatism: William James's Revolutionary Philosophy*. Bloomington: Indiana University Press.

Sullivan, M. (2007). *Legal Pragmatism: Community, Rights and Democracy*. Bloomington: Indiana University Press.

Sullivan, S. (2001). *Living Across and Through Skins: Transactional Bodies, Pragmatism and Feminism*. Bloomington: Indiana University Press.

Sullivan, W.M. (1986). *Reconstructing Public Philosophy*. Berkeley: University of California Press.

Sumner, W.G. (1963). *Social Darwinism: Selected Essays*. Englewood, NJ: Prentice Hall.

Thompson, P.B. and Hilde, T.C. (2000). *The Agrarian Roots of Pragmatism*. Nashville: Vanderbilt University Press.

Thoreau, H.D. (1971). *Great Short Works*. New York: Harper and Row.

Tillich, P. (1966). *On the Boundary*. New York: Scribner and Sons.

Tinbergen, N. (1951, 1969). *The Study of Instinct*. Oxford: Oxford University Press.

Tocqueville, A. de (1830, 1945). *Democracy in America*. New York: Vintage Press.

Todes, D.P. (1989). *Darwin without Malthus*. Oxford: Oxford University Press.

Todes, D.P. (1997). Pavlov's physiology factory. *History of Science Society*, 88: 205–246.

Todes, D.P. (2002). *Pavlov's Physiology Factor*. Baltimore: Johns Hopkins University Press.

Tomasello M. (2009). *Why We Cooperate*. Cambridge, MA: MIT Press.

Tomasello, M., and Call, J. (1997). *Primate Cognition*. Oxford: Oxford University Press.

Tomasello, M., Kruger, A.C., and Ratner, H.H. (1993). Cultural learning. *Behavioral and Brain Sciences*, 16: 495–552.

Toulmin, S. (1977). *Human Understanding*. Princeton: Princeton University Press.

Toulmin, S. (2001). *Return to Reason*. Cambridge, MA: Harvard University Press.

Turing, A. (2004). *The Essential Turing*. Oxford: Clarendon Press.
Ulett, G.A., Han, J., and Han, S. (1998). Traditional and evidence-based acupuncture: History, mechanisms, and present status. *Southern Medical Journal*, 91: 1115–1127.
Ullman, M.T. (2001). A neurocognitive perspective on language: The declarative procedural model. *Nature Neuroscience*, 9: 266–286.
Uvnäs-Moberg, K. (1998). Oxytocin may mediate the benefits of positive social interaction and emotions. *Psychoneuroendocrinology, 23*, 819–835.
Varela, F., Thompson, E., and Rosch, E. (1991). *The Embodied Mind*. Cambridge: Cambridge University Press.
Veatch, R.M. (1981). *A Theory of Medical Ethics*. Boston: Jones and Bartlett.
Vico, G. (1744, 1970). *The New Sciences*. Ithaca: Cornell University Press.
Wallace, A.F. (2002). *A Selection of Writings from the Field*. Baltimore: Johns Hopkins University Press.
Wallace, A.F.C. (1993) *The Long Bitter Trail*. New York: Hill and Wang.
Walls, L.D. (1995). *Seeing New Worlds: Henry David Thoreau and 19th Century Natural Science*. Madison: University of Wisconsin Press.
Walls, L.D. (2003). *Emerson's Life in Science: The Culture of Truth*. Ithaca: Cornell University Press.
Walls, L.D. (2009). *The Passage to Cosmos: Alexander Von Humboldt and the Shaping of America*. Chicago: University of Chicago Press.
Warner, R. (1980). *Morality in Medicine*. Sherman Oaks, CA: Alfred Publishing.
Watson, J.B. (1930, 1961). Autobiography. In: *A History of Psychology in Autobiography*. New York: Russell and Russell.
Watson, J.D. (2001). *A Passion for DNA*. Oxford: Oxford University Press.
Waxman, S. (1999). The dubbing ceremony revisited. In: *Folk Biology* (edited by D.L. Medin and S. Atran). Cambridge, MA: MIT Press.
Weaver, I.C., Cervoni, N., Champagne, F.A., D'Alessio, A.C., Sharma, S., Seckl, J.R., Dymov, S., Szyf, M., and Meaney, M.J. (2004). Epigenetic programming by maternal behavior. *Nature Neuroscience*, 7: 847–854.
Weaver, I.C., Champagne, F.A., Brown, S.E., Dymov, S., Sharma, S., Meaney, M.J., and Szyf, M. (2005). Reversal of maternal programming of stress responses in adult offspring through methyl supplementation: altering epigenetic marking later in life. *Journal of Neuroscience*, 25: 11045–11054.
Weber, M. (1904, 1949). *The Methodology of the Social Sciences* (translated by E. Shils and H. Finch). New York: Free Press.
Weber, M. (1905, 1958). *The Protestant Ethic and the Spirit of Capitalism* (transl. T. Parkson). New York: Scribners.
Weidman, N.M. (1999). *Constructing Scientific Psychology*. Cambridge: Cambridge University Press.
Weiss, P. (1947, 1967). *Nature and Man*. Carbondale: Southern Illinois University Press.
Weiss, P. (1965). Charles S. Peirce: philosopher. In: *Perspectives on Peirce* (edited by R.J. Bernstein). New Haven: Yale University Press.
Weissman, D. (1989). *Hypothesis and the Spiral of Reflection*. Albany, NY: SUNY Press.
Weissman, D. (1993). *Truth's Debt to Value*. New Haven: Yale University Press.
Weissman, D. (2000). *A Social Ontology*. New Haven: Yale University Press.
West, C. (1989). *The American Evasion of Philosophy*. Madison: University of Wisconsin Press.

Wheeler, M., and Clark, A. (1999). Genic representation: reconciling content and causal complexity. *British Journal for the Philosophy of Science*, 50: 103–135.
Wheeler, M., and Clark, A. (2008). Culture, embodiment, and genes: unravelling the triple helix. *Philosophical Transactions of the Royal Society, B*, 373: 3563–3575.
White, M. (1963). *Toward Reunion in Philosophy*. Cambridge, MA: Harvard University Press.
White, M. (1973). *Pragmatism and the American Mind*. Oxford: Oxford University Press.
Whitehead, A.N. (1919, 1982). *An Enquiry Concerning the Principles of Natural Knowledge*. New York: Dover Press.
Whitehead, A.N. (1925, 1953). *Science and Modern World*. New York: Free Press.
Whitehead, A.N. (1927, 1953). *Symbolism*. New York: Macmillan Company.
Whitehead, A.N. (1929, 1958). *The Function of Reason*. Boston: Beacon Press.
Whitehead, A.N. (1929, 1978). *Process and Reality*. New York: Free Press.
Whitehead, A.N. (1933, 1961). *Adventures of Ideas*. New York: Free Press.
Whitehead, A.N. (1938, 1967). *Modes of Thought*. New York: Free Press.
Whitehead, A.N. (1939, 1989). John Dewey and his influence. In: *The Philosophy of John Dewey*. La Salle, IL: Open Court.
Whiten, A., and Byrne, R.W. (1997, editors). *Machiavellian Intelligence II: Extensions and Evaluations*. Cambridge: Cambridge University Press.
Whitman, W. (1855, 1965). *Leaves of Grass*. New York: Norton.
Whitman, W. (1871, 1973). Democratic vistas. In: *The Portable Walt Whitman* (edited by M. Van Doren). New York: Penguin.
Whorton, J.C. (2002). *Nature Cures: The History of Alternative Medicine in America*. Oxford: Oxford University Press.
Wiener, P.P. (1949). *Evolution and the Founders of Pragmatism*. Cambridge, MA: Harvard University Press.
Williams, M. (1996). *Unnatural Doubts*. Princeton: Princeton University Press.
Williams, W.C. (1974). *Selected Poems of Wordsworth*. New York: New Directions.
Williamson, T. (2000). *Knowledge and Its Limits*. Oxford: Oxford University Press.
Wilshire, B. (2000). *The Primal Roots of American Philosophy*. University Park: Penn State University Press.
Wilson, E.B., and Hilferty, M.M. (1929). Note on C.S. Peirce's experimental discussion of the law errors. *Proceedings of the National Academy Sciences*, 15: 120–125.
Wilson, E.O. (1992). *The Diversity of Life*. Cambridge, MA: Harvard University Press.
Wilson, E.O. (1994). *Naturalist*. Washington, DC: Island Press.
Wilson, J.Q. (1993). *The Moral Sense*. New York: Free Press.
Wilson, R.A. (2004). *Boundaries of the Mind*. Cambridge: Cambridge University Press.
Wilson, R.A. (2005). *Genes and the Agents of Life*. Cambridge: Cambridge University Press.
Wirth, J.B. (1989). *Richter and Magendie*. The First International Cyberconference on the Psychobiology of Curt P. Richter.
Witkowski, J.A. (2008). *Davenport's Dream*. Woodbury, NY: Cold Spring Harbor Laboratory Press.
Wittgenstein, L. (1953, 1968). *Philosophical Investigations*. New York: Macmillan Publishing.

Wolin, S.S. (2008). *Managed Democracy and the Specter of Inverted Totalitarianism*. Princeton: Princeton University Press.
Wollstonecraft, M. (1792, 2004). *A Vindication of the Rights of Woman*. New York: Penguin Classics.
Woodbridge, F.J.E. (1940). *An Essay on Nature*. New York: Columbia University Press.
Woodward, C. Vann. (1986). *Thinking Back*. Louisiana: Louisiana State University.
Worster, D. (1977, 1991). *Nature's Economy*. Cambridge: Cambridge University Press.
Worster, D. (2008). *A Passion for Nature: The Life of John Muir*. Oxford: Oxford University Press.
Wright, C. (1877). The evolution of self-consciousness. In: *Chauncey Wright, Philosophical Discussions*. New York: Henry Holt and Co., 199–266.
Wright, C. (1878). Letter to Charles Darwin: August 1, 1871. In: *Letters of Chauncey Wright, with some account of his life by James Bradley Thayer*. Cambridge, MA: 231–234.
Wright, C. (1878). Letter to Charles Darwin: August 29, 1872. In: *Letters of Chauncey Wright, with some account of his life by James Bradley Thayer*. Cambridge, MA: 240–246.
Wright, C. (1878). Letter to Charles Darwin: February 24, 1875. In: *Letters of Chauncey Wright, with some account of his life by James Bradley Thayer*. Cambridge, MA: 330–338.
Wright, C. (1958). *Philosophical Writings*. New York: Liberal Arts Press.
Young, L., and Koenigs, M. (2007). Investigating emotion in moral cognition: a review of evidence from functional neuroimaging and neuropsychology. *British Medical Bulletin*, 84: 69–79.
Zak, P.J., Kurzban, R., and Matzner, W.T. (2005). Oxytocin is associated with human trustworthiness. *Hormones and Behavior*, 48: 522–527.
Zammito, J.H. (2002). *Kant, Herder and the Birth of Anthropology*. Chicago: University of Chicago Press.

Index